Density Functional Methods in Chemistry

Jan K. Labanowski   Jan W. Andzelm
Editors

Density Functional Methods in Chemistry

With 68 Figures

Springer-Verlag
New York Berlin Heidelberg London
Paris Tokyo Hong Kong Barcelona

Jan K. Labanowski
Ohio Supercomputer Center
1224 Kinnear Road
Columbus, OH 43212
USA

Jan W. Andzelm
Industry, Science,
  and Technology Department
Cray Research, Inc.
655 East Lone Oak Drive
Eagan, MN 55121
USA

Library of Congress Cataloging-in-Publication Data
Labanowski, Jan K.
  Density functional methods in chemistry / Jan K. Labanowski, Jan
W. Andzelm.
    p.   cm.
  Includes bibliographical references and index.

  1. Density functionals—Congresses.   2. Quantum chemistry—
Congresses.   3. Electronic structure—Congresses.   I. Andzelm, J.
(Jan)   II. Title.
QD462.6.D45L33   1991
541.2′8—dc20   90-25253

Cover illustration: Electron density from HOMO orbital around hydrogen bond in the formic acid: methylamine complex derived from DMol calculations. The density values were visualized with the apE software, version 2.0.

Printed on acid-free paper.

Camera-ready copy prepared by the editors.

9 8 7 6 5 4 3 2 1

ISBN-13: 978-1-4612-7809-2   e-ISBN-13: 978-1-4612-3136-3
DOI: 10.1007/ 978-1-4612-3136-3

Preface

Predicting molecular structure and energy and explaining the nature of bonding are central goals in quantum chemistry. With this book, the editors assert that the density functional (DF) method satisfies these goals and has come into its own as an advanced method of computational chemistry. The wealth of applications presented in the book, ranging from solid state systems and polymers to organic and organo-metallic molecules, metallic clusters, and biological complexes, prove that DF is becoming a widely used computational tool in chemistry. Progress in the methodology and its implementation documented by the contributions in this book demonstrate that DF calculations are both accurate and efficient.

In fact, the results of DF calculations may pleasantly surprise many chemists. Even the simplest approximation of DF, the local spin density method (LSD), yields molecular structures typical of *ab initio* correlated methods.

The next level of theory, the nonlocal spin density method, predicts the energies of molecular processes within a few kcal/mol or less. Like the Hartree-Fock (HF) and configuration interaction (CI) methods, the DF method is based only on fundamental physical constants. Therefore, it does not require semiempirical parameters and can be applied to any molecular system and to metallic phases. However, DF's greatest advantage is that it can be applied to much larger systems than those approachable by traditional *ab initio* methods, especially when compared with correlated *ab initio* methods.

Although density functional calculations have been performed since the early days of quantum theory, it was the work of Hohenberg, Kohn, and Sham in the mid-1960s that laid the foundation for modern density functional theory (DFT). They proved that the ground state of a many-electron system is completely determined by its electron density.

In the early 1970s, Slater and others started to apply the Xα method to chemical systems. Those efforts, which might be considered as precursors of today's DFT, provided useful results, particularly for transition metal clusters and organo-metallic molecules. However, at that time the method was unable to predict both structure and accurate energies of molecules, mostly because of severe approximations involved in its implementations.

During the same period, analytic first and second derivatives were developed for the Hartree-Fock theory. The Møller-Plesset perturbation theory and CI methods enabled systematic improvements to the calculation of HF energies. Sophisticated *ab initio* programs like GAUSSIAN had laid the basis for routine molecular calculations. As a result, most chemists were not interested in using DFT. Only a small group of researchers pursued this area of research.

The 1980s brought significant improvements in both theory and software for DFT. Accurate calculations of DF energies became possible, and gradient geometry optimization techniques were developed. Recently proposed nonlocal DF potentials provide accurate energetics of chemical reactions. Because vibrational spectra are already being calculated, it is only a matter of time until analytical second derivatives are introduced. Efficient and versatile DF software packages are being used increasingly by academic and industrial researchers.

As we enter the 1990s, we can expect DF software to achieve most of the functionality of the present *ab initio* programs. Since the method has proven to be computationally efficient and reliable and allows the study of relatively large chemical systems, the editors are certain that DF will open new frontiers for chemists in the years to come.

This book is the result of the workshop "Theory and Applications of Density Functional Approaches to Chemistry" held from May 7-9, 1990, in Columbus, Ohio, under the auspices of the Ohio Supercomputer Center. The workshop attracted participants from all over the world, and leading scientists in this field contributed to this book. The organizers want to express their warmest thanks to all those who contributed to the workshop and to this book.

The participation of so many scientists from universities and industry at the workshop gave a clear indication of the growing interest in chemical applications of DF and demonstrated that the discipline has evolved as one of the modern methods of computational chemistry.

We would like to thank Dr. Brett Dunlap, who wrote the introduction to this volume.

We are indebted to many people for making the workshop a success. We particularly thank Charlie Bender, the director of the Ohio Supercomputer Center, for his encouragement and help. David Heisterberg of the Ohio Supercomputer Center gave invaluable assistance in organizing the workshop from the very beginning and in editing this book.

We are grateful to the following organizations for their financial support: BIOSYM Technologies, Bristol-Myers-Squibb Company, Cray Research, Inc., Merck Sharp & Dohme Research Labs, and the Statewide Users Group of the Ohio Supercomputer Center.

The editorial help of Ashley Burns, Lin Daniels, Elizabeth Timmons, and Barbara Woodall of the Ohio Supercomputer Center and the organizational efforts of Sarah A. Sieling from the Department of Conferences & Institutes of The Ohio State University are greatly appreciated.

Jan K. Labanowski, Ohio Supercomputer Center
Jan W. Andzelm, Cray Research, Inc.

October 15, 1990

Contents

Contents   ix

List of Contributors

J.A. Alonso, Departamento de Física Teórica, Universidad de Valladolid, Spain

C. Amador, Física y Química Teórica, Facultad de Química, Universidad Nacional Autónoma de México, Ciudad Universitaria, México D.F. 04510, México

Jan Andzelm, Industry, Science, and Technology Department, Cray Research, Inc., Eagan, MN 55121 USA

A. Ayuela, Departamento de Física Teórica, Universidad de Valladolid, Spain

Evert Jan Baerends, Scheikundig Laboratorium, Vrije Universiteit, De Boelelaan 1083, 1081 HV Amsterdam, The Netherlands

Bruce E. Bursten, The Ohio State University, Columbus, OH 43210 USA

Dennis J. Caldwell, Hercules Aerospace, Physics Division, P.O. Box 98, Bacchus Works, A2, Magna, UT 84044-0098 USA

David A. Case, Department of Molecular Biology, Research Institute of Scripps Clinic, La Jolla, CA 95616 USA

Eduardo A. Castro, Instituto de Investigaciones Físicoquimicas Teóricas y Aplicadas (INIFTA), Division Química Teórica, Casilla de Correo 16, Sucursal 4, (1900)-La Plata, Argentina

Miguel Castro, Sección de Química Teórica, Facultad de Química, Universidad Nacional Autónoma de México, Ciudad Universitaria, México D.F. 04510, México

Andrés Cedillo, Departamento de Química, Universidad Autónoma Metropolitana-Iztapalapa, A.P. 55-534, Mexico D.F. 09340, Mexico

S.H. Chou, Department of Physics and Astronomy, Northwestern University, Evanston, IL 60208-3112, USA

Pietro Cortona, Dipartimento di Fisica, Universitá di Genova, 16146 Genova, Italy

P. Decleva, Dipartimento Scienze Chimiche, Via A. Valerio 22, Universitá di Trieste, I-34127 Trieste, Italy

B. Delley, Paul Scherrer Institut, c/o RCA Laboratories, Ltd., Badener-strasse 569, CH-8048 Zurich, Switzerland

David A. Dixon, Central Research & Development Department, Du Pont, Experimental Station, P.O. Box 80328, Wilmington, DE 19880-0328 USA

Brett I. Dunlap, Theoretical Chemistry Section, Code 6119, Naval Research Laboratory, Washington, DC 20375-5000 USA

Donald E. Ellis, Departments of Physics and Chemistry, Northwestern University, Evanston, IL 60201 USA

Fernando Estrada, Sección de Química Teórica, Facultad de Química, Universidad Nacional Autónoma de México, Ciudad Universitaria, México D.F. 04510, México

George Fitzgerald, Department of Chemistry, University of San Diego, San Diego, CA 92182 USA

René Fournier, Department of Chemistry, Iowa State University, Ames, IA 50011 USA

A.J. Freeman, Department of Physics and Astronomy, Northwestern University, Evanston, IL 60208-3112 USA

G. Fronzoni, Dipartimento Scienze Chimiche, Via A. Valerio 22, Universitá di Trieste, I-34127 Trieste, Italy

Marcelo Galván, Departamento de Química, Division de Ciencias Basicas e Ingenieria, Universidad Autónoma Metropolitana-Iztapalapa, A.P. 55-534, México D.F. 09340, México

José Gázquez, Departamento de Química, Universidad Autónoma Metropolitana-Iztapalapa, A.P. 55-534, México D.F. 09340, México

M. Daniel Glossman, Universidad de Valladolid, Facultad de Ciencias, Departamento de Física Teórica, 47011 Valladolid, Spain

M. Grodzicki, Department of Chemistry, University of New Orleans, New Orleans, LA 70148 USA

Joseph G. Harrison, Department of Physics, University of Alabama at Birmingham, Birmingham, AL 35294 USA

David J. Heisterberg, Ohio Supercomputer Center, 1224 Kinnear Rd., Columbus, OH 43212-1163 USA

Ronald A. Hill, College of Pharmacy, Ohio State University, Columbus, OH 43210-1291 USA

Koblar A. Jackson, Department of Physics, George Mason University, Fairfax, VA 22030 USA

Paul Jasien, Department of Chemistry, University of San Diego, San Diego, CA 92182 USA

Jan K. Labanowski, Ohio Supercomputer Center, 1224 Kinnear Rd., Columbus, OH 43212-1163 USA

Mel Levy, Department of Chemistry and Quantum Theory Group, Tulane University, New Orleans, LA 70118 USA

Ye Ling, Physics Department, Fudan University, Shanghai, People's Republic of China

A. Lisini, Dipartimento Scienze Chimiche, Via A. Valerio 22, Universitá di Trieste, I-34127 Trieste, Italy

J.M. Lopez, Departamento de Física Teórica, Universidad de Valladolid, Spain

M.A. Martínez-Carrillo, Física y Química Teórica, Facultad de Química, Universidad Nacional Autónoma de México, Ciudad Universitaria, México D.F. 04510, México

Francisco Méndez, Departamento de Química, Division de Ciencias Basicas e Ingenieria, Universidad Autónoma Metropolitana-Iztapalapa, A.P. 55-534, México D.F. 09340, México

Duane D. Miller, College of Pharmacy, Ohio State University, Columbus, OH 43210-1291 USA

J.W. Mintmire, Theoretical Chemistry Section, Code 6119, Naval Research Laboratory, Washington, DC 20375-5000 USA

Piotr Młynarski, Quantum Chemistry Group, Institute of Chemistry, University of Lódź, PL 90-136 Lódź, Narutowicza 68, Poland

Louis Noodleman, Department of Molecular Biology, Research Institute of Scripps Clinic, La Jolla, CA 95616 USA

Imre Papai, Institute of Isotopes, Hungarian Academy of Sciences, H-1525 Budapest, P.O.B. 77, Hungary

Mark R. Pederson, Complex Systems Theory Branch, Naval Research Laboratory, Washington, DC 20375-5000 USA

A. Pisanty, Física y Química Teórica, Facultad de Química, Universidad Nacional Autónoma de México, Ciudad Universitaria, México D.F. 04510, México

P. Politzer, Department of Chemistry, University of New Orleans, New Orleans, LA 70148 USA

Patrick K. Redington, Hercules Aerospace, Physics Division, P.O. Box 98, Bacchus Works, A2, Magna, UT 84044-0098 USA

Dennis R. Salahub, Département de chimie, Université de Montréal, C.P. 6128, Succursale A, Montréal, Québec H3C 3J7, Canada

Andreas Savin, Institut für Theoretische Chemie der Universität Stuttgart, Pfaffenwaldring 55, D-7000 Stuttgart 80; and Max-Planck-Institut für Festkörperforschung, Heisenbergstr. 1, D-7000 Stuttgart 80, Germany

William F. Schneider, The Ohio State University, Columbus, OH 43210 USA

J.M. Seminario, Department of Chemistry, University of New Orleans, New Orleans, LA 70148 USA

Vicente Soria, Sección de Química Teórica, Facultad de Química, Universidad Nacional Autónoma de México, Ciudad Universitaria, México D.F. 04510, México

Alain St-Amant, Département de chimie, Université de Montréal, C.P. 6128, Succursale A, Montréal, Québec H3C 3J7, Canada

Richard J. Strittmatter, The Ohio State University, Columbus, OH 43210 USA

D.-R. Su, National Taiwan University, Department of Physics, Taipei, Taiwan 10764, Republic of China

S. Tang, Department of Physics and Astronomy, Northwestern University, Evanston, IL 60208-3112 USA

Vincenzo Tschinke, Department of Chemistry, University of Calgary, Calgary, Alberta T2N 1H4, Canada

Jiro Ushio, Production Engineering Research Laboratory, Hitachi Ltd., 292 Yoshida-machi, Totsuka-ku, Yokohama 244, Japan

Alberto Vela, Departamento de Química, Universidad Autónoma Metropolitana-Iztapalapa, A.P. 55-534, México D.F. 09340, México

Erich Wimmer, Cray Research, Inc., Eagan, MN 55121 USA

Tom Ziegler, Department of Chemistry, University of Calgary, Calgary, Alberta T2N 1H4, Canada

1
Introduction

BRETT I. DUNLAP

This book contains the work of people who spoke and presented posters at the *Ohio Supercomputer Center Workshop on Theory and Applications of Density Functional Theory in Chemistry* 7-9 May 1990, in Columbus Ohio that was organized by the editors, Jan Labanowski and Jan Andzelm. A large fraction of those attending the conference were people from different corners of chemistry who were simply curious about density functional theory in chemistry, Jan Labanowski being one example. That point was underlined for me shortly before my talk. I was holding a molecular model of what I had assumed would be universally recognized as the icosahedral sixty-carbon-atom Buckminsterfullerene[1] molecule. The person sitting next to me interpreted that model as the rhino virus, a very much larger molecule. As this book demonstrates, there was and is the expectation that density functional methods will enable first principles treatment of chemical systems that are significantly larger and more complex than those accessible by *ab initio* methods.

If the field of computational density functional chemistry is new to you, I thank you for letting your curiosity lead you to read at least a portion of this book. The field is rather new compared to most semiempirical and *ab initio* methods of quantum chemistry. Density functional theory has its origins in solid state physics, and that origin can be traced by simply considering the titles of two related earlier conference proceedings,[2,3] which include work by many of the contributors to this volume. The field has evolved sufficiently to focus exclusively on the chemical aspects of density functional theory now.

In fact, my view of density functional theory has changed almost completely three times in the fifteen years that I have worked in this field. At first I was strongly influenced by John Connolly's assessment of early attempts to improve the first local density functional, $X\alpha$, "When one tries to superimpose subtle corrections on a crude approximation, one is left with a slightly altered, but still crude, approximation".[4] In $X\alpha$ the computationally demanding exchange term of Hartree-Fock is replaced by a computationally less demanding function of the density obtained from consideration of the homogeneous electron gas.[5] If it is viewed as an approximation to Hartree-Fock, $X\alpha$ is indeed a very crude approximation. At that time I searched for problems that, first, could be treated by density functional theory and, second, for which the intrinsic inaccuracies of density functional theory would not doom the project. Another view of $X\alpha$ is that it was simply the first local density functional method to be improved

1

by newer local density functionals that are not approximations but direct implementations of the exact Kohn-Sham construction[6] for nondegenerate ground states. In these new methods exact homogeneous electron gas calculations were mapped onto highly degenerate (metallic), noninteracting electron-gas results. These newer local density functionals showed early promise that is reviewed by Salahub, et al., herein and others,[7] but these functionals are not as much of an improvement over the Xα local density functional as they perhaps should be. In the early 1980's I thought that the answer to this puzzle required understanding symmetry-required degeneracy from a density functional point of view, i.e. that density functionals had to be constructed of quantities that are invariant under all the symmetries of the external fields and the density-functional community had to find a consistent way of attaching an irreducible represention label to the density rather than the total wavefunction[8]. Such an understanding is necessary if one wants to follow entirely the breaking of a chemical bond in the case where radicals are formed. On the other hand, a large part of chemistry is concerned with singlet electronic states and potential energy surfaces. Furthermore, as he reviews in this book, Levy formally avoids this degeneracy problem by reformulating density functional theory as a constrained search over all wavefunctions that have the density in question to find the one or more that have the lowest energy as computed by the Schrödinger equation. Although this approach is not computationally advantageous if applied directly, it leads to numerous theorems, which constrain and bound approximate density-functional expressions. Most recently, Becke[9] has chosen to parameterize nonlocal density functionals to atomic data. This new functional of the density and the gradient of the density is quite accurate as demonstrated by the contribution by Ziegler and Tschinke. Results using this and another gradient method are given in contributions by Andzelm, Dixon et al., and Hill et al. Gradient methods could be further improved through the scaling arguments that are also examined by Levy. Currently I want to investigate problems that cannot be treated by density functional theory, such as the avoided crossing of multiple potential energy surfaces discussed herein, and ask why not. Another long-standing problem with local density functionals is their treatment of hydrogen, the atom for which the local density approximation is least appropriate. This problem comes to the fore in treating hydrogen bonding, and example of which is analyzed herein by Hill et al.

Having suggested that the most chemically meaningful definition of density functional theory, if it exists at all, is quite time dependent, I will sketch major views of density functional theory in chemistry that currently coexist. Density functional theory is based on the theorem of Hohenberg and Kohn[10] that a functional of the density exists that, when minimized for a given number of electrons and a given external electric potential, gives the ground-state density and energy. Parr and Yang[11] have reviewed how major chemical concepts follow from the existence of such a functional. Some of these concepts are discussed by Méndez and Galván herein. The Hohenberg-Kohn theorem leads to one view of density functional theory which is that it should be pure and thus the ground-state density is the only independent variable of the theory. Pure density functional theories include the Thomas-Fermi method and its extensions.[12] Such a view is sound for molecules, as can be seen by an algorithm described first to me by Connolly (which he credits to E. B. Wilson): Take the ground-state density and integrate it to find the total number of electrons. Find the cusps in the density to locate all nuclei, and then use the cusp condition—that the radial derivative of the density at the cusp is minus twice the nuclear charge times the density at each cusp—to determine the charges on each nucleus.[13] Finally solve Schrödinger's equation for the ground-state density or any other property that is desired. That algorithm was related to me in a very skeptical tone and has been presented in an optimistic tone by Kohn.[14] This algorithm is not a complete implementation of the Hohenberg-Kohn theorem because it does not include all external electric fields and does not directly address variationality. On the other hand, it is applicable to all densities that have some s component on each nucleus, not just the ground-state density. This includes most chemically important densities, and could, in principle, be extended to all densities by also examining the zeros of the density.

If one considers applying density functional theory to excited states orthogonality becomes a problem. Most of the density functional quantum chemical computer codes described in this volume evaluate the kinetic energy quantum-mechanically using one-electron orbitals derived from a local one-electron potential. A local potential, one that when expressed as a function of position is the same for all like-spin one-electron orbitals, need not be a local density functional—local potential methods can treat the Becke correction as demonstrated in the contributions by Ziegler and Tschinke, Andzelm, Dixon et al., and Hill et al. In the simplest case of all orbitals being occupied or unoccupied, such a construction solves the orthogonality problem and satisfies the Pauli principle if these orbitals are associated with a single-determinant wavefunction. Kohn and Sham[6] related the ground states calculated by such methods to the Hohenberg-Kohn theorem and pure density functional theory by mapping the real interacting electronic system of a given ground-state density onto a noninteracting electronic system of the same density. A still less demanding view of density functional theory is that it should be used only to compute all the correlation energy missing from a Hartree-Fock or limited configuration-interaction calculation. This view is represented by Savin in this volume.

All-purpose density functional quantum chemistry was pioneered by Johnson and coworkers from Slater's group in the late 1960's with the advent of the Scattered-Wave computer code for finite systems.[15] (In this volume Wimmer reviews the relationship between local density functional molecular calculations and corresponding band-structure calculations, which are appropriate for periodically infinite systems and for which local density functional methods were developed.) That computer code[16] gets scant mention in this book or elsewhere, because the code requires the user to choose the size of the sphere centered about each atom in which to integrate the local one-electron potential over solid angle in order to obtain an average radial potential. Unfortunately, it was originally assumed that these spheres should touch and not be allowed to overlap. The touching spheres criterion can be met in most, particularly heteronuclear, molecules in a continuum of ways spanning a wide range of total energy. On the other hand, if spheres are allowed to overlap and are chosen to enclose roughly the chemically correct amount of charge, the method becomes much more stable.[17] Perhaps a resurgence of the method will occur if enough readers of this book are influenced by Noodleman, Case, and Baerends through their chapter, which shows that Scattered-Wave calculations compare favorably to calculations using a newer, more precise, but computationally more demanding method. In any event, muffin-tin orbitals provide the preferred starting point for numerious applications in chemistry and physics. One such application is the problem of embedding a cluster into a solid, which is discussed in a chapter by Pisanty, Amador, and Martínez-Carillo.

All of the more recent all-purpose density functional quantum chemistry computer codes replace the muffin-tin approximation which requires only one-dimensional numerical integration with three-dimensional numerical integration of at least the density functional expression for the correlation energy. The first such code[18] is based on similiar band-structure computer code.[19] It uses three-dimensional numerical integration to construct the one-electron secular equation as well as to compute the total energy. Thus any type of orbital basis function can be used. In this volume Delley describes the use of numerical atomic basis functions while Ziegler and Tschinke describe calculations using Slater-type orbitals in current computer codes based on this discrete variational method. The use of numerical basis functions has the advantage that a minimal basis set is exact in the separated-atom limit within the atomic central-field approximation. On the other hand, there is more collective exprerience throughout quantum chemistry in augmenting Slater-type orbital basis sets. Relativistic discrete-variational calcuations are reviewed by Schneider et al. herein. Inspired by *ab initio* Gaussian-type-orbital quantum chemistry computer codes, Sambe and Felton[20] proposed using Gaussian basis functions in local density functional calculations on molecules. Using Gaussian basis functions and fitting the potential to appropriate functional forms lead to a secular equation that can be treated analytically. The total energy in this Gaussian approach becomes numerically stable if the Coulomb potential of the electrons is treated variationally.[21] Extensions and uses of this molecular density-functional method are described in contributions whose first authors are Andzelm, Caldwell, Dixon, Dunlap, Reddington, and Salahub. Pederson and Jackson describe a Gaussian method relying on numerical integration of the secular matrix. In quite detailed fashion, Mintmire tells how to extend the Gaussian approach to helical polymers, using screw symmetry. Results from commercial versions of numerical-basis-set and Gaussian-basis-set molecular local density functional computer codes are compared to each other and *ab initio* results in the contributions by Hill *et al.* and Dixon *et al.*

For me, the most exciting new development in these nonmuffin-tin local density functional methods is the advent of gradients of the total energy using various methods, which are described throughout this volume. The rapid growth in the number of people considering and using density functional methods in chemistry suggests that there are major advances yet to come.

Several times throughout this introduction I have included quite critical statements of John Connolly. These insights have served chemical density functional very well. First, they have taught me and not a few others to be self critical. Second, they perhaps contributed to John Connolly's leaving the field and creating the National Science Foundation supercomputer centers. These centers were in turn a stimulus and example for the formation of the Ohio Supercomputer Center, which hosted this first conference on density functional methods in chemistry.

REFERENCES

1. R.F. Curl and R.E. Smalley, *Science* **242**, 1018 (1988).

2. J.P. Dahl and J. Avery (eds.), *Local Density Approximations in Quantum Chemistry and Solid State Physics* (Plenum 1984).

3. R.M. Dreizler and J. da Provindencia, *Density Functional Methods in Physics* (Plenum 1985).

4. J.W.D. Connolly, in *Modern Theoretical Chemistry*, Vol. 7, ed. G.A. Segal (Plenum, New York, 1977) p. 105.

5. J.C. Slater, *Quantum Theory of Molecules and Solids*, Vol. 4 (McGraw-Hill, New York, 1974).

6. W. Kohn and L.J. Sham, *Phys. Rev.* **140**, A1133 (1965).

7. R.O. Jones and O. Gunnarsson, *Rev. Mod. Phys.* **61**, 789 (1989).

8. B.I. Dunlap, *Adv. Chem. Phys.* **69**, 287 (1987).

9. A.D. Becke, *Phys. Rev. A* **38**, 3098 (1988).

10. P. Hohenberg, and W. Kohn, *Phys. Rev.*, **136**, B864 (1964).

11. R.G. Parr and W. Yang, *Density Functional Theory of Atoms and Molecules* (Oxford, New York, 1989).

12. L.R. Pratt, G.G. Hoffman, and R.A. Harris, *J. Chem. Phys.* **88**, 1818 (1988).

13. E. Steiner, *J. Chem. Phys.* **39**, 2365 (1963).

14. W. Kohn, in *Density Functional Methods in Physics*, eds. R.M. Dreizler and J. da Provindencia (Plenum, New York, 1985) p. 1.

15. K.H. Johnson, *Adv. Q. Chem.* **7**, 143 (1973).

16. M. Cook and D.A. Case, Quantum Chemistry Program Exchange #465, QCPE, Bloomington, IN 47405.

17. J.G. Norman, Jr., *Mol. Phys.* **31**, 1191 (1976).

18. E.J. Baerends, D.E. Ellis, and P. Ros, *Chem. Phys.* **2**, 41 (1973).

19. G.S. Painter and D.E. Ellis, *Phys. Rev. B* **1**, 4747 (1970).

20. H. Sambe and R.H. Felton, *J. Chem. Phys.* **62**, 1122 (1975).

21. B.I. Dunlap, J.W.D. Connolly, and J.R. Sabin, *J. Chem. Phys.* **71**, 3396; 4993 (1979).

2
Density Functional Theory for Solids, Surfaces, and Molecules: From Energy Bands to Molecular Bonds

ERICH WIMMER

Introduction

In the 1950's, Slater introduced the concept of the electron-gas approximation as a simplification of the Hartree-Fock equations (Slater, 1951). The goal was to make energy band structure calculations practical for crystalline solids such as bulk Cu. The exchange term of the Hartree-Fock equations needed to be approximated while the kinetic energy and the Coulomb repulsion terms of the Fock operator could be treated rather well within a periodic lattice, particularly if the so-called muffin-tin approximation (Slater, 1937) was made.

It was found that the leading term in the exchange operator is proportional to $\rho^{1/3}$ with ρ being the electron density at a given point in real space. However, the pre-factor became a matter of controversy. Depending on the way the electron gas approximation was introduced into the Hartree-Fock equations, the results differed between 2/3 and 1. If a pre-factor of 2/3 was used (Gáspár, 1954), the one-particle wave functions calculated from this approximation resembled rather closely the Hartree-Fock results, yet the one-particle eigenvalues were found to be significantly different (their absolute values were markedly smaller than the Hartree-Fock eigenvalues). On the other hand, using a pre-factor of 1 yielded eigenvalues closer to the Hartree-Fock results, but the wave functions (and hence the electron density) were too contracted compared with the Hartree-Fock results.

This dilemma lead to the idea of introducing a variable parameter, α, which could be adjusted between 2/3 and 1 (Slater, 1974). One way to determine this parameter was to choose a value for each atom so that the atomic $X\alpha$ calculations fulfilled the virial theorem (Schwarz, 1972). The resulting values for α were found to be typically around 0.7. Combined with the muffin-tin approximation, this so-called $X\alpha$ method was widely used in band structure calculations on a number of systems including refractory metal compounds such as TiC (Neckel et al., 1976).

While the muffin-tin approximation works surprisingly well for closed-packed metallic systems (Moruzzi et al., 1978) and other solids, the attempt to use this concept for molecular calculations turned out to introduce an element of arbitrariness such as the choice of the muffin-tin radii. Although the results from this scattered-wave $X\alpha$ method were very useful in many cases, the approach was finally unacceptable compared with rigorous molecular ab initio calculations which used an implementation of Hartree-Fock theory without any approximation except in the basis set. The development of analytic first and second derivatives in the Hartree-Fock theory during the 1970's and 1980's and the systematic incorporation of electronic correlation effects via Møller-Plesset perturbation theory, through configuration interactions and multi-reference methods combined with the availability of programs such as Gaussian, Hondo, Gamess, and Gradscf established these ab initio methods as the standard for molecular calculations, especially in the realm of organic chemistry (cf. Hehre et al., 1986; and references therein).

In the mid 1960's, W. Kohn, P. Hohenberg, and L. Sham (Hohenberg and Kohn, 1964; Kohn and Sham, 1965) laid the foundation of today's density functional theory (DFT) by showing that the electron density could be used as fundamental quantity to develop a rigorous many-body theory, applicable to any atomic, molecular, or solid state system. Although the resulting equations looked very much like Slater's $X\alpha$ equations, the theoretical foundation had changed fundamentally: no longer were these equations an approximation to Hartree-Fock theory, but they resulted from a many-body theory which stands logically on the same level as the Hartree-Fock picture. In the derivation of the density functional equations, no empirical or adjustable parameters occur and therefore DFT should be considered as an *ab initio* theory.

In the early 1970's, the muffin-tin approximation was widely used in band structure calculations based on the augmented-plane-wave (APW) method (Slater, 1937), the Korringa-Kohn-Rostoker (KKR) method (Korringa, 1947; Kohn and Rostoker, 1954), and the linearized muffin-tin orbital (LMTO) method (Andersen, 1975) with impressive accuracy for densely packed solids (Moruzzi et al., 1978). It is interesting to note that the theoretical framework of these methods such as the APW and KKR methods had been formulated several decades earlier, but their applications had to await the availability of electronic computers in the 1970's with adequate computational speed and memory. The desire to treat systems of lower symmetry such as surfaces lead to efforts in the late 1970's and early 1980's to eliminate the muffin-tin approximation and "warped muffin-tin" (Herzig, 1974; Krakauer et al., 1978) and "full potential" methods (Wimmer et al., 1981) were introduced.

By the early 1980's, full-potential local density functional calculations represented the state-of-the-art in solid state electronic structure theory. At the same time, Hartree-Fock methods on the single reference SCF level as well as correlated methods represented the equivalent state-of-the-art in molecular calculations, particularly for organic and small inorganic molecules. During that time, Hartree-Fock programs offered analytic first and second derivatives of the total energy allowing geometry optimizations and vibrational frequency calculations in an elegant and powerful way while solid state programs just started to have reliable single-point total energies. Nevertheless, even with single point energies, geometry optimization started to become possible and showed very promising results (Weinert et al., 1982).

During the same time, substantial progress was made in the development of molecular density functional methods using Slater-type orbitals and numerical atomic basis functions (Baerends et al., 1973; Averill and Ellis, 1973; Rosén and Ellis, 1976; Delley and Ellis, 1982), numerical functions generated from a muffin-tin potential (Andersen and Woolley, 1973; Gunnarsson et al., 1977ab) and Gaussian-type orbitals (Sambe and Felton, 1974, 1975; Dunlap et al., 1979). The results indicated a great promise for the use of density functional theory in molecular calculations. For example, spectroscopic constants calculated for first-row diatomics (Gunnarsson et al., 1977b) gave consistently better results than Hartree-Fock theory and were found to be comparable with those of configuration interaction calculations. These and similar results encouraged the further developments of molecular density functional methods (Ravenek and Baerends, 1984; Dunlap, 1986; Salahub, 1987; Ziegler and Versluis, 1988; Andzelm et al., 1989; Delley, 1990).

It is somewhat surprising that it took relatively long for density functional methods to find a wider acceptance among chemists (Borman, 1990). Perhaps some of the reasons were the lack of analytic gradient methods and automated geometry optimizations in molecular density functional programs, the lack of widely available and easy-to-use computer programs, and the lack of a sufficient number of typical chemical applications and systematic comparisons with results from Hartree-Fock and correlated calculations on typical organic and inorganic molecules. During the last five years, these obstacles have gradually been removed so that today molecular density functional methods, implemented in the form of efficient computer programs with analytic gradient methods start to become more widely available and the method is applied to a variety of chemically and technologically interesting systems including organic and inorganic molecules, fluorocarbons, organometallic compounds, and clusters (Wimmer et al., 1987; Andzelm et al., 1989; Dixon et al., 1988; Dixon et al., this volume). There is now rapidly increasing evidence that molecular density functional calculations give remarkably accurate results for molecular structures and electronic properties. Furthermore, very recent studies seem to indicate that even the prediction of reaction energies, at least those of isodesmic reactions, can be predicted from DFT within a chemically meaningful accuracy (Dixon et al., this volume). Density functional theory can describe a great variety of molecular bonds including those of highly correlated systems such as the FOOF molecule (Dixon et al., to be published). For increasing system size, the computational effort in density functional calculations scales between N^2 and N^3 (with N being the number of basis functions) while correlated Hartree-Fock methods giving results of similar quality scale at least with a fifth power. These features of molecular density functional theory opens exciting opportunities for the study of large and complex molecular systems which are out of reach for other ab initio molecular orbital methods.

In the following sections, the basic concepts of density functional theory are reviewed and practical implementations for both solid state and molecular systems are discussed. Applications of the method using different implementations are then shown for characteristic examples ranging from transition metal surfaces to biologically active molecules and metal-protein complexes.

Concepts of Density Functional Theory

In order to appreciate the concepts of density functional theory, it is useful to compare the derivation of the Hartree-Fock equations with that of the one-particle density functional or Kohn-Sham equations (eqs. 1-5). The starting point in the derivation of Hartree-Fock (HF) theory (Hartree, 1928; Fock, 1930ab) is the total energy expressed as a function of the total wave function (1a). The total energy also depends, of course, on the positions of the atoms, R_α .

HF (1928, 1930)	DFT (1964, 1965)	

$$E = E[\Psi, R_\alpha] \qquad\qquad E = E [\rho, R_\alpha] \qquad (1ab)$$

$$E = \int \Psi^*[\Sigma_i h_i + \Sigma_{i>j} 1/r_{ij}]\Psi \, d\tau \qquad E = T[\rho] + U[\rho] + E_{xc}[\rho] \qquad (2ab)$$

$$\Psi = |\psi_1(1), \psi_2(2), ..., \psi_n(n)| \qquad \rho(r) = \Sigma_{occ} |\psi_i(r)|^2 \qquad (3ab)$$

$$\partial E / \partial \Psi = 0 \qquad\qquad \partial E / \partial \rho = 0 \qquad (4ab)$$

$$[-1/2\nabla^2 + V_C (r) + \mu^i_x(r)] \, \psi_i = \varepsilon_i\psi_i \quad [-1/2\nabla^2 + V_C (r) + \mu_{xc}(r)] \, \psi_i = \varepsilon_i\psi_i \quad (5ab)$$

In DFT, the derivation begins with the total energy written as a functional of the total electron density for given positions of the atomic nuclei (1b). In contrast to HF theory, DFT uses a physical observable, the electron density, as fundamental quantity. In HF theory, the total energy is expressed as an expectation value of the exact non-relativistic Hamiltonian (2a) using a Slater determinant (3a) as an approximation for the total wave function. In DFT, the total energy is decomposed in a formally exact way into three terms (2b), a kinetic energy term, $T[\rho]$, an electrostatic or Coulomb energy term, $U[\rho]$, and a many-body term, $E_{xc}[\rho]$, which contains all exchange and correlation effects. This decomposition is constructed in such a way that $T[\rho]$ corresponds to the kinetic energy of a system of non-interacting particles that yield the same density as the original electron system.

The total density is decomposed into single-particle densities which originate from one-particle wave functions, ψ_i (3b). DFT requires (4b) that, upon variation of the total electron density, the total energy assumes a minimum (Kohn and Sham, 1965). This leads to conditions for the one-particle wave functions in the form of effective one-particle Schrödinger equations or Kohn-Sham equations (5b). In HF theory, a variational principle applied to the Slater determinant (4a) leads to one-particle eigenvalue equations which are known as the Hartree-Fock equations (5a).

The HF equations can be written in a form that resembles very closely that of the Kohn-Sham equations (Slater, 1974; Wimmer, 1979). In fact, the only formal difference is in the term $\mu(r)$. In HF theory, this term, by definition, only describes exchange effects and depends on the actual orbital, ψ_i, the Fock operator is acting on. In contrast, the corresponding term in DFT contains, by construction, all many-body effects and does not depend on the orbital index, i. The Hamiltonian in the DFT equations is an effective one-electron operator. It contains a one-electron kinetic energy operator, a Coulomb potential operator, V_C, which includes all electrostatic interactions (electron-electron, electron-nuclei, nuclei-nuclei), and the exchange-correlation potential operator. The DFT orbitals are the eigenfunctions of this operator with the eigenvalues ε_i. The kinetic energy and Coulomb potential operator are identical in HF theory and DFT. Approximations in DFT are introduced in the exchange-correlation potential operator while, in principle, there are no conceptual approximations made in the wave functions or any other place. One could say that the Hartree-Fock equations solve the exact Hamiltonian with approximate many-body wave functions while the DFT equations solve an approximate many-body Hamiltonian with exact wave functions. Hartree-Fock based methods converge to the exact solution of the many-body Schrödinger equation through systematic improvements in the form of the many-body wave functions, for example by configuration interaction expansions. DFT theory converges to the exact solution by improving the effective exchange-correlation potential operator, $\mu_{xc}(r)$. This is equivalent to improving the description of the total exchange-correlation energy, E_{xc}, because of the relation

$$\mu_{xc}(r) = \delta E_{xc}[\rho] / d\rho(r). \tag{6}$$

In other words, if the exchange-correlation functional of a system would be known, the DFT equations would provide the exact solution of the many-body problem. Currently, the most common approximation to the exchange-correlation energy is the so-called local density approximation (LDA) (Hohenberg and Kohn, 1964; Kohn and Sham, 1965):

$$E_{xc}[\rho] \approx \int \rho(r)\, \varepsilon_{xc}[\rho(r)]\, dr. \tag{7}$$

In eq. (7), $\varepsilon_{xc}[\rho]$ is the exchange-correlation energy per electron in an interacting electron system of constant density ρ. This quantity is known from many-body theory (Wigner, 1934, 1938; Hedin and Lundqvist, 1972; Ceperley and Alder, 1980). For metallic and strongly delocalized systems with fairly constant electron density, the LDA comes very close to the exact solution whereas this approximation can be expected to be less accurate for systems with strongly varying electron densities.

It is a fundamental assumption in the LDA (7) that the length-scale associated with exchange and correlation effects is small compared with the length scale on which the electron density varies significantly. The surprisingly good results obtained for a number of solid state and molecular systems seem to indicate that it is justified to assume a rather local nature of many-body effects in solid state and molecular systems. In contrast to this approach, configuration interaction expansions use molecular orbitals, which extend spatially over an entire molecule, to capture correlation effects. The notoriously slow convergence of CI expansions may actually be an indication that the length-scale implicit in molecular orbitals may not properly match the length-scale of exchange-correlation effects. In other words, perhaps the density functional picture, at least for systems with many electrons, is more "natural" for expressing exchange and correlation effects than a wave function approach.

The non-local nature of the Hartree-Fock exchange term, in contrast to the local characteristics of the exchange-correlation term in the local density functional (LDF) method can also be seen from a comparison of the corresponding terms in the HF and LDF equations. To this end, the exchange term in the HF equations is written in the form of a potential energy operator (Slater, 1974; Wimmer, 1979):

$$\mu^i_x(r) = -\sum_j \delta(\sigma_i, \sigma_j) \, [\int \psi_i^*(r) \, \psi_j^*(r') \, 1/|r-r'| \, \psi_j(r) \, \psi_i(r') \, dr'] \, / \, [\psi_i^*(r)\psi_i(r)] \quad (8)$$

This HF exchange operator, which includes only interactions between electrons with the same spin, depends on the orbital, ψ_i, and contains information of all wave functions from the entire system. Thus, the operator is highly non-local. The HF exchange operator (8) does not contain any Coulombic screening which causes a well-known qualitative breakdown of HF theory in the limit of the homogeneous electron gas (Pines, 1955).

In LDF theory, the corresponding term can be written in the form

$$\mu_{xc}(r) = \varepsilon_{xc}[\rho(r)] + \rho(r)\{\delta\varepsilon_{xc}[\rho(r)] \, / \, \delta\rho(r)\} \quad (9)$$

This exchange-correlation operator contains information of just the electron density at point r and the exchange correlation energy, ε_{xc}, associated with that density. The essence of the LDA is the assumption that the exchange-correlation energy arising from a volume element, dr, in an inhomogeneous system is the same as if that volume element would be embedded in a homogeneous electron gas of that same density. Furthermore, it is assumed that the total exchange-correlation energy of a system can be obtained by integrating the local contributions over all space.

For computational purposes it is convenient to represent the results for ε_{xc} and μ_{xc} for various densities in an analytic form. The most commonly used forms are those by Hedin and Lundqvist (1972) and Vosko, Wilk, and Nusair (VWN) (Vosko et al., 1980). It is found that such a representation can be written as

$$\mu_{xc}(r) = \alpha \, \rho(r)^{1/3} + \ldots \tag{10}$$

in which the leading term has the form which was originally proposed by Slater (1974) in the $X\alpha$ method. Thus, the $X\alpha$ method can be seen as an approximation to the LDA. It should be noted again, though, that the LDA does not contain any empirically adjustable parameters, but is based on a rigorous first-principles many-body approach.

Self-consistent Solution of the DFT Equations

The density functional one-particle equations (5b) have to be solved in an iterative self-consistent-field (SCF) procedure (cf. Fig. 1) similar to that used in Hartree-Fock theory. The procedure begins with a starting density corresponding to a certain geometric arrangement of the nuclei. Usually, the starting density is constructed from a superposition of atomic densities. The Coulomb operator of the DFT equations can be evaluated either explicitly by solving Poisson's equation or by calculating Coulomb integrals for a given basis set using a fitted density. The former procedure is most commonly used in solid state calculations where the calculations of the Coulomb potential involves summations over the entire lattice. It is also employed in the DMol program (Delley, 1990) in which numerical basis functions are used. In the case of Gaussian-type basis functions, it is more convenient to follow the latter procedure. The calculation of the exchange-correlation potential is typically done on a numerical grid in real space and the results are then fitted or expressed in a convenient analytic form.

In the standard procedure, the DFT equations are solved by expanding the one-particle wave functions (molecular orbitals) variationally into a basis set as will be discussed below. Once the Hamiltonian and overlap matrix elements are calculated, the eigenvalues and eigenvectors are found from the diagonalization of the matrix $H-\varepsilon S$. Following an Aufbau principle (i.e. using Fermi-Dirac statistics), the orbitals are occupied and a new density is formed. This completes one cycle in the SCF procedure. At this point, convergence acceleration schemes are used to create a new input density from the previous input and output densities. After self-consistency is achieved, the total energy for this geometry with the corresponding molecular properties can be calculated. If the goal is a geometry optimization, the energy gradients are evaluated and used in the choice of a new geometry. A new SCF cycle is then

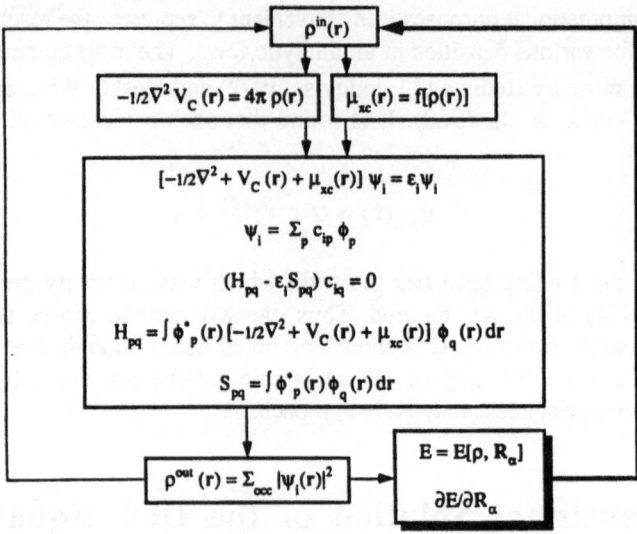

Figure 1. Self-consistent field (SCF) scheme used in the solution of the Kohn-Sham equations. The fundamental quantity that governs the iterative process is the electron density, which defines both the starting and the end points of the SCF procedure. A description of this figure is given in the text.

started with the new geometry. At this point, the SCF cycle can be started with the SCF density from the previous geometry, provided that the geometry has not changed too much.

In general, DFT SCF procedures seem to converge somewhat slower than corresponding Hartree-Fock calculations. Typical closed-shell organic molecules do not pose any serious challenges for the SCF convergence. However, for organometallic complexes and metallic systems with a high density of states near the Fermi level such as Ni and Pd, SCF convergence can become a very serious problem. Carefully chosen convergence schemes such as attenuated feedback with small mixing parameters, Broyden methods, level shifting, and fractional occupation numbers around the highest occupied level need to be used in order to achieve convergence.

Implementations of DFT

Table 1 provides an overview of various choices in the implementation of density functional theory for the calculations on solids, surfaces, and molecules. The choice of the exchange-correlation form is essentially independent of the boundary conditions, the shape of the potential, and the choice of variational basis set. On the other hand, the treatment of the potential is fairly closely related to the choice of the basis set. For example, a pure plane-wave basis is meaningful only in connection with a pseudopotential.

Augmented-plane-wave basis sets are used for all-electron treatments, typically with periodic boundary conditions. Localized orbital basis sets (numerical, Slater-type, and Gaussian-type) can be chosen for periodic as well as non-periodic systems.

Plane-waves have a systematic convergence behavior. In fact, there is only one parameter, the maximum k-vector, that determines the quality of the basis set. The maximum k-vector also defines the quality of an augmented-plane-wave basis set together with the highest angular momentum number for the spherical harmonics expansion within the muffin-tin spheres. In plane-wave basis sets the number of basis functions per atom is typically much higher than for localized basis sets. However, the construction of localized basis sets, which includes the determination of Gaussian expansion coefficients and contraction schemes, is a tedious task and there is, a priori, no "best" solution. Thus the compactness of localized basis sets is somewhat off-set by the difficulty in establishing the optimal basis set. Pure plane-wave basis sets, in conjunction with soft pseudopotentials, lead to computationally highly efficient procedures. Therefore, even very large numbers of basis functions may not pose a serious computational obstacle (Teter et al., 1989). In order to illustrate the issues involved in the choice of a certain implementation, two cases will be discussed below in more detail: the full-potential linearized-augmented-plane-wave (FLAPW) method and a Gaussian-type orbital scheme.

Table 1. Implementation choices for density functional calculations.

Potential	Exchange-Correlation	Basis Sets
Boundary conditions	Local	O plane-waves
O periodic	O Vosko-Wilk-Nusair	O augmented-plane-waves
O non-periodic	O Hedin-Lundqvist	O numerical atomic orb.
	O von Barth-Hedin	O Slater-type orbitals
Shape	O Kohn-Sham + Wigner	O augmented Slater-type
O muffin-tin	O $X\alpha$	O Gaussian-type
O full potential	O other	O Gaussians + plane-waves
		O other
Treatment of	Non-local	
core electrons	O Becke-Stoll	
O all-electron	O Becke-Perdew	O basis-set free
O pseudopotential	O other	

Full-potential Linearized Augmented Plane Wave (FLAPW) Method

In the full-potential linearized augmented plane wave (FLAPW) method (Wimmer, 1981 and references therein) the real space is partitioned into different regions according to the shape of the potential. In each of these regions, the appropriate form of the wave function is chosen. For surface calculations using a single-slab (or thin film) geometry, three regions are necessary: spheres around the atoms, an interstitial region between these spheres, and a vacuum regions on both sides of the slab. The "natural" form of the wave functions in these regions are (i) radial functions, multiplied by spherical harmonics inside the atomic spheres, (ii) plane waves for the interstitial region, and (iii) functions that depend on the normal distance (z-direction) from the surface, modulated with plane waves parallel to the surface. On the boundaries, the basis functions are matched such that the values and normal derivatives across all boundaries are continuous. The radial functions and the z-dependent functions are recalculated in each cycle of the SCF iteration. In other words, the basis functions adapt to the changes in the potential during the SCF procedure. This is in contrast to localized Gaussian, Slater-type, or numerical basis functions which remain rigid throughout the SCF process. In order to match the plane waves on the sphere boundaries, angular momentum components up to $l=8$ are typically included. The radial functions for each of these l-values consists of two linearly independent functions. This gives the FLAPW basis a remarkably high degree of variational flexibility and a benchmark-type quality. In fact, the two linearly independent radial functions per angular momentum quantum number, which are recalculated in each iteration, are reminiscent of the a split-valence basis used in molecular calculations.

In the FLAPW method Poisson's equation is solved by the following procedure (Weinert, 1981; Wimmer et al. 1981): the actual electron density is replaced by a smooth pseudo-density which is constructed in such a way that the multipole moments outside the atomic spheres are identical to those of the actual density. The smooth pseudo-density is Fourier-transformed. In this representation, the solution of Poissons's equation is very simple yielding the correct electrostatic potential everywhere outside and on the surface of the atomic spheres. Using the correct potential on the surface of the atomic spheres and the original electron density inside each sphere, the correct potential can subsequently be calculated also inside the spheres by a Green's function method.

The exchange-correlation potential is evaluated on a real space grid and then fitted to analytic representations in each of the regions corresponding to the form of the wave functions. In this way, all integrals in the Hamiltonian and overlap matrices can be evaluated analytically.

In band structure calculations, the Kohn-Sham equations have to be solved for a number of k-points in the irreducible part of the Brillouin zone. Typically, between about 3 and 100 k-points are used as sampling points for the integration over the Brillouin zone. The actual number depends on the type of system and the size of the unit cell. In fact, for very large unit cells of an insulating material, even one k-point can be adequate.

The computational efficiency, as in many other variational electronic structure methods, depends on the number of basis functions. For a typical closely-packed material with fairly similar sizes of atomic-spheres, about 50 basis functions per atom are needed. The number of basis functions in this method is determined by the shape of the electron density (or the potential) on the surfaces of the atomic spheres. As long as all spheres have a similar behavior (as is the case in a bulk metal) the non-local nature of a plane-wave basis is able to provide enough variational flexibility throughout the system. If certain atomic spheres, for example in a molecule with short bonds adsorbed on a metal surface, require additional variational freedom, the number of basis functions in the entire system has to be increased since plane waves cannot be targeted at specific atoms. This problem is aggravated by increasing certain interatomic distances while retaining short bond lengths in other regions, as might be the case in the study of chemical reactions on surfaces. In such systems, the use of localized basis sets, as discussed in the following section, appears to be more appropriate.

Gaussian-type Orbital Method

In search of efficient methods to solve the molecular DFT problem, it turned out that Gaussian-type orbitals are appealing for the same reasons as they are in Hartree-Fock theory. The wealth of experience gained from the use of Gaussian-type basis sets in Hartree-Fock theory (Dunning and Hay, 1977; Hehre et al., 1986 and references therein) can thus be transferred to DFT calculations, especially in the construction of basis sets (Huzinaga et al., 1984), the evaluation of two-electron integrals (King et al., 1976; Dupuis et al, 1976; Rys et al., 1983; Amos, 1984; Obara and Saika, 1986; Head-Gordon and Pople, 1988), and the evaluation of analytic derivatives (Stevens et al., 1963; Pulay, 1969; McIver and Komornicki, 1971; Pople et al., 1979; Schlegel, 1982). Molecular Gaussian-type DFT methods were developed by several research groups (Sambe and Felton, 1974; Dunlap et al., 1979; Andzelm and Salahub, 1986; Andzelm et al., 1989). In contrast to Hartree-Fock calculations, molecular DFT calculations based on Gaussian-type orbitals provide an opportunity to employ Gaussians not only for the expansion of the orbitals, but also for expressing the electron density and the exchange-correlation potential. This allows to make full use of the N^3 characteristic of density functional theory. Following Sambe and Felton (1974) the electron density is approximated by

$$\rho(r) \approx \rho'(r) = \Sigma_r \, \rho_r \, g_r \tag{11}$$

with $\{g_r \, ; \, r=1,...,N_\rho\}$ being an auxiliary Gaussian basis for the electron density. Similarly, the exchange-correlation potential, $\mu_{xc}(r)$, is expanded in another set of auxiliary Gaussian-type functions, $\{g_s \, ; \, s=1,...,N_\mu\}$ in the form

$$\mu_{xc}(r) = \Sigma_s \, \mu_s \, g_s. \tag{12}$$

The molecular orbitals, ψ_i, are represented by Gaussians in the same way as in the Hartree-Fock method by

$$\psi_i = \Sigma_p \, c_{ip} \, g_p \tag{13}$$

with $\{g_p \, ; \, p=1,...,N\}$ being a set of contracted Gaussian basis functions.

Using expressions (2) through (9) in eq. (1), the expansion coefficients, c_{ip}, are determined by a variational procedure analogous to that used in Hartree-Fock theory leading to the generalized eigenvalue problem

$$(H_{pq} - \varepsilon_i S_{pq}) \, c_{iq} = 0 \tag{14}$$

with the following matrix elements

$$H_{pq} = h_{pq} + \Sigma_r \, \rho_r \, [pq\|r] + \Sigma_s \, \mu_s \, [pqs] \tag{15}$$

and

$$S_{pq} = [pq] \tag{16}$$

with the definitions

$$h_{pq} \equiv \int g_p(r) \, (-1/2\nabla^2 - \Sigma_\alpha \, Z_\alpha / \, |R_\alpha - r| \,) \, g_q(r) \, dr \tag{17}$$

$$[pq\|r] \equiv \int\int g_p(r) \, g_q(r) \, (1/|r-r'|) \, g_r(r') \, dr \, dr' \tag{18}$$

$$[pqs] \equiv \int g_p(r) \, g_q(r) \, g_s(r) \, dr \tag{19}$$

$$[pq] \equiv \int g_p(r) \, g_q(r) \, dr \, . \tag{20}$$

The expression for the total energy follows as

$$E_{LDA} = \Sigma_{pq} \, P_{pq} \{h_{pq} + \Sigma_r \, \rho_r \, [pq\|r] + \Sigma_s \, \varepsilon_s \, [pqs] \, \} - 1/2 \, \Sigma_{rr'} \, \rho_r \, \rho_{r'} \, [r\|r'] + U_N. \tag{21}$$

Here, P_{pq} is the density matrix

$$P_{pq} = \Sigma_i c_{ip} c_{iq} \tag{22}$$

with the summation extending over all occupied molecular orbitals. Similar to the expansion of the exchange-correlation potential given in eq. (12), the exchange-correlation energy needed for the total energy expression (21), is expanded in a set of auxiliary Gaussian-type functions in the form

$$\varepsilon_{xc}(r) = \Sigma_s \varepsilon_s g_s. \tag{23}$$

In fact, the same set of functions $\{g_s\}$ is used as in the expansion of μ_{xc} given in eq.(12). The notation [r||r'] is an abbreviation for the integrals

$$[r||r'] = \int\int g_r(r) \ (1/|r\text{-}r'|) \ g_{r'}(r') \ dr \ dr' \tag{24}$$

and

$$U_N = 1/2 \ \Sigma_{\alpha\alpha'} \ Z_\alpha \ Z_{\alpha'} / \ |R_\alpha - R_{\alpha'}| \tag{25}$$

denotes the Coulomb repulsion energy between all nuclei.

In contrast to the usual implementations of the Hartree-Fock method, the present DFT method requires the evaluation of only two and three-index integrals. A recurrence scheme, originally developed by Obara and Saika (1986) for the computation of four-index integrals over Cartesian Gaussian functions has been reformulated for three-index integrals to meet the needs of DFT (Andzelm et al., 1989 and references therein). The resulting scheme turns out to be computationally highly efficient on vector and parallel computers.

Following Dunlap et al. (1979) the fitting coefficients for the electron density, ρ_r, are defined such that the residual Coulomb energy,

$$\Delta \equiv \int\int \delta\rho(r) \ (1/|r\text{-}r'|) \ \delta\rho(r') \ dr \ dr' \tag{26}$$

arising from the difference between the fitted and original density,

$$\delta\rho(r) \equiv \rho(r) - \rho'(r) \tag{27}$$

is minimized while maintaining charge conservation. Using the definition of the original density and that of the density matrix (22), we find the coefficients ρ_r to be determined by

$$\rho_r = \Sigma_{r'} \ C_{rr'}^{-1} \ \{ \ \Sigma_{pq} P_{pq} \ [pq||r'] - \Lambda(r') \} \tag{28}$$

The Lagrange multipliers, $\Lambda(r')$, guarantee charge conservation. The matrix C is defined by its elements

$$C_{rr'} \equiv \int \int g_r(r) \ (1/|r-r'|) \ g_{r'}(r') \ dr \ dr' \tag{29}$$

It can be seen that all the integrals necessary to calculate the density fitting coefficients, ρ_r , can be obtained from analytic, Coulomb-type integrals.

The fitting coefficients for the exchange-correlation potential of eq. (12) are given by the following relations:

$$\mu_s = \Sigma_{s'} \ S_{ss'}{}^{-1} \int g_{s'}(r) \ \mu_{xc}(r) \ dr \tag{30}$$

with $S_{ss'}$ being the usual overlap matrix element defined in eqs. (16) and (20). The evaluation of the integrals in eq. (30) is done numerically using a grid which is constructed in an adaptive way as to take into account variations in the electron density.

The first derivatives of the energy with respect to nuclear displacements can be evaluated analytically (Fournier et al., 1989; Andzelm et al., 1989) using the expression

$$\partial E_{LSD}/\partial x = F_{HFB} + F_D \tag{31}$$

with F_{HFB} being the Hellman-Feynman force with a correction arising from the incompleteness of the orbital basis

$$F_{HFB} = \Sigma_{pq} P_{pq} \ \{\partial h_{pq}/\partial x + \Sigma_r \rho_r \ [\partial(pq)/\partial x||r] + \mu_{xc}(r) \ [\partial(pq)/\partial x] \\ + \partial U_n/\partial x - \Sigma_{pq} W_{pq} \ \partial[pq]/\partial x \tag{32}$$

and F_D being a term due to the incompleteness of the density fit,

$$F_D = \Sigma_r \rho_r \ [\partial(r)/\partial x \ || \ (\rho - \rho')] = \Sigma_r \rho_r \ [\partial(r)/\partial x \ || \ (\Sigma_{pq} P_{pq} \ pq - \Sigma_t \rho_t \ t)] \tag{33}$$

It is important to note that expressions (32) and (33) contain in essence derivatives of three-index integrals, which can be evaluated analytically in the same way as the integrals in the SCF calculations. The exchange-correlation term in eq. (32) is evaluated by a numerical integration which turns out to be more accurate than using the fitted form of μ_{xc}, provided a highly accurate grid is used. The matrix W is an energy weighted density matrix similar to that used in the gradients in Hartree-Fock theory.

Illustrative Examples

Cs Adsorption on a Mo(100) Surface

Interactions of alkali metals with transition metal surfaces play a fundamental role in a variety of technologically important phenomena including catalysis, electron emission, and ion formation (Wimmer et al., 1983). Density functional theory has proven to be a valuable tool for the quantitative description of these interactions and have lead to a new understanding of alkali adsorption on transition metal surfaces (Wimmer et al.,1982, 1983; Soukiassian et al., 1985; Chubb et al., 1987).

Since the pioneering work of Kingdon and Langmuir (1923) it had been assumed that alkali metals such as Cs are essentially completely ionized when adsorbed on a transition metal surface. Density functional calculations revealed a more complex picture. In these calculations (Chubb et al., 1987), the adsorption was modeled by a thin film geometry with five layers of Mo and Cs mono-layers on both sides of the film. The DFT equations were solved using the all-electron FLAPW method as described above. It turns out that the valence s-electron of the alkali metal form a covalent-polarized bond with characteristic surface states of the Mo(100) surface which is very similar to that on the W(100) surface (Wimmer et al., 1982, 1983).

This is illustrated by the surface energy band structure of the clean and cesiated Mo(100) surface (cf. Fig. 2). On a clean Mo(100) as well as on a W(100) surface, occupied surface states of d_z2-character exist just below the Fermi level. The orbitals associated with these states project into the vacuum region. Upon alkali metal adsorption, bonding and anti-bonding states are formed between these transition metal d-states and the alkali s-states. The calculations predict an alkali-induced shift of the surface states by about 1 eV (cf. Fig. 2). Independently, such a shift of 1 eV has been observed in angle-resolved photoemission experiments (Soukiassian et al., 1985 and references therein). The new bonding state has its strongest transition metal d-character at the point Γ of the surface Brillouin zone. Away from Γ, a partially occupied band with strong Cs-s character disperses upward which is characteristic for s-bands (cf. Fig. 2). Thus, the energy band structure demonstrates that Cs, in the high coverage regime, is not fully ionized.

The re-distribution of electronic charge upon adsorption of the alkali metals on a transition metal surface can be visualized by electron density difference maps. In Fig. 3, such a map is shown for the system Cs/Mo(100). In this analysis, the electron densities of a clean Mo(100) surface and that of a free-standing Cs monolayer are subtracted from the self-consistent charge of the Cs/Mo system. It can be seen (cf. Fig. 3) that the major redistribution of charge occurs in the region between the Cs and surface Mo atoms, i.e. the Cs valence s-electrons are polarized toward the surface, but stay in the vicinity of the Cs atoms. This leads to a built-up of charge above the surface Mo atoms, in a region where the Cs-6s functions would have their outer maximum, and a depletion of electronic charge between and above the Cs atoms. This charge rearrangement induces are strong electric dipole within the Cs overlayer which is the primary cause of the work function lowering upon alkali metal adsorption. It is interesting to note that the Cs-5p semi-core electrons are counter-polarized leading to an increase of electronic charge directly above the Cs atoms and a depletion just below the Cs atoms.

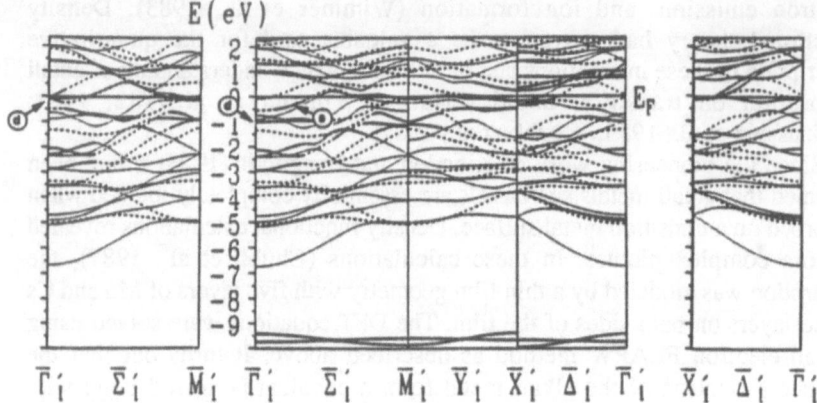

Figure 2. Energy band structure of a five-layer Mo(100) film, covered with Cs monolayers on both sides (Chubb et al, 1987). Only states with even symmetry with respect to mirror reflection on the central plane of the film are shown. The outer panels show the energy band structure for the clean Mo(001) film. The label 'd' refers to a characteristic Mo d-like surface state before and after cesiation and 's' marks a Cs-6s like band which hybridizes with the Mo surface states.

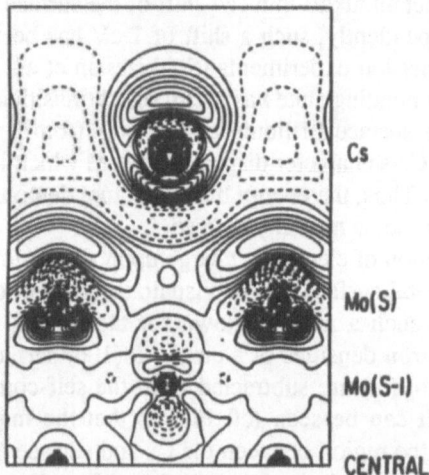

Figure 3. Electron density difference for a cesium monolayer adsorbed on a Mo(100) surface in a plane perpendicular to the surface. The map is obtained by subtracting the self-consistent electron density of a clean five-layer Mo(100) film and that of a free-standing Cs monolayer from the self-consistent electron density of the Cs/Mo system. Contours are evenly spaced with increments of 5×10^{-4} electrons/(a.u.)^3. Solid contours mark regions of increased electron density while dashed lines refer to a loss of electron density (Chubb et al. 1987).

CO Adsorption on a Ni(100) Surface

The interaction of CO with a Ni(100) surface represents a widely studied idealized model for heterogeneous catalysis. To this end, the electronic structure of CO adsorbed on a Ni(100) surface in the form of an ordered c(2x2) overlayer has been investigated using density functional theory and the FLAPW method (Wimmer et al., 1985). FLAPW calculations for a clean Ni(100) surface, represented by a five-layer film, and unsupported layers of CO molecules are used as reference systems. It is interesting to see that the FLAPW band structure approach, which was originally designed to describe energy band structures, also captures the electronic structure of molecular orbitals, as can be seen from a plot of the electron density of the highest occupied molecular orbital of the CO molecule (cf. Fig. 4).

The analysis of the electron density, similar to that performed for the Cs/Mo system, reveals that upon chemisorption of the CO molecule on a Ni(100) surface, σ-like charge is lost, especially in the C-O bond while the anti-bonding $2\pi^*$-state becomes partially occupied through back-donation of electrons from the transition metal surface to the adsorbed CO molecules. This change in the electronic structure, induced by the transition metal surface, causes a weakening of the C-O bond which can be seen as the first step in the catalytic breaking of the C-O bond.

Figure 4. Density of the highest occupied molecular orbital, 5σ, of a CO molecule in an unsupported monolayer geometry, obtained by the FLAPW band structure method (Wimmer et al., 1985).

Graphite Monolayer

One of the most important topics in the molecular sciences is the exploration of the total energy as a function of geometric conformations. Thus, the value of a theory hinges on its ability to predict accurately and reliably the energy hypersurfaces of molecules, molecular assemblies, and condensed phases. To this end, one of the first cases after the development of total energy capabilities in the thin-film FLAPW method (Weinert et al., 1982) was the investigation of the structural properties of a graphite monolayer. Because of the layered structure in a graphite crystal with its large inter-layer separations, the in-plane C-C bond distances can be expected to be very close to those obtained for a single graphite monolayer. In fact, upon thermal expansion of graphite, the in-plane C-C bond distances slightly decrease as the inter-layer distances increase. This suggests that the C-C bond length in a free-standing graphite monolayer is slightly shorter than in bulk graphite.

An optimization of the C-C bond length in a graphite monolayer through a series of single-point total energy calculations yields a C-C bond length of 2.450 Å. This is in excellent agreement with the experimental value of 2.461 Å. Such a good performance of the local density approximation is, to some extent, surprising, since the electron density in the C-C bond of graphite is far from being uniform. It should be pointed out, though, that the delocalized nature of the π-electrons in graphite lends itself for a density functional picture.

It was found (Weinert et al., 1982) that the predicted C-C bond distance is rather insensitive to the quality of the basis set whereas convergence in the value of the binding energy requires a fairly large basis set. Using a fully converged basis set, the cohesive energy of a graphite monolayer was calculated to be almost 1 eV too large compared with experiment. This over-binding in LDA calculations has been found for many cases and appears to be intrinsic to this approximation. It could be argued that the larger error actually occurs in the free atoms which define the reference for the cohesive energy. In fact, the inclusion of the most important configuration-interaction terms for the C atom lowers its total energy by about 1 eV (Weinert et al., 1982) compared with the atomic state in the LDA picture.

Conformations of a Biologically Active Molecule

Knowledge of the molecular conformations accessible under biological conditions is of fundamental importance in structure-based drug design. To investigate the importance of an adequately accurate level of theory, Wimmer and Delley (unpublished) have studied the conformational properties of compound 1, which belongs to the class of imidazolinone herbicides (Los, 1984) . For simplicity, R=H was assumed in the calculations.

1

One factor that determines the conformation of compound **1** is the electrostatic repulsion between the nitrogen and oxygen atoms in the five-membered ring and the carboxylic acid residue. This repulsion drives the planes of the five-membered ring as well as that of the carboxy group out of the plane of the pyridine ring. On the other hand, one can expect some small π-bonding components between the two ring systems, which would energetically favor a co-planar conformation. Given this electronic effect, a simple force-field can be expected to be unreliable in predicting the ground state structure of this compound. Optimization of the structure on the semi-empirical level using the MOPAC program (Stewart, 1988) with an AM1 Hamiltonian yields dihedral angles τ_1 and τ_2 of 35 and 54 degrees, respectively. Optimization at the HF/STO-3G level using the Gaussian program (Frisch et al., 1988) yields values of 17 and 84 degrees, whereas 16 and 74 degrees are obtained with a 3-21G basis. Going from semi-empirical quantum mechanics to ab initio theory, the computing time increases from a few minutes to several hours. It is interesting to note that local density functional theory, using a precursor of the the DMol program (Delley, 1990) with a DZ-like basis gave values of 20 and 69 degrees which are in fairly good agreement with the experimental values of 17 and 67 degrees obtained from x-ray data (Los, 1984) on a crystal of an ester of this acid (with R=isopropyl). From these findings it can be concluded that Hartree-Fock theory as well as density functional theory agree to within about 5 degrees for the two sensitive dihedral angles, whereas the semi-empirical result differs more substantially. The good agreement between the Hartree-Fock, density functional, and x-ray results indicate that crystal packing and substitution effects in the R-group seem to play a minor role in determining the structure of this herbicide.

LDF Approach to Metal-Protein Interactions: Zn-Insulin

Despite the importance of metal ions in biological systems, many structural and electronic properties related to their interactions with their biochemical

Table 2. Computational parameters and requirements for Gaussian-type orbital DFT calculations on Zn-insulin fragments of increasing size. Im denotes an imidazole ring, His refers to the complete amino acid histidine, and HSG represents the amino acid sequence histidine-serine-glycine. TZd denotes a triple zeta fitting set including d-functions. A direct scheme has been used for the integral evaluation. The computing time refers to 20 iterations in the SCF procedure and has been measured as wall-clock time on an 8-processor CRAY Y-MP system with 32 MW of memory.

	Zn-Im$_3$	Zn-His$_3$	Zn-(HSG)$_3$
Atoms	28	58	112
Electrons	136	250	478
Basis functions (6-31G*)	278	533	1031
Fitting functions (TZd)	665	1295	2513
Grid points	15900	32700	62900
Computing time (minutes)			
1 Y-MP processor	6	21	85
8 Y-MP processors	1	4	14

environments remain open. One of the reasons, so far, has been the lack of theoretical and computational approaches that would provide an adequate level of accuracy for such complex systems. While Hartree-Fock theory has proven to be valuable for predicting structural and vibrational properties of small organic molecules, the investigation of larger molecular structures and the occurrence of metal atoms presents a considerable challenge to this approach. It is well known that the SCF level of Hartree-Fock theory is not sufficient for systems involving metals and the inclusion of electron correlation is needed. However, correlated methods are computationally extremely demanding and thus are restricted to only small molecules which may not adequately represent the structural complexity of biomolecular systems.

In order to test the applicability of DFT calculations for large systems such as the binding of metal ions in proteins and to probe the computational requirements for this kind of systems, fragments of a Zn-insulin complex were investigated using the Gaussian-type orbital approach (Andzelm, unpublished). In the Zn-insulin crystal (Shields, private communication), the Zn ions are coordinated by three N atoms and three water molecules to form a distorted octahedral complex. The N atoms are part of imidazole rings which belong to histidine residues of three different insulin molecules.

In the present computational study, three models of these Zn-insulin complexes have been considered: (A) Zn-Im$_3$, where Im represents an imidazole molecule, (B) Zn-His$_3$, where His is the complete histidine amino acid , and (C) Zn-(HSG)$_3$, where HSG denotes a sequence of the three amino acids histidine, serine, and glycine. Model (C) contains a total of 112 atoms. No symmetry was used in the calculations. Table 2 summarizes the computational aspects for these calculations.

The analysis of the electrostatic potential around the Zn ion reveals that the qualitative features of electropositive and electronegative regions near the Zn ion are present already in model (A). However, a quantitative description of the electrostatic potential, as might be expected, requires a fragment at least as big as model (C). Table 2 shows that fairly accurate DFT calculations for this size of systems is now becoming possible, provided that adequate computational power and efficient algorithmic implementations are used.

Summary and Outlook

Density functional theory, originally intended for band structure calculations on crystalline solids, is now becoming an increasingly important tool in the investigation of molecular systems. In this contribution, the development and concepts of density functional theory were reviewed and compared with those of Hartree-Fock theory. Different algorithmic and computational implementations of density functional theory were discussed, especially the full-potential linearized augmented plane-wave method and a Gaussian-type orbital approach. Solid state, surface, and molecular applications of density functional calculations were presented to illustrate the strengths and limitations of the various implementations. The examples included the adsorption of atoms and molecules on metal surfaces (Cs on Mo and CO on Ni surfaces), the geometry of a graphite monolayer, the conformation of organic molecules with biological activity, and calculations on large Zn-insulin fragments. The results indicate that the local density functional approximation reproduces geometries to within about 0.02 Å and dihedral angles to within a few degrees. This accuracy is comparable with the Hartree-Fock MP2 level of theory. As might be expected, DFT describes rather well the charge distributions and electrostatic potentials for solid state systems as well as organic and biological molecules.

Three major factors make the DFT approach appealing: (i) the agreement of calculated DFT geometries and electronic properties are comparable with traditional correlated molecular orbital methods, (ii) DFT has been demonstrated to be applicable to all atoms of the periodic table and to various types of bonding with consistent accuracy, and (iii) the scaling of the computational effort with the molecular size is equal or less than a third power, which is in marked contrast to traditional correlated molecular orbital methods.

However, currently there are several major shortcomings of practical DFT implementations: (i) there is no systematic way to improve the local density approximation (LDA) towards the exact solution, (ii) in the LDA, binding energies are substantially over-estimated, (iii) there are no implementations of analytic second derivatives as necessary for accurate molecular vibrational analysis and transition state searches, (iv) there is a lack of experience for molecular systems that would allow to establish error bars for the prediction of molecular geometries and energies.

It can be expected that the current successes of DFT will stimulate research in these areas so that these shortcomings can be eliminated. One of the most critical, but also most difficult obstacles is the lack of systematic improvements beyond the LDA. Perhaps detailed comparisons between highly precise correlated methods such as full CI calculations and DFT calculations could provide at least a heuristic guideline on how to improve the LDA. To this end, it will be necessary to eliminate all computational errors not connected to the LDA originating, for example, from basis functions and numerical grids. Clearly, intense research in this important area is called for.

Given the computational speed and efficiency of today's DFT programs on present computational hardware, it is intriguing to speculate on the size and kind of systems that might become tractable in the future. If we assume that during the 1990's hardware performance will increase by two orders of magnitude using advanced parallel architectures and if we assume that improvements in algorithmic implementations will lead to another order of magnitude in gain of computational efficiency, then by the end of this decade, DFT calculations on systems with several thousand atoms might become possible. Perhaps more intriguingly, it might be possible to study dynamic phenomena including chemical and enzymatic reactions in a way similar to current molecular dynamics simulations, but employing DFT forces rather than empirical force-fields. Using today's technology, an energy and force evaluation for a given geometry for a system consisting of 100 atoms can be done within about 1 hour of computing time. A gain of three orders of magnitude would reduce this to the time scale of seconds. If time intervals of 1 femto second are used, which is typical of today's molecular dynamics simulations, the evolution of a 100 atom system over 100 ps could be calculated in less than 100 hours. Thus, molecular dynamics simulations on the DFT level might become practical within this decade. In fact, ideas by Car and Parrinello (Car and Parrinello, 1985) already point in this direction.

Acknowledgements

The author is grateful to A. Neckel and his group in Vienna for sparking the interest in density functional theory in the early 1970's. He would also like to thank A. J. Freeman for his continuous encouragements and enthusiasm, B.

Delley and J. Andzelm for many fruitful conversations on the implementations of DFT methods, D. A. Dixon for his critical and constructive comments as well as his industrial perspective of DFT applications, and all colleagues at Cray Research, Inc. who have created an exciting and stimulating environment.

References

Amos, R., 1984, *Chem. Phys. Lett.* 108: 347

Andersen, O. K., and Woolley, R. G., 1973, *Mol. Phys.* 26:905

Andersen, O. K., 1975, *Phys. Rev.* B12:3060

Andzelm, J., Wimmer, E., and Salahub, D. R., 1989, *The Challenge of d and f Electrons. Theory and Computation*, ACS Symposium Series 394:228

Averill, F. W., and Ellis, D. E., 1973, *J. Chem. Phys.* 59:6412

Baerends, E. J.,Ellis, D. E., and Ros, P., 1973, *Chem. Phys.* 2:41

Borman, S., 1990, *Chemical and Engineering News*, April 9, p.22

Car, R., and Parrinello, 1985, *Phys. Rev. Lett.* 55:2471

Ceperley, D. M., and Alder, B. J.,1980, *Phys. Rev. Lett.* 45:566

Chubb, S. R., Wimmer, E., Freeman, A. J., Hiskes, J. R., and Karo, A. M., 1987, *Phys. Rev.* B36:4112

Delley, B., and Ellis, D. E., 1982, *J. Chem. Phys.* 76:1949

Delley, B., 1990, *J. Chem. Phys.* 92:508

Dixon, D. A., Capobianco, P. J., Mertz, J. E., and Wimmer, E., 1988, *Science and Engineering on Cray Supercomputers*, Proceedsings of the Fourth International Symposium, Cray Research, Minneapolis, MN, p.189

Dixon, D. A., Andzelm, J., Fitzgerald, G., Wimmer, E., and Delley, B, 1990, this volume

Dixon, D. A., Andzelm, J., Fitzgerald, G., and Wimmer, E., to be published

Dunlap, B. I., Connolly, J. W. D., and Sabin, J. R., 1979, *J. Chem. Phys.* 71:4993

Dunlap, B. I., 1986, *J. Phys. Chem.* 90:5524

Dunning, T. H. Jr., 1970, *J. Chem. Phys.* 53:2823

Dunning, T. H. Jr., and Hay, P. J., 1977, *Methods of Electronic Structure Theory*, Ch. 1., (Edited by Schaefer, H. F. III), Plenum Press, New York

Dupuis, M., Rys., J., and King, H. F., 1976, *J. Chem. Phys.* 65:111

Fock, V., 1930, *Z. Phys.* 61:126; ibid. 62: 795

Fournier, R., Andzelm, J., and Salahub, D. R., 1989, *J. Chem .Phys.* 90:6371

Frisch, M. J., Head-Gordon, M., Schlegel, H. B., Raghavachari, K., Binkley, J. S., Gonzalez, C., DeFrees, D. J., Fox, D. J., Whiteside, R. A., Seeger, R., Melius, C. F., Baker, J., Martin, R. L., Kahn, L. R., Stewart, J. J. P., Fluder, E. M.,Topiol, S., and Pople, J. A., 1988, Gaussian 88, (Gaussian Inc., Pittsburgh, PA)

Gáspár, R., 1954, *Acta Phys. Acad. Sci. Hung.* 3: 263

Gunnarsson, O., Harris, J., and Jones, R. O., 1977, *Phys. Rev.* **B15**:3027

Gunnarsson, O., Harris, J., and Jones, R. O., 1977, *J. Chem. Phys.* **67**:3970

Hartree, D. R., 1928, *Proc. Cambridge Phil. Soc.* **24**: 89

Head-Gordon, M., and Pople, J. A., 1988, *J. Chem. Phys.* **89**:5777

Hedin, L., and Lundqvist, S., 1972, *J. Phys. (France)* **33**:C3-73

Hehre, W. J., Radom, L., Schleyer, P., and Pople, J. A., 1986, *Ab initio molecular orbital theory*, John Wiley & Sons, NewYork

Herzig, P., 1974, *Ph.D. Thesis, University of Vienna*

Hohenberg, P., and Kohn, W., 1964, *Phys. Rev.*, **136**: B 864

Huzinaga, S., Andzelm, J., Klobukowski, M., Radzio, E., Sakai, Y., and Tatasaki, H., 1984, *Gaussian Basis Sets for Molecular Calculations*, Elsevier, Amsterdam

King, H. F., and Dupuis, M., 1976, *J. Comp. Phys.* **21**:144

Kingdon, K. H., and Langmuir, I., 1923, *Phys. Rev.* **21**:380

Kohn, W., and Rostoker, N., 1954, *Phys. Rev.* **94**:1111

Kohn, W., and Sham, L. J., 1965, *Phys. Rev.* **140**: A1133

Korringa, J. 1947, *Physica* **13**:392

Krakauer, H., Posternak, M., and Freeman, A. J., 1978, *Phys. Rev. Lett.* **41**:1072

Los, M., 1984, *ACS Symposium Series* **225**:29

McIver, J. W., and Komornicki, A., 1971, *Chem. Phys. Lett.* **10**:303

Moruzzi, V. L., Janak, J. F., and Williams, A. R., 1978, *Calculated Electronic Pro-perties of Metals*, Pergamon, New York

Neckel, A., Rastl, P., Eibler, R.,Weinberger, P., and Schwarz, K., 1976, *J. Phys. C:*

Solid St. Phys. **9**:579

Obara, S., and Saika, A., 1986, *J. Chem. Phys.* **84**:3963

Pines, D., 1955, *Solid State Physics* **1**:367

Ravenek, W., and Baerends, E. J., 1984, *J. Chem. Phys.* **81**:865

Pople, J. A., Krishnan, R., Schlegel, H. B., and Binkley, J. S., 1979, *Int. Quant. Chem. Symp.* **13**:225

Pulay, P., 1969, *Mol. Phys.* **17**:197

Rosén, A., and Ellis, D. E., 1976, *J. Chem. Phys.* **65**:3629

Rys, J., Dupuis, M., and King, H. F., 1983, *J. Comp. Chem.* **2**:154

Salahub, D. R., 1987, in *Ab Initio Methods in Quantum Methods in Quantum Chemistry-II*, ed. Lawly, K.P, Wiley & Sons, New York, p.447

Sambe, H., and Felton, R. H., 1974, *J. Chem. Phys.* **61**:3862

Sambe, H., and Felton, R.H., 1975, *J. Chem. Phys.* **62**:1122

Schlegel, B., 1982, *J. Comp. Chem.* **3**:214

Schwarz, K., 1972, *Phys. Rev.* **B5**:2466

Shields, J., private communication

Slater, J. C., 1937, *Phys. Rev.* **51**:846

Slater, J. C., 1951, *Phys. Rev.* **81**: 385

Slater, J. C., 1974, *Quantum Theory of Molecules and Solids Vol. 4*, McGraw-Hill, New York

Soukiassian, P., Riwan, R.,Lecante, J., Wimmer, E., Chubb, S. R., and Freeman, A. J., 1985, *Phys. Rev.* B31:4911

Stevens, R. M., Pitzer, R. M., and Lipscomb, W. N., 1963, *J. Chem. Phys.* 38:550

Stewart, J. J. P., 1988, MOPAC, A General Molecular Orbital Package, QCPE program #455

Teter, M. P., Payne, M. C., and Allan, D. C., 1989, *Phys. Rev.* B40:12255

Vosko, S. H., Wilk, L., and Nusair, M., 1980, *Can. J. Phys.* 58:1200

Weinert, M., 1981, *J. Math. Phys.* 22:2433

Weinert, M., Wimmer, E., and Freeman, A. J., 1982, *Phys. Rev.* B26:4571

Wigner, E. P., 1934, *Phys. Rev.* 46:1002

Wigner, E. P., 1938, *Trans. Faraday Soc.* 34:678

Wimmer, E., 1979, *Theoret. Chim. Acta (Berl.)* 51:339

Wimmer, E., Krakauer, H., Weinert, M., and Freeman, A. J., 1981, *Phys. Rev.* B24:864

Wimmer, E., Freeman, A. J., Weinert, M., Krakauer, H., Hiskes, J. R., and Karo, A. M.; 1982, *Phys. Rev. Lett.* 48:1128

Wimmer, E., Freeman, H., Hiskes, J. R., and Karo, A. M.; 1983, *Phys. Rev.* B28:3074

Wimmer, E., Fu, C.L., and Freeman, A. J., 1985, *Phys. Rev. Lett.* 55:2618

Wimmer, E., Freeman, A. J., Fu, C.-L., Cao, P.-L., Chou, S.-H., and Delley, B., 1977, *ACS Symposium Series* 353, Jensen, K. F.; Truhlar, D. G., editors, p.49

Ziegler, T., and Versluis, L.,1988, *J. Chem. Phys.* 88:322

3
Benchmark and Testing of the Local Density Functional Method for Molecular Systems

DAVID A. DIXON, JAN ANDZELM, GEORGE FITZGERALD, ERICH WIMMER, AND PAUL JASIEN

Introduction

The prediction of molecular properties from computer simulations is playing an increasingly important role in the chemical and pharmaceutical industry (Dixon, 1987; Dixon et al., 1988a; 1988b). Although molecular modeling groups were already popular in the 1970's with the promise of solving the rational drug design problem in a quantitative way, the available computational resources and software were not adequate for the task. With the more ready access by chemists to supercomputers in the 1980's and the revolutions in theoretical developments, algorithm design and software implementation, it is now possible for computational science to have a quantitative impact on molecular design. The primary areas that have been responsible for this renaissance have been the development of force field methods including molecular dynamics and the more routine application of quantum chemical methods (Hehre et al., 1986) to increasingly more complex molecular systems of interest to the bench chemist. This chapter will focus on the latter application area.

The workhorse for quantum chemical applications has been the molecular orbital method with the largest number of calculations being done at the Hartree-Fock (HF) self-consistent field (SCF) level. In order to obtain good energetics of the accuracy required to make reasonable predictions for industrial applications, some form of correlation energy correction is usually required. Because of the size of the systems under consideration which often involve many electron pairs, correlation is usually included at the level of second order Møller-Plesset perturbation theory, MP-2 (Møller et al., 1934; Pople et al., 1976). The usual type of calculation is first to perform a Hartree-Fock geometry optimization with an adequate basis set, usually double-zeta augmented by polarization functions, and then to calculate a correlation energy correction at the optimum Hartree-Fock geometry (Dixon et al., 1990). The molecular force field can be calculated analytically at the SCF level at the optimum geometry. On occasion, one can use correlated methods to calculate the geometry and the frequency, but these are quite expensive computationally even at the MP-2 level. Higher order correlation corrections are much more expensive computationally and thus can be done only for smaller molecules. The computational expense for the HF calculations comes from its scaling as N^4 where N is the number of basis functions. Although this scaling can be smaller for extended molecules because of integral truncation, it is still a limiting factor. The correlation calculations scale as N^m where the value of m is ≥ 5 because even for the simplest correlation treatment such as MP-2, a sort and transformation of the integrals is required. Higher order correlation treatments have an even worse scaling factor.

Computational resources are not improving as the fourth power, and there is a search on for new computational methods that do not scale with such a large exponent but are based on rigorous theory with no empirical parameters. One potential method is the local density functional (LDF) method (Parr et al., 1989) which has been used with great success in solid state and surface physics. It has the advantage of scaling as N^3 and, as in the case of the Hartree-Fock method, may have a lower exponent depending on implementation. It is based on rigorous physics and contains no empirical parameters. Because of its form, it does include correlation effects in its calculation of the energy. A disadvantage at present is that it has not been widely applied to a broad range of chemical systems, and so there is no history of experience on molecular systems as has been built up from traditional molecular orbital methods. Furthermore, it is only very recently that it has been implemented in software that is becoming more widely available (Salahub, 1987; Wimmer et al., 1987; Jones et al., 1989; Dixon et al., 1990). It also has suffered because only recently (Delley, private communication; Versluis et al., 1988; Fournier et al., 1989) have analytic derivative methods become available so geometry optimization of asymmetric molecules has been very difficult and time-consuming. Yet chemists are very interested in molecular structure and are interested in obtaining the best energies as possible. (An error of a few kcal/mol due to errors in the geometry can change whether a chemical process is of technological interest to pursue experimentally or not.)

In this chapter, we focus on applications of two approaches to solving the LDF equations and compare these results to other calculations and experiment where possible. We are interested in providing some benchmark results to demonstrate the ability of the LDF method to treat complex chemical systems for a variety of properties including geometries, force fields and reaction energies. It is well established that non-local corrections need to be applied if bond-breaking occurs in a chemical reaction as discussed elsewhere in this book. We present the first examples of the application of non-local corrections to the calculation of isodesmic reaction energies. Because the methods are well-discussed in other chapters, we will focus on the results of our simulations.

Methods

The calculations described below were done with the program systems DMol (Delley, 1990) and DGauss (Andzelm et al., 1989; Andzelm et al. 1990).

DMol employs numerical functions for the atomic basis sets. The atomic basis functions are given numerically on an atom-centered, spherical-polar mesh. The radial portion of the grid is obtained from the solution of the atomic LDF equations by numerical methods. The radial functions are stored as sets of cubic spline coefficients so that the radial functions are piece-wise analytic, a necessity for the evaluation of gradients. The use of exact spherical atom results offers some advantages. The molecule will dissociate exactly to its atoms within the LDF framework, although this does not guarantee correct dissociation energies. Furthermore, because of the quality of the atomic basis sets, basis set superposition effects should be minimized and correct behavior at the nucleus is obtained.

Since the basis sets are numerical, the various integrals arising from the expression for the energy equation need to be evaluated over a grid. The integration points are generated in terms of angular functions and spherical harmonics. The number of radial points N_R is given as

$$N_R = 1.2 \cdot 14(Z + 2)^{1/3} \qquad (1)$$

where Z is the atomic number. The maximum distance for any function is 12 a.u. The angular integration points N_O are generated at the N_R radial points to form shells around each nucleus. The value of N_O ranges from 14 to 302 depending on the behavior of the density (This grid can be obtained by using the FINE parameter in DMol.) The Coulomb potential corresponding to the electron repulsion term could be solved by evaluation of integrals. However, since the method is based on the density, it was found to be more appropriate to determine the Coulomb potential directly from the electron density by solving Poisson's equation.

$$- \nabla^2 V_e(r) = 4\pi e^2 \rho(r) \qquad (2)$$

In DMol, the form for the exchange-correlation energy of the uniform electron gas is that derived by von Barth and Hedin (von Barth et al., 1972).

All of the DMol calculations were done with a double numerical basis set augmented by polarization functions. This can be considered in terms of size as a polarized double zeta basis set. However because we are using exact numerical solutions for the atom, this basis set is of significantly higher quality than a normal molecular orbital double zeta basis set. The fitting functions were all done using an angular momentum number one greater than that of the polarization function. For hydrogen with a p polarization function, a value of $l = 2$ was used in the fitting functions. For all of the other atoms which had d polarization functions, a value of $l = 3$ was used for the fitting function.

An efficient, analytical approach to variationally fit the density has been developed previously (Dunlap et al., 1979; Andzelm et al., 1989). The analytical implementation of the LDF method requires that Gaussian basis sets be used. Additionally, this allows one to build on the wealth of experience gained from Hartree-Fock molecular orbital calculations. In the DGauss program (Andzelm et al., 1989; Andzelm et al., 1990), one can use the same type of basis sets as used in modern Hartree-Fock methods. One can also apply analytical techniques from ab initio methods such as those employed in evaluating the gradient of the energy for geometry optimizations in the LDF approach. To this end, Cartesian Gaussians are used as primitives and are contracted in the same way as in Hartree-Fock methods. In fact, even the familiar basis sets developed by the Pople group can be used (Hehre et al., 1986). The Gaussian basis sets employed in the DGauss calculations are summarized in Table 1 together with the grid information. Although the experience about basis sets gained from Hartree-Fock molecular calculations is invaluable in terms of defining the basis set size and the need for additional polarization and diffuse functions, Hartree-Fock basis sets used in an LDF calculation may exhibit large basis set superposition errors (BSSE). In order to minimize BSSE, LDF optimized basis sets were developed (Andzelm et al., 1985; Dunning et al, 1977; Huzinaga et al., 1989) and these are the ones presented below.

The approach used in DGauss is an analytical implementation of the LDF method as opposed to the purely numerical approach taken in DMol. In the analytical approach, the density can be variationally fit (Dunlap et al., 1979) leading to exact Coulomb forces (Andzelm et al, 1989; Fournier et al., 1989) The remaining exchange-correlation

Table 1. Basis Sets for DGauss Calculations

Basis	Size	Contraction	Fitting Set	Grid[a]
DG-1	9/5/1	621/41/1	TZ	M
DG-2	9/5/1	621/41/1	TZ	F
DG-3	9/5/2	621/41/2	TZ	M
DG-4	9/5/2	621/41/2	QZ	M
DG-5	10/6/1	721/51/1	TZ	M
DG-6	10/6/1	721/51/2	QZ	M
DG-7	10/6/2	721/51/2	QZ	F
DG-8	11/7/1	821/61/1	TZ	M
DG-9	10/6/2	7111/411/2	QZ	M

[a] M = MEDIUM grid. F = FINE grid

energy term is a smooth function of the density and can be accurately fit on a small, adaptive set of grid points. It is this numerical evaluation of the exchange-correlation energy term that leads to the differences in the geometries obtained from the minimum in the energy and the one with zero gradient as discussed below. A new method for variational exchange-correlation fitting has recently been developed (Dunlap et al., 1990) which may alleviate the inherent errors in the gradient calculations of the exchange-correlation forces although this method is not implemented yet in DGauss.

An additional fitting basis set is required to fit the electron density as obtained from the molecular orbitals and the exchange-correlation potential and energy. This additional basis set is expanded in a set of atom-centered Gaussian type basis functions with s,p and d character. It is found that even-tempered expansions of triple-zeta(TZ) and quadruple-zeta(QZ) quality are adequate to reproduce the electron density. The coefficients in this density expansion are determined from a variational condition that requires that the residual (second order) Coulomb energy term arising from the difference between the exact and the fitted density be minimized. In practice, this leads to analytic expressions for the fitting coefficients involving Coulomb-type three-index two-electron integrals.

The exchange-correlation potential is also expanded in Gaussian-type basis functions. However, the determination of the expansion coefficients is carried out numerically using a grid similar to the one described above for the DMol program. The grid selection in DGauss is derived from an adaptive procedure based on the approximation of the exchange-correlation energy in the angular shell around each atom. The radial distribution of points is accomplished by using the method of Becke (Becke, 1988). Once the Gaussian expansions of the exchange-correlation potential are determined, the matrix elements involving the three-index, one-electron integrals are calculated. To this end, the DGauss approach leads to an N^3 method which in practice can be made close to N^2 by using sparse, direct matrix algorithms for the integral calculations, numerical integration and density synthesis. This can be compared to the four-index, two-electron integrals required for the Hartree-Fock method which formally scale as N^4. All three-index, two-electron integrals are calculated by using an algorithm based on the work of Obara and Saika (Obara et al., 1986). In DGauss, the form for the exchange-correlation energy of the uniform electron gas is that derived by Vosko, Wilk and Nusair (Vosko et al., 1980).

Geometries were determined by optimization using analytic gradient methods. First derivatives in the LDF framework can be calculated efficiently and only take on the order of 3-4 SCF iterations or 10-25% of the calculation of the energy. In the DGauss implementation, the evaluation of the gradient requires about 5 SCF iterations or about 30-40% of the time required for the energy evaluation. This is in contrast to the evaluation of the derivatives in traditional molecular orbital methods which usually takes at least a factor of 1-2 times the evaluation of the energy (Komornicki et al, 1977; Pulay, 1977; Jorgenson et al. 1986). There are two problems with evaluating gradients in the LDF framework which are due to the numerical methods that are used. The first is that the energy minimum does not necessarily correspond exactly to the point with a zero derivative. The second is that the sum of the gradients may not always be zero as required for translational invariance. These tend to introduce errors on the order of 0.001 Å in the calculation of the coordinates if both a reasonable grid and basis set are used. This gives bond lengths and angles with reasonable error limits. The difference of 0.001 Å is about an order of magnitude smaller than the accuracy of the LDF geometries as compared to experiment.

Results and Discussion

The first application is the calculation of the structures and energetics of the fluorinated methanes (Dixon, 1988) The structural data for the bond lengths is given in Table 2 and the energetics for a series of isodesmic reactions are given in Table 3. The general trends that can be observed in Table 2 for the C-F bond lengths are that increasing fluorination leads to a decrease in the bond length. The two theoretical methods agree with experiment in this with the experimental shortening of 0.064 Å going from CFH$_3$ to CF$_4$ bracketed by the LDF shortenings of 0.055 Å (Gaussian) and 0.058 Å (numerical) and the HF shortening of 0.066 Å. The numerical LDF values are longer than the experimental values by 0.007 to 0.013 Å and the Gaussian LDF values are longer by 0.001 to 0.011 Å. By comparison, the HF values are 0.016 to 0.018 Å shorter than experiment. The longer LDF values are typical of other work in fluorocarbons which shows that correlation leads to bond lengths longer than experiment (Dixon et al., 1986). The C-H bond lengths show different trends. There is no regularity to the experimental measurements due in part to the different types of experiments in use and to the difficulty in measuring the exact position of the light H atom. The SCF values are shorter than experiment and exhibit the trend of a decrease in the bond length with increasing fluorination. The LDF values are longer than experiment and exhibit a different trend with an increase in the C-H bond length with increasing fluorination up to 2 fluorines. This difference in predicted behavior awaits experimental confirmation.

Although good geometrical predictions might be expected from the LDF method, it was not at all clear if one could use the total energies in the prediction of reaction energies. A set of isodesmic reactions involving the fluoromethanes is given in Table 3. For this series of reactions, the molecular orbital method at both the HF and correlated level with a triple zeta basis set augmented with two sets of

Table 2. Bond Lengths for the Fluorinated Methanes (A)

Molecule	EXPT	r(C-F) LDF DMol	LDF DG-5	HF
CH$_3$F	1.383	1.390	1.384	1.367
CH$_2$F$_2$	1.357	1.366	1.362	1.339
CHF$_3$	1.332	1.346	1.343	1.316
CF$_4$	1.319	1.332	1.329	1.301
		r(C-H)		
CH$_4$	1.092	1.097	1.100	1.084
CH$_3$F	1.100	1.102	1.106	1.082
CH$_2$F$_2$	1.078	1.104	1.109	1.080
CHF$_3$	1.098	1.104	1.108	1.077

a See (Dixon, 1988) for details of experimental and HF (Hartree-Fock) results. HF geometries calculated with the polarized double-zeta basis set. Bond distances in Å.

polarization functions gives good agreement with experiment if the results are corrected for zero point energy effects which are on the order of 2 kcal/mol for some of these reactions. The SCF results with a basis set including only polarization functions on C do not agree as well with experiment although if polarization functions are added to all atoms good agreement can be obtained. The numerical LDF results are within 2 kcal/mol of all of the reaction energies and are within the error bars for all of the reactions except for Reaction 1 which has a very small error. The Gaussian LDF results show slightly worse agreement but still yield reasonable values. They show slightly worse agreement than the HF or correlated results with the TZ+2P basis set but are significantly better than the SCF results with the DZ+Dc basis set.

In order to provide more information about the behavior of the numerical LDF method, we varied both the number of grid points and the degree of angular truncation (l') in the evaluation of the Coulomb potential. The results are shown in Table 4 for the bond distances and in Table 5 for the isodesmic reaction energies. As shown in Table 4, within 0.001 Å there is convergence in the grid when comparing the FINE and MEDIUM grids for the same value of l' for the bond

Table 3. Fluorinated Methanes - Isodesmic Reaction Energies ΔH (kcal/mol) [a]

Reactions

(1)	$CH_4 + CF_4$	---->	$2CH_2F_2$
(2)	$CH_4 + CF_4$	---->	$CHF_3 + CH_3F$
(3)	$CH_4 + CH_2F_2$	---->	$2CH_3F$
(4)	$CH_3F + CHF_3$	---->	$2CH_2F_2$
(5)	$CH_2F_2 + CF_4$	---->	$2CHF_3$
(6)	$CH_2F_2 + CH_3F$	---->	$CH_4 + CHF_3$

RXN	LDF DMol	LDF DG-5	LDF DG-5/B-P	HF (DZD$_c$)	HF (TZ+2P)	HF (TZ+2P) +MP-2	EXPT
1	26.0	27.8	21.0	17.5	23.7	24.2	24.5±1.0
2	18.8	20.1	15.1	13.7	16.9	17.5	17.3±2.7
3	14.6	15.3	12.3	10.7	11.5	12.9	12.5±3.5
4	7.2	7.7	5.8	4.7	6.7	6.6	7.2±4.1
5	-2.9	-2.8	-3.5	-2.5	-1.3	-2.0	-2.4±1.8
6	-21.7	-23.0	-18.1	-15.5	-18.3	-19.6	-19.9±2.6

[a] See (Dixon, 1988) for experimental and HF values. MP-2 calculated at the valence level.

Table 4. Bond Distances (Å) as a Function of Grid and Fit for DMol

Molecule	FINE $l' = 1 + 1$	FINE $l' = 1$	MEDIUM $l' = 1 + 1$	MEDIUM $l' = 1$
		r(C-F)		
CH_3F	1.390	1.383	1.390	1.386
CH_2F_2	1.366	1.362	1.365	1.361
CH_3F	1.346	1.342	1.346	1.342
CF_4	1.332	1.329	1.333	1.330
		r(C-H)		
CH_4	1.097	1.097	1.097	1.097
CH_3F	1.102	1.097	1.102	1.098
CH_2F_2	1.104	1.100	1.104	1.101
CHF_3	1.104	1.100	1.104	1.10

distances. However there is clearly a difference in the bond distances which is dependent on the value of l'. A value of $l' = 1$ where 1 is the angular momentum of the polarization function gives shorter bond distances than do values of $l' = 1 + 1$. This is most pronounced for the C-F bond distance in CH_3F.

In Table 5, the reaction energies (not the reaction enthalpies, ΔH, given in Table 3) are compared. Here one can see that the reaction energies are converged in terms of l' for a given grid but that the grid fit can yield differences of up to 1 kcal/mol. The lower quality grid predicts reaction energies that are worse when compared to experiment than found with the higher quality grid.

Table 5. Reaction Energies (ΔE in kcal/mol) as a Function of Grid and Fit for DMol

Reaction	EXPT [a]	FINE $l' = 1 + 1$	FINE $l' = 1$	MEDIUM $l' = 1 + 1$	MEDIUM $l' = 1$
1	22.1	23.6	23.6	24.6	24.6
2	15.5	17.0	17.0	17.4	17.4
3	11.9	14.0	14.0	13.8	13.8
4	6.6	6.6	6.6	7.2	7.2
5	-3.0	-3.5	-3.5	-3.6	-3.7
6	-18.7	-20.5	-20.6	-21.0	-21.0

[a] Dixon, 1988

A variety of basis sets were tested with the Gaussian LDF code in order to determine the optimum compromise between speed and accuracy. The first basis sets that were tested were the standard 6-31G** and 6-311G** basis sets developed by Pople and coworkers.(Hehre et al., 1986). Since the first reaction in Table 2 is the most difficult to calculate, we focussed on optimizing the geometries for the molecules in this reaction and predicting the energy of the reaction. As shown in Table 6, the methods all predict about the same bond distance for the C-H bonds. The C-F bond is a little more sensitive to improvement in the quality of the basis set with the 6-31G** and 6-311G** basis sets giving shorter bond lengths than the other optimized basis sets. Only basis set DG-9 shows any difference in comparison to the other LDF optimized basis sets giving a slightly longer bond. Although the 6-31G** and 6-311G** basis sets do an adequate job of predicting the molecular geometries, they fail in predicting the energy for reaction 1 giving a result that is too positive. In contrast, the LDF optimized basis sets show much better agreement with the experimental value. The results are most sensitive to the number of contracted functions in the core orbitals. The smaller basis sets contracted from 6 primitives in the core show an error of about 4 kcal/mol whereas an improved treatment of the core by just one additional function decreases this error by more than 1 kcal/mol. The reaction energy is not strongly dependent on the choice of fitting functions or on the quality of the grid as long as a good grid is chosen as the starting point. The behavior of improving the grid with the Gaussian LDF slightly worsens the agreement with experiment which is opposite to the result found with numerical LDF. The results in Table 6 indicate at the level of a double-zeta basis augmented by polarization functions that convergence of the results has been achieved within an accuracy of 0.5 kcal/mol and 0.005 Å in the bond distances. Within these limits the results are not sensitive to the choice of grid, fitting set or size of the core representations. The results found for basis set DG-9 which is triple zeta show that care may need to be taken in the design of a more uncontracted basis set and that a larger primitive set may be required.

We have employed a set of non-local corrections from the work of Becke (Becke, 1989a, b) and Perdew (Perdew, 1986) to test the performance of these corrections in a series of isodesmic reactions. These gradient-corrected, exchange-correlation potentials have been used in the DGauss program with basis set DG-5 in the same way that we use MP-2 corrections for correcting the energetics from Hartree-Fock calculations. The optimized geometry from the LDF calculations is used and the Becke-Perdew (B-P) correction is applied to the energy at this optimum geometry. The results in Table 3 shows

Table 6. Test of Gaussian Basis Sets

Basis	r(C-H) CH$_4$	r(C-F) CF$_4$	r(C-H) CF$_2$H$_2$	r(C-F) CF$_2$H$_2$	ΔE
6-31G**	1.102	1.322	1.112	1.347	38.4
6-311G**	1.099	1.319	1.108	1.349	33.5
DG-1	1.100	1.325	1.108	1.357	26.1
DG-2	1.101	1.325	1.108	1.356	26.7
DG-3	1.109	1.325	1.108	1.357	26.2
DG-4	1.100	1.324	1.108	1.355	26.1
DG-5	1.100	1.329	1.109	1.361	24.9
DG-6	1.100	1.326	1.108	1.358	24.4
DG-7	1.100	1.326	1.108	1.358	24.8
DG-8	1.100	1.332	1.109	1.363	24.8
DG-9	1.098	1.321	1.105	1.352	26.2
EXPT	1.092	1.319	1.098	1.357	22.1±1.0

that in general the B-P correction leads to too large a decrease in the reaction energy. The correction, except for reaction 5 for which no effect is predicted, leads to magnitudes of the reaction energies that are smaller than the experimental values. The correction is in the right direction but the experimental value is overshot and the results are not really improved over the simple LDF values.

The dipole moments for the three polar methanes are given in Table 7. The Hartree Fock values are greater than the experimental values and an improvement in the basis set leads to a decrease in the dipole moment. The best calculations give dipoles that are 7% too high. The numerical LDF dipole moments are 8-10% less than the experimental values. It is well-known that the Hartree-Fock approximation can build too much ionic character into the wave function and hence gives dipole moments that are too large. In contrast, the LDF method will have more covalent character from the form of the Hamiltonian that is used and for these small systems will not have sufficient ionic character. This is consistent with the smaller dipoles found at the LDF level.

One of the great successes of Hartree-Fock theory besides the prediction of molecular geometries has been its ability to deal with the energetics associated with conformational change, for example rotation and inversion barriers (Payne et al., 1977). In Table 8, we compare the geometry of C$_2$F$_6$ with calculated and experimental values (Dixon et al., 1988b). There is excellent agreement between both theoretical methods and experiment for the C-C bond distance.

Table 7. Fluorinated Methanes - Dipole Moments(Debye) [a]

Molecule	Expt	LDF DMol	HF(TZ+2P)	HF(DZP)	HF(DZ+Dc)
CH_3F	1.86	1.67	1.97	2.08	2.38
CH_2F_2	1.97	1.79	2.11	2.18	2.52
CHF_3	1.65	1.52	1.77	1.84	2.13

[a] See Ref (Dixon, 1988)

Both theoretical methods predict the C-C bond length to increase on rotation. The C-F bond lengths show the same behavior as found in the fluoromethanes. The SCF rotation barrier is in excellent agreement with the experimental value and has no correlation correction. The LDF barrier on the other hand is too low by about 1 kcal/mol. This could arise from the lower ionic character of the LDF density. The C-F bonds are quite polar and in the eclipsed form the interaction of the bond dipoles contributes significantly to the rotation barrier. Since the LDF electron density tends to have less polar character, the bond dipoles will be smaller and hence the barrier will be smaller, exactly as found. Another possible explanation for the low barrier at the LDF level is that there is too much attraction between the fluorines that are eclipsed as it has been found previously that LDF overemphasizes the attraction between closed shell atoms (Andzelm, unpublished work).

Conclusions

The above results demonstrate that the LDF method can be reliably used in the study of molecular systems. When adequate representations of the density are provided, good molecular geometries can be obtained. The total energies from the LDF calculations can be used to predict relative energies based on isodesmic reactions. This is an important result as it broadens the use of the LDF method for chemical systems. The molecular dipole moments are too small at the LDF level and provide some insight into the nature of the chemical effects of the LDF calculation. The rotational barrier in C_2F_6 is too small by 1 kcal/mol as compared to the experimental value. This result is consistent with dipole moments calculated at the LDF level being less than the

Table 8. Geometries and Rotation Barriers for C_2F_6 [a]

BOND DISTANCES Å

Symmetry	Expt.	LDF DMol	6-31G* HF
		R(C-C)	
D_{3d}	1.545	1.535	1.528
D_{3h}		1.563	1.556
		R(C-F)	
D_{3d}	1.326	1.339	1.310
D_{3h}		1.338	1.309

ROTATION BARRIER (kcal/mol)

LDF (DMol)	2.81
HF/6-31G*	3.88
Expt.	3.92

[a] See (Dixon et al., 1988b) for HF and experimental values.

experimental values. The use of non-local corrections in the prediction of isodesmic reaction energies shows that unlike the prediction of bond dissociation energies, the non-local corrections did not improve agreement with experiment, and, in fact, may give worse agreement. This is however a new research area and more testing of the methods is clearly required. There is also a significant amount of work going on at the present in deriving improved non-local correction terms such as the work of Levy (Wilson et al., 1990) Final conclusions will have to await more extensive testing of both present and future correction terms.

Acknowledgment

We thank Nathalie Godbout for contributions to the development of the basis sets for C and F.

References

Andzelm, J., Radzio, E., and Salahub, D. R., 1985, *J. Comput. Chem.* 6:520.

Andzelm, J., Wimmer, E., and Salahub, D. R., 1989, *The Challenge of d and f Electrons: Theory and Computation,* ACS Symposium Series, No. 394 (Edited by Salahub, D. R, and Zerner, M. C.), p.228. American Chemical Society, Washington, D.C.

Andzelm, J.,and Wimmer, E., this volume, Chapt.

Becke, A. D., 1989a, *Inter. J. Quantum Chem. Quantum. Chem. Symp.* 23:599.

Becke, A. D., 1989b, *The Challenge of d and f Electrons: Theory and Computation*, ACS Symposium Series No. 394, (Edited by Salahub, D. R., and Zerner, M. C.), p. 166. American Chemical Society, Washington, D.C.

Becke, A., 1988, *J. Chem. Phys.* 88:2547.

Delley, B., 1990, *J. Chem. Phys.* 92:508. DMol is available commercially from BIOSYM Technologies, San Diego, CA.

Dixon, D. A., Fukunaga, T., and Smart, B. E., 1986, *J. Am. Chem. Soc.* 108:1585.

Dixon, D. A., 1987, *Science and Engineering on Cray Supercomputers, Proceedings of the Third International Symposium*, p. 169. Cray Research, Minneapolis, MN.

Dixon, D. A., 1988, *J. Phys. Chem.* 92:86.

Dixon, D. A., Capobianco, P. J., Mertz, J. E., Wimmer, E., 1988a, *Science and Engineering on Cray Supercomputers, Proceedings of the Fourth International Symposium*, p. 189. Cray Research, Minneapolis, MN.

Dixon, D. A., and Van-Catledge, F. A., 1988b, *Int. J. Supercomputer Appli.* 2, No. 2: 62.

Dixon, D. A., Andzelm, A., Fitzgerald, G., Wimmer, E., and Delley, B., 1990, *Science and Engineering on Cray Supercomputers, Proceedings of the Fifth International Symposium*. Cray Research, Minneapolis, Mn.

Dixon, D. A., and Smart, B. E., in press 1990, *Chemical Engineering Communications*.

Dunlap, B., Connolly, J., and Sabin, J., 1979, *J. Chem. Phys.* **71**:3396.

Dunlap, B., Andzelm, J., and Mintmire, J., in press 1990, *Phys. Rev. A.*

Dunning , T. H., Jr., and Hay, P. J., 1977, *Methods of Electronic Structure Theory* , Ch. 1, (Edited by Schaefer, H. F., III). Plenum Press, New York.

Fournier, R., Andzelm, J., and Salahub, D. R., 1989, *J. Chem. Phys.* **90**:6371.

Hehre, W. J., Radom, L., Schleyer, P. vR., and Pople, J. A., 1986, *Ab Initio Molecular Orbital Theory*. John Wiley & Sons, New York.

Huzinaga, S., Andzelm, J., Klobukowski, M., Radzio, E., Sakai, Y., and Tatasaki, H., 1984, *Gaussian Basis Sets for Molecular Calculations*. Elsevier, Amsterdam.

Jones, R. O., and Gunnarsson, O., 1989, *Rev. Mod. Phys.* **61**:689.

Jorgenson, P., and Simons, J. Eds., 1986c, *Geometrical Derivatives of Energy Surfaces and Molecular Properties*, NATO ASI Series C. Vol. 166, p. 207. D. Reidel, Dordrecht.

Komornicki, A., Ishida, K., Morokuma, K., Ditchfield, R., and Conrad, M., 1977, *Chem. Phys. Let.* **45**:595.

Møller, C., and Plesset, M. S., 1934, *Phys. Rev.* **46**:618.

Obara, S., and Saika, A., 1986, *J. Chem. Phys.* **84**:3963.

Parr, R. G., and Yang, W., 1989, *Density Functional Theory of Atoms and Molecules*. Oxford University Press, New York, and references therein.

48   D.A. Dixon et al.

Payne and Allen, L. C., 1977, *Applications of Electronic Structure Theory,* (Edited by Schaeffer, H.F., III).

Perdew, J. P., 1986, *Phys. Rev.B* **33**:8822.

Pople, J. A., Binkley, J. S., and Seeger, R., 1976b, I*nt. J. Quantum Chem. Symp.* **10**:1.

Pulay, P. , 1977b, *Applications of Electronic Structure Theory*, (Edited by Schaefer, H. F. III), p. 153. Plenum Press, New York.

Salahub, D. R., 1987, *Ab Initio Methods in Quantum Methods in Quantum Chemistry-II ,* (Edited by Lawley, K. P. J), p. 447. Wiley & Sons, New York.

Versluis, L., and Ziegler, T., 1988, *J. Chem. Phys.* **88**:3322.

von Barth, U., and Hedin, L., 1972, J*. Phys. C* **5**:1629.

Vosko, S. H., Wilk, L., and Nusair, M., 1980, *Can. J. Phys.* **58**:1200.

Wilson, L. C., and Levy M., 1990, *Phys. Rev. B.* **41**:12930.

Wimmer, E., Freeman, A. J., Fu, C.-L., Cao, P.-L., Chou, S.-H., and Delley, B., 1987, *Supercomputer Research in Chemistry and Chemical Engineering* (Edited by Jensen, K. F., and Truhlar, D. G.) ACS Symposium Series, p. 49. American Chemical Society, Washington, D. C.

4
Symmetry and Local Potential Methods

BRETT I. DUNLAP

Introduction

The Xα method[1] and its offspring are becoming more and more important in quantum chemistry for chemical systems that are too large for accurate configuration interaction (CI) calculations. All self-consistent-field (SCF) methods of quantum chemistry method begin with one or more systems of one-electron equations of the form,

$$\epsilon_i \, \phi_i(\mathbf{r}) = \left[-\tfrac{1}{2}\nabla^2 + V \right] \phi_i(\mathbf{r}). \tag{1}$$

Invariably V is separated into its one-electron and two-electron parts, the electron-nuclear interaction potential, V_{en}, and the electron-electron interaction potential, V_{ee}, respectively. Only V_{ee} distinguishes the various SCF methods of quantum chemistry. In what are now called local density functional (LDF) methods the electron-electron potential is written,

$$V_{ee} = V_{ee}(\mathbf{r}_1) = \int \frac{\rho(\mathbf{r}_2)d\mathbf{r}_2}{r_{12}} + V_{xc}(\rho(\mathbf{r}_1)), \tag{2}$$

where the total electron density, ρ, can be divided into spin-up and spin-down components, and the expression for the spin-up component,

$$\rho_\uparrow(\mathbf{r}) = n_{i\uparrow} \, \phi_{i\uparrow}^*(\mathbf{r}) \, \phi_{i\uparrow}(\mathbf{r}), \tag{3}$$

where the n_i are the (possibly fractional) number of electrons occupying each orbital of Eq. 1, contains an implicit summation over orbital index i. These methods are called LDF's because the exchange and correlation potential, $V_{xc}(\rho(\mathbf{r}_1))$, at any point in space is constrained to depend only on the density at the same point. If we constrain V_{xc} to be local and require the system of equations to satisfy the virial theorem at all extrema on all Born-Oppenheimer potential energy surfaces, we are led to the Xα approximation for spin-up orbitals,

$$V_{xc}^\uparrow(\rho(\mathbf{r}_1)) = 3\,\alpha \left(\frac{3}{4\pi} \rho_\uparrow(\mathbf{r}) \right)^{1/3}. \tag{4}$$

This result follows from simply considering the homogeneous electron gas at all densities and requires that LDF expressions not include gradients of the density. The value of α for the homogeneous electron gas based on variation is two-thirds[2] and based on averaging is one.[1] For heteronuclear molecules the best choice is perhaps 0.7.

On the other hand, the Kohn-Sham[2] mapping of the the ground state of the real system onto the ground state of a fictitious noninteracting set of electrons, which has the same density as the real system, provides a different interpretation of these equations. If both the real and noninteracting sets of electrons are nondegenerate, the mapping is one-to-one and the wavefunction of the noninteracting set of electrons is a single determinant of one-electron orbitals likely satisfying Eq. 1 for some V. (The counterexamples of non-V-representability use degeneracy of one-electron orbitals.[3-5]) If we map the energy of a single-determinant of plane-wave orbitals onto the essentially exact energy[6] of the homogeneous electron gas as a function of background positive charge density in the completely spin-paired and completely spin-polarized extremes, we are led to a different set of LDF's that typically overbind molecules compared to $X\alpha$ and experiment.[7] The best of this other set is perhaps the Perdew-Zunger (PZ) parameterization,[8] which, although it is too complicated to write out here, gives, apart from slightly rescaling the total energies, the same description of most nonmagnetic molecules as does $X\alpha$. The reason for almost no difference is that the one-third power and functions similar to it tremendously deamplify variations in the nonnegative density itself.

The computational attraction of Eq. 1 and 2 is that $V(\mathbf{r})$ is a local potential, i.e., it is the same for each orbital.[9] While all the orbital generating equations of quantum chemistry can be rendered in the form of Eq. 1 (if all else fails by using projection) only with a local potential is $V(\mathbf{r})$ a single-valued (orbital independent) function of position. Thus obtaining $V(\mathbf{r})$ is central to all local potential methods and that problem is straightforward. Eqs. 1-3 can be solved for any real vector of occupation numbers, $\mathbf{n}$. All LDF approximations to V_{xc} yield local potentials, but not all local potentials result from LDF approximations, as is obvious from the direct Coulomb term in Eq. 2.

For any density functional expression for V_{xc} the coupled one-electron equations of motion, Eq. 1, can be integrated to yield the total energy expression,

$$E = \langle T \rangle + \langle U \rangle, \tag{5}$$

where the total kinetic energy is given by an expression,

$$\langle T \rangle = -\tfrac{1}{2} n_{is}\, \phi_{is}^*(\mathbf{r})\, \nabla^2 \phi_{is}(\mathbf{r}), \tag{6}$$

containing an implicit summation over the two spin directions through the index s, and

$$\langle U \rangle = \int \rho_s(\mathbf{r})\, U_s(\rho(\mathbf{r}))\, d\mathbf{r}, \tag{7}$$

where the one-electron potential and the density functional potential energy operator, U, are related by an independent variation of $\langle U \rangle$ with respect to each spin density at each point,

$$\int V_\uparrow(\rho)\, \delta\rho_\uparrow(\mathbf{r})\, d\mathbf{r} = \delta\langle U \rangle \tag{8}$$

For any part of V that is approximated as being an LDF, Eq. 8 is precisely a differentiation,

$$V_\uparrow^{LDF}(\mathbf{r}) = \frac{d\left[\rho_s(\mathbf{r})\, U_s^{LDF}(\rho(\mathbf{r}))\right]}{d\rho_\uparrow(\mathbf{r})}. \tag{9}$$

That Eq. 8 does not hold for a nonLDF contribution to V is obvious if one considers multiplying Eq. 2 by $\rho(\mathbf{r}_1)$. In that case the $\rho(\mathbf{r}_1)$ and $\rho(\mathbf{r}_2)$ in the direct Coulomb term are asymetrical; $\mathbf{r}_2$ is an integration variable while $\mathbf{r}_1$ is not.

As is the case with Hartree-Fock-based quantum chemical methods, the analytic basis set approach is the overwhelming favorite in nonmuffin-tin local potential quantum chemical methods, except for lower dimensional problems such as atoms[10] and linear molecules.[11-13] The all-purpose analytic basis set is Gaussians[7,14] in both quantum chemical methods. In contrast to Hartree-Fock based methods, computational efficiency dictates and complicated expressions such as Eq. 4 inspire fitting V in analytic-basis-set local potential methods.[15-16] If the potential is fit, which is indicated by placing a bar over the fitted quantity, then the potential energy should be evaluated,[17]

$$\langle U \rangle \approx \int \rho(\mathbf{r})\, \overline{V}(\rho(\mathbf{r}))\, d\mathbf{r} + \int \overline{\rho(\mathbf{r})\left[U(\rho(\mathbf{r})) - V(\rho(\mathbf{r}))\right]}\, d\mathbf{r}, \tag{10}$$

because then the nonfitted density only occurs multiplied by the local potential. Variation with respect to the orbitals (occurring only in the nonfitted density) gives the local potential term of the one-electron equations, and the total energy is insensitive in first order to changes of the one-electron orbitals, provided the fitting procedure follows variationally from this equation. Eq. 10 yields an energy-stationary way to fit the charge density,[7,18] to fit the direct electron-electron Coulomb potential itself,[19] and to fit the $X\alpha$ exchange potential.[7,20] For the more complicated V_{xc} such as occur in the PZ functional, Eq. 10 yields variational weights for any numerical fitting scheme.[21]

Independent of how V is determined, it is natural to ask what symmetry it has and to what extent symmetry can help in local potential calculations. The system of equations is nonlinear, and therefore its solutions can and do yield densities that break the symmetry of the collection of nuclear charges.[22] There are three approaches to this important problem. First, one can accept broken-symmetry solutions.[23-26] Second, one can restrict one's attention to density functional methods that yield V having the symmetry of the nuclear charges.[3-5] Third, one can explicitly symmetrize V by symmetrizing a broken-symmetry wavefunction before computing the kinetic energy and density during the SCF process.[27]

Empirically, the experimental ionization potential of a core-hole from a set of symmetry-equivalent atoms is better reproduced in local potential methods when symmetry is broken and the hole self-consistently localizes on one of that set of atoms.[23,24] Unfortunately, in contrast to *ab initio* methods, the variational principle does not also suggest this solution. In current local potential methods, a localized core hole results in a higher total energy than a delocalized-hole total energy. When applied to spin, symmetry-restriction leads to a poor description of Cr_2,[25,26] which has a singlet electronic ground state but must dissociate into two septet atoms. All current LDF methods, however, are not invariant under rotations in spin space.[22,28] Therefore they can never yield spin-densities that transform properly for any magnetic system that has any minority-spin electrons, without further assumptions (such as using the same orbitals for the two spins). Nevertheless, significant progress is being made in interpreting solutions in which the density is invariant but the spin densities break the molecular symmetry[29,30] (spin density waves). Apart from the special case of core holes, there has been no general attempt to interpret broken-spatial-symmetry solutions (charge density waves) despite the fact that they they often arise asymptotically in molecular dissociation.[11,31] In cases where the real system does not exhibit charge-density-wave behavior, these broken-spatial-symmetry errors typically make binding energies uncertain at the tenth of an electron volt level.

Symmetrizing V removes this uncertainty. (For atoms this means making the central field approximation.) Furthermore, a symmetrized V can be readily taken advantage of in linear-combination-of-atomic-orbitals

(LCAO) local potential computer codes.[22] Symmetrization without further approximation allows routine LDF calculations on symmetric systems containing tens of transition metal atoms,[32-34] or larger mixed system such as C_{60} and yttrium diphtalocyanine ($YC_{64}H_{32}N_{16}$).[35,36]

In what follows a density functional theory[27,28] of the broken-spatial-symmetry problem for singlet electronic states is reviewed. It relies on the fact the density is to be symmetrized, which, in turn, symmetrizes V provided it is a function of the density. This symmetrization validates the fractional occupation number (FON) description of the ground state.[37] The theory contains the mathematics of *ab initio* configuration interaction. A good introduction to any discussion of correlation and LDF theory is the work of Cook and Karplus.[38]

FON Singlet-State Configuration Interaction

The simplest densities to consider correspond to closed-shell molecules, in which the density clearly has the symmetry of the molecule. For such a molecule consider the case where the highest occupied molecular orbital (HOMO) is antibonding and the lowest-unoccupied molecular orbital (LUMO) is bonding and compress the appropriate internuclear separations until these orbitals become nearly degenerate (or consider the opposite case and expand the bonds). For concreteness, consider a $^1\Sigma_g^+$ state of a homonuclear diatomic molecule and a broken-symmetry HOMO of mixed σ_g and σ_u character,

$$\phi = a\chi_{\sigma_g} + b\chi_{\sigma_u}. \tag{11}$$

The g and u components must be orthogonal from symmetry considerations alone. (If they are also normalized, then the sum of the magnitudes squared of a and b is unity.) Since the kinetic energy operator is invariant under any spatial symmetry operation, ϕ's contribution to the kinetic energy,

$$T_\phi = -a^*a\langle\chi_{\sigma_g}|\tfrac{1}{2}\nabla^2|\chi_{\sigma_g}\rangle - b^*b\langle\chi_{\sigma_u}|\tfrac{1}{2}\nabla^2|\chi_{\sigma_u}\rangle, \tag{12}$$

is diagonal in its components; thus the kinetic energy is the same as that of a symmetry-restricted FON calculation using $n_{\sigma_g} = a^*a$ and $n_{\sigma_u} = b^*b$. In contrast to the situation for the overlap and kinetic energy, there is a difference between the broken-symmetry and symmetry-restricted densities. The difference is the unsymmetrical expression,

$$a^*b\, u_{\sigma_g}^* u_{\sigma_u} + b^*a\, u_{\sigma_u}^* u_{\sigma_g}, \tag{13}$$

that has larger magnitude on either the left-hand atom or on the right hand atom depending on the magnitudes and phases of a and b. If, however, the density is symmetrized by, in this case, averaging the original broken-symmetry orbital density with the same density after inverting the coordinate system, then both orbital densities are the same,

$$a^* a\, u^*_{\sigma_g} u_{\sigma_g} + b^* b\, u^*_{\sigma_u} u_{\sigma_u}. \tag{14}$$

Thus symmetrized-broken-symmetry and the FON density-functional calculations are identical. For the above example, both give a charge density consistent with a $^1\Sigma^+_g$ electronic state.

FON SCF calculations are always slower to converge than fixed-occupation-number LDF calculations. This is precisely because there is at least one more degree of freedom involved. Another way to analyze this slowness is to note that for fixed occupation-number calculations, the relevant gauge of speed of convergence is the gap between the HOMO and LUMO energies,[39] whereas FON convergence is also strongly influenced by the density of states surrounding the HOMO and LUMO energies.[17] (For the FON case, the HOMO must be defined as the highest level that is greater than half occupied and the LUMO is the lowest level that is less than half occupied.) Therefore it is best to avoid direct FON calculations if one can.

To an excellent approximation the total LDF energy is a quadratic function of the occupation numbers,[37] because all of its components except V_{xc} are, and V_{xc} is a very slowly varying function of the density. This quadratic property is also suggested by the fact that the exact total energy depends only on the first and second order density matrices. If the LDF energy is a quadratic function of the occupation numbers, then the LDF one-electron eigenvalues, which are the derivatives of the total energy with respect to the corresponding occupation number,[40]

$$\epsilon_{is} = \frac{\partial E}{\partial n_{is}} \tag{15}$$

must be linear functions of the occupation numbers. In particular, this approximation means that for the σ_g and σ_u orbitals under consideration,

$$\Delta\epsilon = \epsilon_u - \epsilon_g \tag{16}$$

varies linearly with fractional number of electrons, n, transferred from the σ_g orbital to the σ_u orbital. The value of this eigenvalue difference for all numbers of electrons transferred then follows from the lagrange interpolating formula,

$$\Delta\epsilon(n) = \Delta\epsilon(0)\frac{(n-2)}{(0-2)} - \Delta\epsilon(2)\frac{(n-0)}{(2-0)}, \tag{17}$$

which reduces,

$$\Delta\epsilon(n) = \Delta\epsilon(0) - [(\Delta\epsilon(2) - \Delta\epsilon(0)]\frac{n}{2}. \qquad (18)$$

There is an FON solution $\Delta\epsilon(n_f) = 0$, if and only if

$$\Delta\epsilon(0)\,\Delta\epsilon(2) < 0. \qquad (19)$$

In that case, transferring n_f electrons,

$$n_f = \frac{2\Delta\epsilon(0)}{\Delta\epsilon(0) - \Delta\epsilon(2)}. \qquad (20)$$

gives the FON ground state.

Integrating to get the total energy at the FON solution from the $n = 0$ limit,

$$\Delta E = \int_0^{n_f} \Delta\epsilon(n)\,dn \qquad (21)$$

gives one formula for the FON total energy,

$$E_{f0} = E(0) + \frac{\Delta\epsilon(0)^2}{\Delta\epsilon(0) - \Delta\epsilon(2)}, \qquad (22)$$

and integrating from the other limit gives another formula,

$$E_{f2} = E(2) + \frac{\Delta\epsilon(2)^2}{\Delta\epsilon(0) - \Delta\epsilon(2)}. \qquad (23)$$

Lagrange interpolating between these two expressions for the FON total energy gives an FON energy,[28]

$$E_l = \frac{-E(0)\Delta\epsilon(2) + E(2)\Delta\epsilon(0) - \Delta\epsilon(0)\Delta\epsilon(2)}{\Delta\epsilon(0) - \Delta\epsilon(2)}, \qquad (24)$$

that connects continuously to the pure state solutions outside the FON range, geometries for which Eq. 19 is not satisfied.

The FON energy lowering can be viewed as resulting from configuration interaction between the $n = 0$ and the $n = 2$ states, in which the upper FON energy,

$$E_u = \frac{E(0)\Delta\epsilon(0) - E(2)\Delta\epsilon(2) + \Delta\epsilon(0)\Delta\epsilon(2)}{\Delta\epsilon(0) - \Delta\epsilon(2)}, \qquad (25)$$

lies as much higher in energy above the average of $E(0)$ and $E(2)$ as E_l lies below that average. This approach can be extended to the general

$n \times n$ configuration interaction problem. In such an extension the pure state energies become diagonal entries in an $n \times n$ eigenvalue problem. In this eigenvalue problem one adds an off-diagonal matrix between any two pure states,

$$V = \frac{\sqrt{(E_l - E_u)^2 - (E(0) - E(2))^2}}{2}. \tag{26}$$

if and only if the two pure states differ precisely in that two electrons are transferred from one orbital to another and that the eigenvalue difference between these two orbitals change according to Eq. 19 in going from the one state to the other. This postulate gives as its ground state the FON solution when restricted to only allowing a single pair, but any single pair, of states to exchange electrons.

Al$_4$

The Xα ground state of Al$_4$ is a D$_{2h}$-symmetric rhombus with triplet electronic configuration having a spin-up electron in each of the $9a_{1g}$ and $1b_{1u}$ orbitals outside a 50-electron $8a_g^2 2b_{1g}^2 3b_{1u}^2 1b_{2g}^2 5b_{2u}^2 1b_{3g}^2 5b_{3u}^2$ closed-shell configuration. The b_{1g} orbital is a π orbital with nodal plane in the plane of the molecule. This agrees with an *ab initio* description.[41] The triplet-state atomization energy (the energy required to dissociate the molecule into four separated R(3)-symmetrized spin-polarized atoms) is 6.52 eV, its bond distance is 4.99 bohr, and it has a bond angle of 88°.

Much more interesting than the ground state potential energy surface are the singlet potential energy surfaces. Fig. 1a gives the symmetry-restricted low-lying surfaces in square-planar D$_{4h}$ symmetry. These breathing-mode surfaces are derived from the 54-electron $5a_{1g}^2 1a_{2g}^2 2a_{2u}^2 4b_{1g}^2 2b_{2g}^2 1b_{2u}^2 1e_g^4 5e_u^4$ configuration by anhilating two electrons from the valence a_{1g}, a_{2u}, b_{1g}, and b_{2g} orbitals respectively. There are five curve crossings in the figure. This case is ideal for using Eqs. 24-26 in a 4×4 eigenvalue problem. The result is the solid lines of Fig. 1b, where all crossings, of the original, dashed lines, are avoided.

The most interesting avoided crossing in Fig. 1b is the one, at smallest Al-Al bond distance, between the a_{1g}^0 and b_{1g}^0 states. It is the weakest. This is because the a_{1g} orbital, being nodeless, is largely centered on the atomic centers, like the b_{1g} orbital which, being $d_{x^2-y^2}$-like, has nodes at the bond centers for atoms chosen as they were to lie in the $\pm x$ and $\pm y$ directions. Because the densities are similar, transferring two electrons between these two orbitals has the smallest effect on their eigenvalue difference, and the configurations do not repel each other very much.

This is not the whole story for singlet Al$_4$, by any means. Fig. 2a considers D$_{2h}$-symmetric bending motion at the minimum Al-Al bond distance

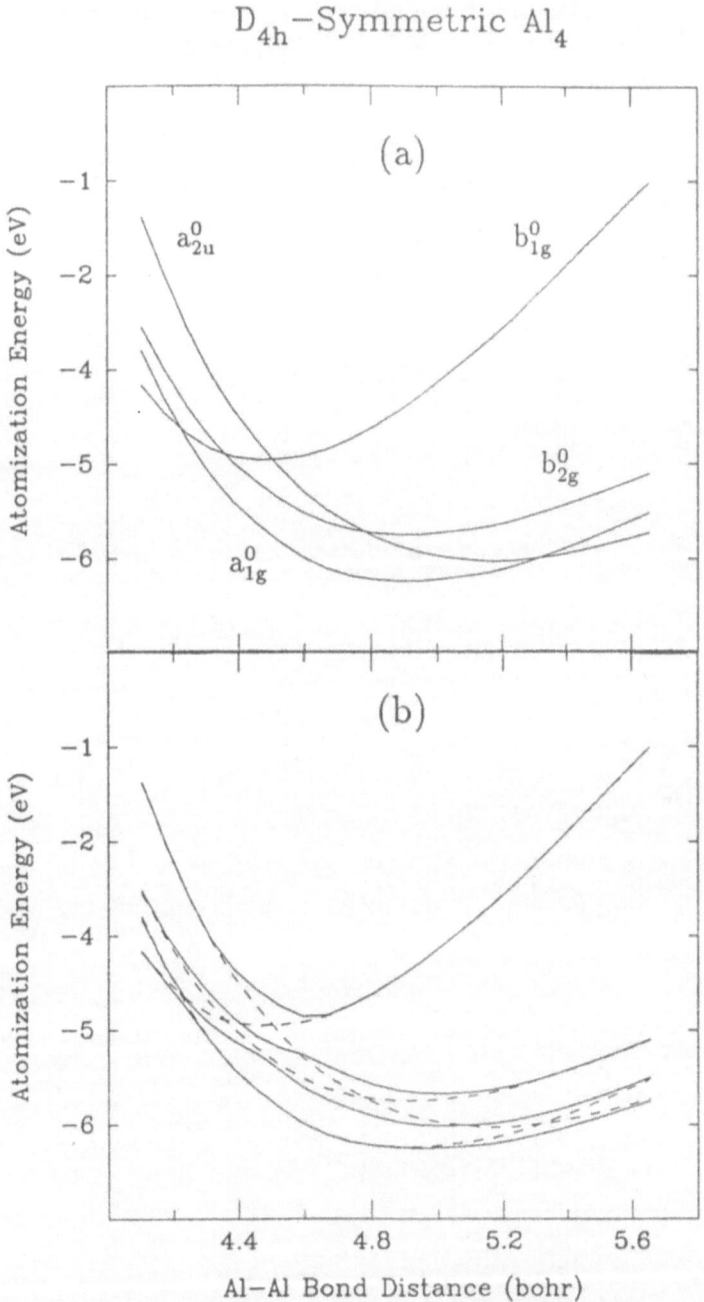

Figure 1. Breathing mode potential energy curves for singlet square-planar Al$_4$.

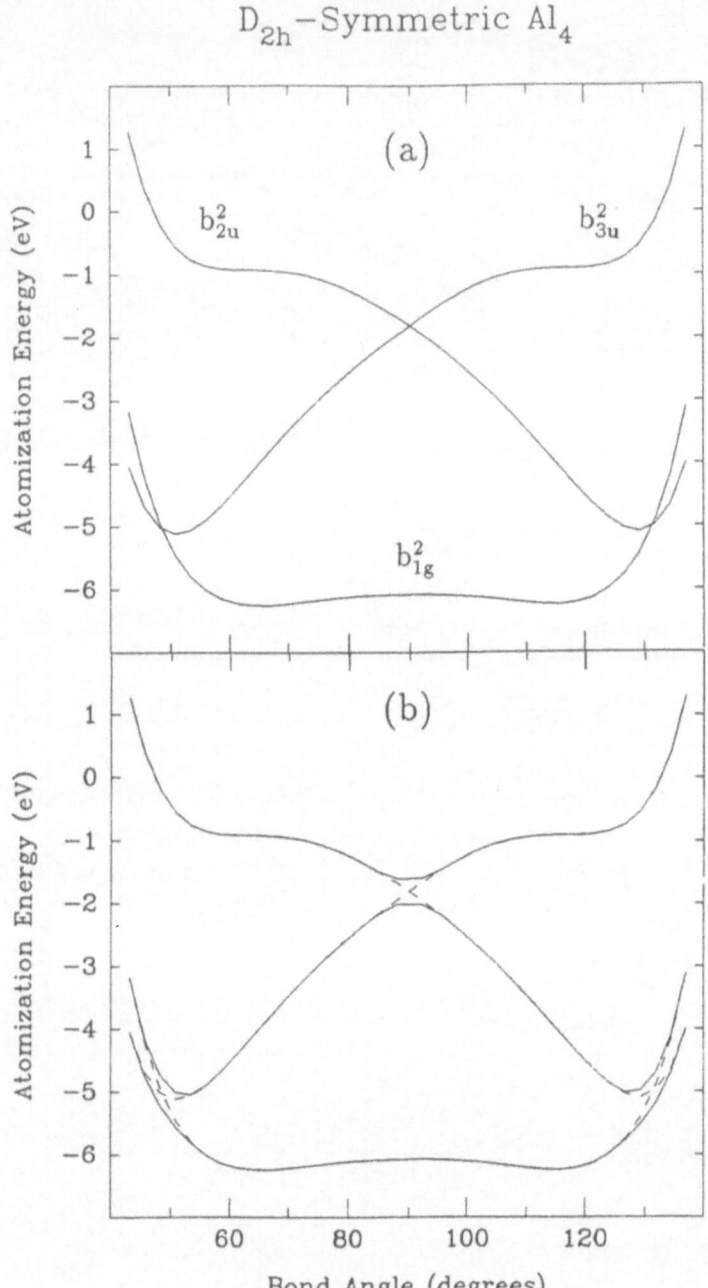

Figure 2. In-plane bending potential energy curves for singlet Al$_4$ for Al-Al bond distances of 4.9 bohr.

of 4.9 bohr in Fig. 1, where the ground state is largely a_{1g}^0. Under this motion, the a_{1g}^0 state interacts with two p-like states that have nodes along the shortest Al-Al second neighbor bond when they cross the a_{1g}^0 state. These two p-like states are degenerate in D_{4h} symmetry and thus are different, higher lying in energy, from the states considered in Fig. 1. Fig. 2b gives the result of using Eqs. 24-26 on the pure states of Fig. 2a.

Conclusions

The work of defining local potential, Xα-like, methods for use in quantum chemisty is not done. Only pure singlet states were addressed in this work, and only methods that resulted in no net spin-polarization anywhere in space is space were used. The problem of coupling spin, fractional occupation numbers, and symmetry in a complete approximate LDF theory[22,27,28] is unfinished.

This work was supported by the Office of Naval Research, contract number N0001490WX24264, and through a grant of computer time by the Research Advisory Committee of the Naval Research Laboratory.

REFERENCES

1. J.C. Slater, *Quantum Theory of Molecules and Solids*, Vol. 4 (McGraw-Hill, New York, 1974).

2. W. Kohn and L.J. Sham, *Phys. Rev.* **140**, A1133 (1965).

3. M. Levy, *Phys. Rev. A* **26**, 1200 (1982).

4. E.H. Lieb, *Int. J. Quantum Chem.* **24**, 243 (1983).

5. H. Englisch and R. Englisch, *Physica* **121A**, 253 (1983).

6. D.M. Ceperley and B.J. Alder, *Phys. Rev. Lett.* **45**, 566 (1980).

7. B.I. Dunlap, J.W.D. Connolly and J.R. Sabin, *J. Chem. Phys.* **71**, 3396; 4993 (1979).

8. J.P. Perdew and A. Zunger, *Phys. Rev. B* **23**, 5048 (1981).

9. J.W.D. Connolly, in *Modern Theoretical Chemistry*, Vol. 7, ed. G.A. Segal (Plenum, New York, 1977) p. 105.

10. F. Herman and S. Skillman, *Atomic Structure Calculations* (Prentice-Hall, Englewood Cliffs, 1963).

11. A.D. Becke, *I. J. Quantum Chem.* **27**, 585 (1985).

12. L. Laaksonen, D. Sundholm and P. Pyykkö, *Comp. Phys. Rept.* **4**, 313 (1986).

13. D. Heinemann, B. Fricke and D. Kolb, *Chem. Phys. Lett.* **145**, 125 (1988).

14. W. Hehre, L. Radom, P.v.R. Schleyer, and J.A. Pople, *Ab Initio Molecular Orbital Theory* (Wiley-Interscience, New York, 1986).

15. E.J. Baerends, D.E. Ellis, and P. Ros, *Chem. Phys.* **2**, 41 (1973).

16. H. Sambe and R.H. Felton, *J. Chem. Phys.* **62**, 1122 (1975).

17. B.I. Dunlap and N. Rösch, in *Density Functional Theory for Many-Fermion Systems*, S.B. Trickey, editor, *Adv. Quantum Chem.*, to be published.

18. J.W. Mintmire and B.I. Dunlap, *Phys. Rev. A* **25**, 88 (1982).

19. B.I. Dunlap, *J. Chem. Phys.* **78**, 3140 (1983).

20. B.I. Dunlap, *J. Phys. Chem.* **90**, 5524 (1986).

21. B.I. Dunlap and M. Cook, *Int. J. Quantum Chem.* **29**, 767 (1986).

22. B.I. Dunlap, *Adv. Chem. Phys.* **69**, 287 (1987).

23. J.W.D. Connolly, H. Siegbahn, U. Gelius, and C. Nordling, *J. Chem. Phys.* **58**, 4265 (1973).

24. B.I. Dunlap, P.A. Mills, and D.E. Ramaker, *J. Chem. Phys.* **75**, 300 (1981).

25. N.A. Baykara, B.N. McMaster, and D.R. Salahub, *Mol. Phys.* **52**, 891 (1984).

26. G.S. Painter, *J. Phys. Chem.* **90**, 5530 (1986).

27. B.I. Dunlap, *Phys. Rev. A* **29**, 2902 (1984).

28. B.I. Dunlap, *Chem. Phys.* **125**, 89 (1988).

29. L. Noodleman, *J. Chem. Phys.* **74**, 5737 (1981).

30. L. Noodleman, D.L. Case, and F. Sontum *J. Chim. Phys. Phys-chim. Biol.* **86**, 743 (1989).

31. F.W. Kutzler and G.S. Painter, *Phys. Rev. Lett.* **59**, 1285 (1987).

32. B.I. Dunlap and N. Rösch, *J. Chim. Phys. Phys-chim. Biol.* **86**, 671 (1989).

33. N. Rösch, P. Knappe, P. Sandl, A. Görling, and B.I. Dunlap, in *The Challenge of d and f Electrons. Theory and Computation*, eds. D.R. Salahub and M.C. Zerner, ACS Symposium Series **394** (Washington 1989) p. 180.

34. B.I. Dunlap, *Phys. Rev. A* **41**, 5691 (1990).

35. B.I. Dunlap, *Int. J. Quantum Chem.* **S22**, 257 (1988).

36. B.I. Dunlap and C.T. White, unpublished.

37. J.C. Slater, J.B. Mann, T.M. Wilson, and J.H. Wood, *Phys. Rev.*, **184**, 672 (1969).

38. M. Cook and M. Karplus, *J. Phys. Chem.* **91**, 31 (1987).

39. B.I. Dunlap, *Phys. Rev. A* **25**, 2847 (1982).

40. J.F. Janak, *Phys. Rev. B* **18**, 7165 (1978).

41. C.W. Bauschlicher, Jr., and L.G.M. Pettersson, *J. Chem. Phys.* **87**, 2198 (1987).

5
Local Density DMOL Studies of Noble and Alkali Metal Adsorption on the Silicon Surface

A.J. FREEMAN, S. TANG, S.H. CHOU, YE LING, AND B. DELLEY

Abstract

Noble and alkali metal adsorption on the Si surface is investigated by means of self-consistent local density molecular cluster calculations using the DMOL method and clusters which range up to 98 atoms. Results are reported for (i) Cu and Ag on the Si(111) surface; (ii) alkali metal adsorption on the (2×1) Si(100) reconstructed surface and (iii) co-adsorption of oxygen and potassium on the (2×1) Si(100) surface.

Introduction

It is now widely recognized that the interaction between metals and silicon is one of the most studied subjects in surface science. The motivation for all this effort comes from both technological and fundamental scientific interest in the metal/silicon interface, especially in the manufacture of electronic devices. It has been shown that silicide-like compounds can be formed at metal/silicon interfaces when metals are deposited onto the clean silicon surface at some adequate temperature. The product at the interface is of great importance in the technology of very-large-scale integrated circuits. The metal–silicon interaction also plays a central role in the catalysis of silicon surface reactions. For example, the noble metals (Au, Ag and Cu) are found to increase the Si oxidation rate when deposited on the Si surface.

Some examples (Chou et al., 1987; Chou et al., 1989; Ye et al., 1989; Ye et al.) of our work on the metal/Si interface are presented in this paper.

Theoretical Approach

We determine the electronic structure by solving the Kohn–Sham equations in the local-density approximation. Now it should be emphasized that in the period of time just after the development of density-functional theory, the numerical methods for solving the Kohn–Sham equations were so crude that resulting errors completely obscured the effects of the physical approximations. Only recently has it become possible, due to enormously increased computer power and sophisticated combined theoretical and computational approaches, to solve these equations in a much more exact fashion.

We use molecular cluster models to simulate the Si(111) and Si(100) substrate using the DMOL method. For Cu, Ag/Si(111) systems, we used 47 atoms with 20 Si atoms, 26 H atoms and one noble atom. For alkali metal/Si(100) systems, we used several models containing up to 89 atoms with 39 Si atoms, 44 H atoms and 6 K atoms.

In this work, DMOL — an all-electron numerical method (Chou et al., 1987, 1989; Delley, 1986, 1990; Delley et al., 1983ab) for solving the local density functional equations — is applied to study the chemisorption systems. For a more realistic simulation of the Si substrate, we have also computed the relaxation effects in the K, Na/Si(100) system with an analytical energy gradient technique (Delley) newly incorporated into DMOL. By adding forces into the calculations, we have successfully established a way of choosing cluster of appropriately smaller size (Tang et al.). This LCAO approach uses numerical atomic wave functions generated from a LDF spherical atom program as basis functions. The Hamiltonian and overlap matrix elements are evaluated by a numerical integration technique on a grid of 20,000 to 40,000 sampling points. The von Barth–Hedin (von Barth and Hedin, 1972) exchange-correlation functionals are used in the calculation. The adsorption energy is determined as the difference in total energies of the clean and adsorbed clusters:

$$\Delta E = E_{tot}(\text{adsorbed cluster}) - E_{tot}(\text{clean surface}) \qquad (1)$$

For the co-adsorption of O_2 molecule and K atom on the Si(100) surface, the adsorption energy is defined as:

$$\Delta E = E_{tot}(\text{adsorbed cluster}) - E_{tot}(\text{clean surface}) - E_{tot}(O_2) \qquad (2)$$

Noble Metal Adsorption on the Si(111) Surface

Although much work has been done on the noble metal/silicon interface, questions concerning structural characterization, the role of metal as a catalyst and the bonding between metal and silicon are not well known. For example, various experimental studies, which aimed at determining the Ag adsorption sites in the Ag/Si(111)$\sqrt{3} \times \sqrt{3}$ system, have led to controversial results. Copper exhibits high catalytic activity in the synthesis of methylchlorosilanes via the Direct Process, but the major function of Cu in this process is still not fully understood. Therefore, a detailed comparison between Cu and Ag, interacting with a Si surface, is expected to provide new insights into this catalytic system. There is experimental evidence showing several significant differences between Cu and Ag on a Si(111) surface: low energy electron diffraction (LEED) studies have shown a (3×1) pattern at high temperatures, at coverages between $\frac{1}{3}$ and $\frac{2}{3}$ of a Ag monolayer, while at a coverage between $\frac{2}{3}$ and 1 monolayer, a $(\sqrt{3} \times \sqrt{3})$ pattern is observed (Le Lay, 1983a; Yokotuka *et al.*, 1983). The LEED structure for the unannealed Cu/Si(111) surface exhibits a (1×1) pattern, which changes into (5×5) after a few minutes of annealing at 600° C (Chambers *et al.*, 1985; Ringersen *et al.*, 1983). In contrast to Ag, Cu forms stable silicides, Cu_3Si (Le Lay, 1983b). Our studies based on simple cluster models (Chou *et al.*, 1987, 1989) do provide some insight regarding the nature of the adsorption of Cu and Ag on Si(111), which can't be obtained by experimental methods. We will briefly describe our theoretical models and results for these systems in this section.

The clean Si(111) surface is known to reconstruct into (2×1) and (7×7) structures (for a review, see van der Veen, 1985). These reconstructions are driven by the presence of dangling-bond Si surface states. Since the same states can also be expected to dominate the interactions between surface and adsorbate atoms, the metal–silicon interaction is in competition with the surface reconstruction. However, the theoretical studies of Redondo *et al.* (Redondo *et al.*, 1982) conclude that the mechanism responsible for the surface reconstruction is not due to nearest-neighbor interactions. They also suggest that such reconstruction probably results from collective interactions in which large numbers of surface and subsurface atoms are involved. Now, since the dominant interaction between Cu or Ag with Si(111) is local, the use of a finite cluster without surface reconstruction to simulate the noble metal–silicon surface interactions should be a reasonable first step approach.

The DMOL calculations reveal that in their ground states both Cu and Ag are adsorbed in the threefold hollow position as shown in Fig. 1 with equilibrium heights of 0.74 Å and 1.48 Å, and binding energies of 92 and 72 kcal/mol for Cu and Ag, respectively (Chou *et al.*, 1987, 1989). Assuming a rigid substrate, the vibrational mode perpendicular to the surface is very soft for Cu ($\omega^{-1} = 58$ cm^{-1}), but almost twice as large for Ag ($\omega^{-1} = 90$ cm^{-1}).

Our results (Chou *et al.*, 1987, 1989) and those of other studies (Barone *et al.*, 1980; Julg and Allouche, 1982; Zheng *et al.*, 1988; Stöhr and Jaeger, 1982; Stöhr *et al.*, 1983; Le Lay *et al.*, 1978) of the adsorption heights and energies of Cu/Si(111) and Ag/Si(111) are presented in Table 1. To our knowledge, no experimental data for adsorption geometries, binding energies and vibrational frequencies of Cu atoms on an Si(111) surface are available for the unannealed Cu/Si(111) system. Lewis *et al.* (Lewis *et al.*, 1988) estimate that the Cu–Si bond energy is 30.6 ± 2.4 kcal/mol. Our calculations show that each Cu atom bonds to three surface silicon atoms

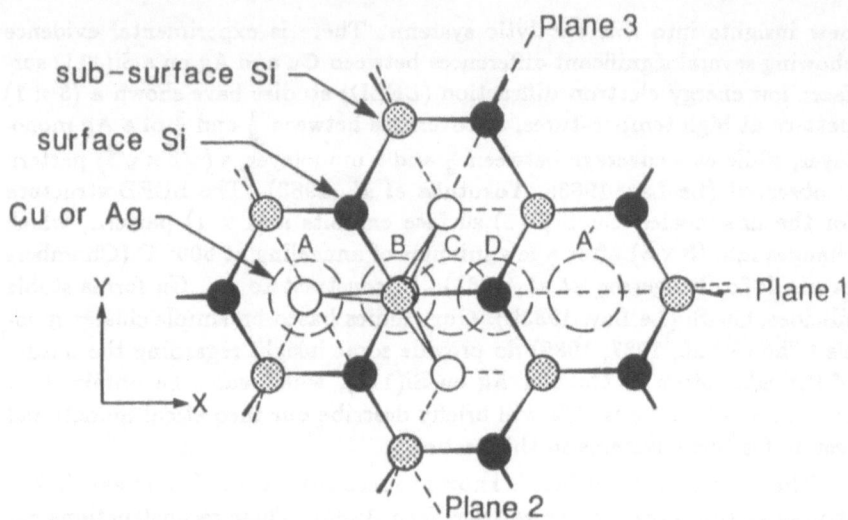

Figure 1: Si(111) cluster model. The four different adsorption sites are labeled as **A** (threefold hollow), **B** (on top of fourfold coordinated subsurface Si), **C** (bridge), **D** (on top of threefold coordinated surface Si), and **A'** (threefold hollow with missing third layer Si).

with a total binding energy of 92 kcal/mol. This is equivalent to 30.7 kcal/mol for each Cu–Si bond, and is in excellent agreement with the value obtained in Lewis *et al.*, 1988.

From isothermal desorption spectroscopy (Le Lay *et al.*, 1978), the binding energies of Ag/Si(111) (3 × 1) and Ag/Si(111) ($\sqrt{3} \times \sqrt{3}$) were measured to be 68 and 65 kcal/mol, and the vibrational frequencies are 77 and 60 cm^{-1}, respectively. The binding energies were measured at a much higher coverage than our theoretical isolated atom model. Our theoretical binding energy for an isolated Ag atom on the unreconstructed and unrelaxed Si(111) surface is 72 kcal/mol. Julg *et al.* (Julg and Allouche, 1982) found that the binding energy per Ag atom would increase by about 0.2 to 0.3 eV as the coverage decreases from 1 to 1/3. This amounts to a change of 5 to 7 kcal/mol in binding energy. Thus, our theoretical results are in agreement with low coverage (or submonolayer coverage) experimental data.

The lateral diffusion barriers, assuming an unreconstructed rigid Si(111) surface, are found to be 12 and 8 kcal/mol for Cu and Ag, respectively. Calculations for Cu and Ag atoms moved towards the interior of the cluster, including the geometric relaxation of the nearest neighbor Si atoms, demonstrate that Cu has a much lower vertical penetration barrier than does Ag (4 vs. 53/kcal/mol). This theoretical result reveals a qualitative differ-

Table 1: Equilibrium heights h (Å), adsorption energies E_b (kcal/mole) for Cu and Ag atoms on an unreconstructed Si(111) surface.

	Cu^{Theory}	Ag^{Theory}				Ag^{Exp}
h	0.74^a	1.48^a	$1.50 - 1.93^b$	$2.18 - 2.19^c$	0.05^d	0.7^e
E_b	92^a	72^a	$23 - 42^b$	$69 - 120^c$	—	$65 - 68^f$

a From (Chou et al., 1987, 1989)
b For AgSi₃ (Barone et al., 1980)
c Single Ag in three-fold site in (Julg and Allouche, 1982)
d From (ZHeng et al., 1988)
e From (Stöhr and Jaeger, 1982; Stöhr et al., 1983)
f From (Le Lay et al., 1978)

ence between Cu and Ag on the Si(111) surface: at elevated temperatures, Cu can be expected to penetrate into the Si crystal, whereas Ag remains above the surface Si atoms and can be expected to evaporate into the gas phase as the temperature is raised further. Our calculations also reveal that Cu weakens the backbonds between the surface and the underlayer silicon atoms, while Ag has a significantly smaller effect.

Experimental studies of the desorption of Cu and Ag on Si(111) surface exhibit marked differences in these two systems as a function of temperature. Auger electron spectroscopy and thermal desorption studies of Cu/Si(111) and Ag/Si(111) show that at submonolayer coverages Cu diffuses into the bulk, whereas Ag remains on the surface and thermally desorbs (Gentle and Owen). Thus, our theoretical conclusions are in agreement with the experimental findings.

Our theoretical results also suggest that the catalytic activity of Cu and the absence of activity of Ag in the synthesis of methylchlorosilanes ("direct process") is possibly due to the ability of Cu to penetrate into the surface thus forming the initial stages of a copper–silicide, whereas Ag stays at the surface and desorbs at higher temperatures.

Alkali Metal Adsorption on the (2×1) Si(100) Reconstructed Surface

Alkali metals deposited on the silicon surface represents another trend in the study of metal/Si interfaces. The changes of electronic properties, especially the work-function lowering effect, make these systems interesting for technical applications, e.g., as electron emitters or as spin-polarized electron guns. There are also applications in negative-electron-affinity devices and unique catalytic reactions mediated by alkali–metal agents. Alkali metals have rather simple electronic properties; their adsorption on the Si surface can be treated as simple chemisorption systems which are attractive for theoretical consideration. However, the nature of the alkali metal/silicon bonding is not well established nor is the geometry of the adsorption site. We shall discuss the K and Na/Si(100) adsorption system in this section.

It is generally accepted that the Si(100) surface reconstructs into a (2 × 1) structure by forming surface dimers that are arranged in parallel rows (Levine, 1983). In this work, we first assume that the substrate is described by a simple dimer model where only the surface atoms form a symmetric dimer with dimer bond length equal to the bulk bond length, in order to find out possible adsorption site. Then we use the optimised dimer structure (Tang *et al.*) as the substrate, and calculate the adsorption energy at cave and pedestal site. Finally, the adsorption cluster is optimised according to the force and total energy. We assume that the alkali metal adsorption only affects the first four layer of Si; the layers below fifth layer are kept in the ideal position. Since our purpose is to know the relative stability of the adsorption site rather than to get the exact information on the structural changes upon metal adsorption, we did not optimize all the atoms in the first four layers, instead we optimized the nine silicon atoms nearest to the adsorption atom within the first to fourth layers.

1   K Atom Adsorption on Si(100) Surface Based on a Simple Dimer Model

Four possible adsorption sites were studied with only one K atom on the surface. The sixfold cave site was found to have the lowest binding energy (−3.7 eV), whereas the pedestal site, the bridge site and the valley bridge site have −3.0 eV, −2.95 eV and −3.4 eV, respectively. The K–Si bond length for the corresponding four clusters are 3.22 Å, 3.44 Å, 3.11 Å and 3.54 Å. Although the valley bridge site has relatively low binding energy, this chemisorption site is ruled out since the K–Si bond length does not agree with the experimental value (Kendelewicz *et al.*, 1988). Other theoretical calculations have also found that the valley bridge site is less binding than the cave and pedestal site (Ramires, 1989; Batra, 1990). In addition, they also showed that the bridge site is less stable than the cave and pedestal site. Hence, in the following the emphasis will be focussed on the cave and pedestal site.

In simulating the higher coverage case, it is generally assumed that a one-dimensional array of alkalis is formed on the surface. Thus K-chain structure on cave and pedestal sites have been studied using several different clusters as illustrated in Ye *et al.*, 1989. The resulting K–Si bond lengths for cave and pedestal chains are 3.22 Å and 3.21 Å which agree with the observed value of 3.14 ± 0.10 Å(Kendelewicz *et al.*, 1988). It is interesting to find out that the K–Si bond length at the cave site remain fixed when coverage changes whereas that at the pedestal site changes with the coverage. For K chain chemisorption on the cave and pedestal sites, the charge transfer from K to the surface is found to be 0.53 e and 0.63 e, respectively. Hence, the bonding is not simply ionic or covalent, but is instead of a mixed type.

The above results were obtained assumming a silicon substrate described with a simple dimer model (Levine, 1983). Experiments and theoretical studies have revealed that the Si(100)2 × 1 surface has relaxations down to several layers. We have recently studied the dimer models of the Si(100)2 × 1 clean surface and have worked out an optimized dimer structure using the DMOL method (Tang *et al.*). In the following section, single K and Na atoms adsorption on cave and pedestal sites will be given using the optimized dimer structure as the silicon substrate.

2   K and Na Atoms Adsorption on Si(100) Surface Based on Optimized Dimer Structure

Four different size clusters have been used for the study of the cave site in order to have a knowledge of cluster size effects on the result. The smallest cluster consists of 24 atoms, whereas the biggest one has 70 atoms (see Fig.2(a)–Fig.2(d)). Each cluster contains at least 4 layers of silicon atoms. The calculated adsorption energy, the K–Si bond length and the vertical distance of K from the first Si layer are shown in Table 2. It can be seen that the bond length and the vertical distance of K from the Si surface are almost unchanged for the clusters of different size. This implies that the interaction between the K atom and the Si surface is mainly short-ranged in nature. The calculated adsorption energies are not consistent among the four clusters chosen. However, when the cluster size is over 50 atoms, the adsorption energy converges. It is suggestive from these results that clusters smaller than 50 atoms are not large enough to get correct adsorption energies for the study of the K/Si(100) adsorption system.

Table 2: The adsorption energy, ΔE, the K–Si bond length, d_{K-Si}, and the vertical distance between K atom and the silicon surface $d_\perp$ for K adsorbed on cave site of Si(100) surface. Energies are in eV unit, distances in Å.

	$K_1Si_9H_{14}$	$K_1Si_{17}H_{22}$	$K_1Si_{23}H_{30}$	$K_1Si_{31}H_{38}$
ΔE	−2.36	−2.42	−2.20	−2.20
d_{K-Si}	3.21	3.21	3.21	3.21
$d_\perp$	1.69	1.70	1.70	1.70

For the pedestal site, a cluster which contains 1 potassium, 29 silicon and 28 hydrogen atoms has been chosen (Fig.2(e)). The resulting adsorption energy and the K–Si bond length are −2.26 eV and 3.34 Å. It seems that the pedestal site has lower binding energy (by only 0.06 eV) and thus should be the stable adsorption site. However, as demonstrated by Batra, (Batra, 1990) by including the substrate relaxation in the alkali–metal/Si system, the results can be different from those obtained without relaxation. Although experimentalists do not find any significant changes with alkali metal chemisorption on Si, it is worthwhile to do the substrate relaxation since the exact structure of the clean surface itself is not known clearly (Chadi, 1979; Yin and Cohen, 1981; Holland et al., 1984; Bechstedt and Reichardt, 1988; Haneman, 1987). The results including lattice relaxation will be given in the following.

The calculation of sodium adsorption on Si(100) was carried out in a similar way to that for K/Si(100). The $Na_1Si_{23}H_{30}$ and the $Na_1Si_{29}H_{28}$ clusters are chosen for the cave and pedestal site, respectively. The calculated binding energies for cave and pedestal site are -1.90 eV and -2.04 eV. Since the energy difference between cave and pedestal site is relatively large, the lower energy site (pedestal) can be considered to be the preferred adsorption site.

3   K and Na Adsorption on Si(100) with Substrate Relaxation

For the cave site, we start with the $K_1Si_{17}H_{22}$ cluster (cf., Fig.2(b)), and optimized only the nine atoms (as shown in Fig. 2(a)) in the first four layers which are nearest to the adsorbed atom. Convergence is achieved when all the forces acting on the atom assigned are less than 4.0×10^{-3} (Ry/a.u.). The optimized coordinates obtained are then used in larger clusters, $K_1Si_{23}H_{30}$ and $K_1Si_{31}H_{38}$ (cf., Fig. 2), and the self-consistent procedure repeated. The results for the three clusters are shown in Table 3. Note that the small cluster gains more energy than the larger one after lattice relaxation. This is due to the effect of the boundary which can only be reduced in the larger sized cluster. It can be seen again from the Table that the two clusters which have more than 50 atoms give the same results for adsorption energy and bond length. The bond length becomes smaller after relaxation, but agrees well with the experimental value (Kendelewicz et al., 1988).

Table 3: Results of K adsorption on cave site of Si(100) surface with lattice relaxation, ΔE, d_{K-Si} and $d_\perp$ have been described in Table 2.

	$K_1Si_{17}H_{22}$	$K_1Si_{23}H_{30}$	$K_1Si_{31}H_{38}$
ΔE	-2.72	-2.38	-2.38
d_{K-Si}	3.15	3.16	3.16
$d_\perp$	1.70	1.71	1.71

As to the pedestal site, the optimization done on the $K_1Si_{29}H_{28}$ cluster gives an adsorption energy of -2.39 eV and a bond length of 3.33 Å,

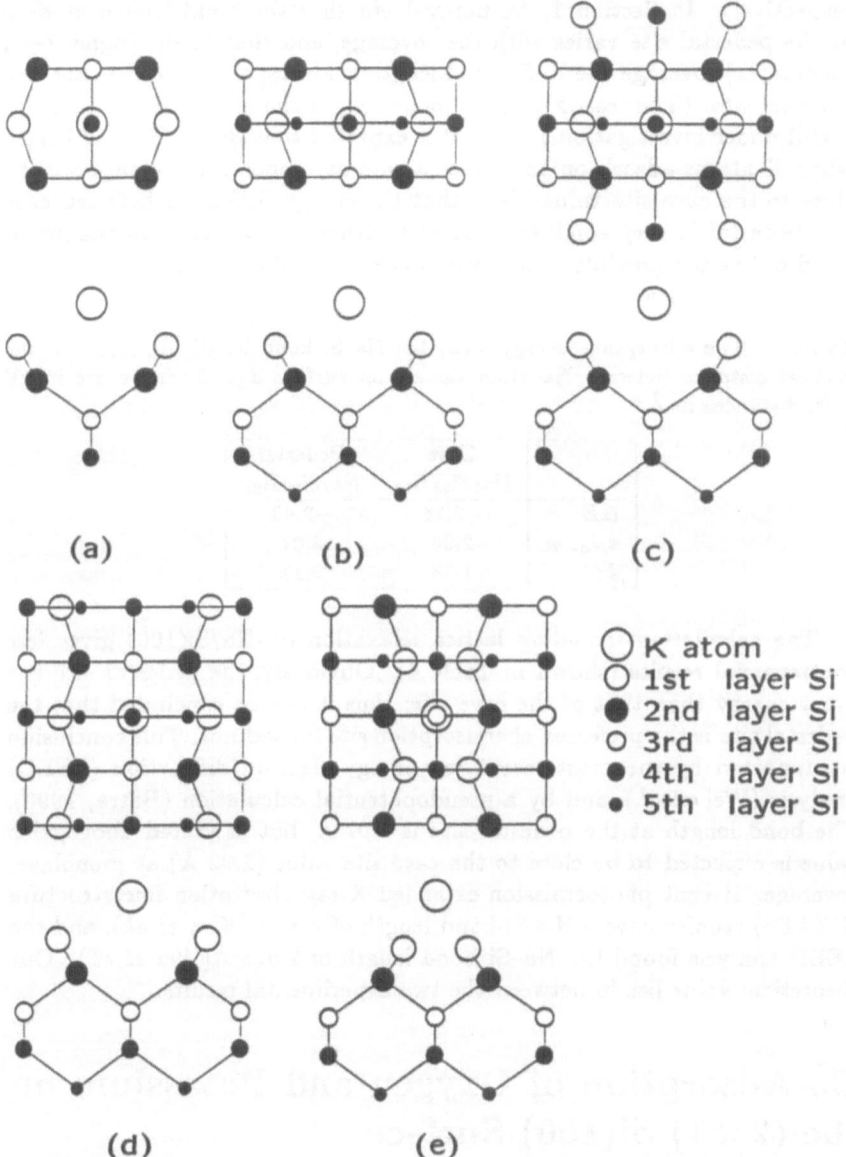

Figure 2: Atomic cluster models simulating the cave ((a)–(d) and the pedestal site (e)) of Si(100) 2×1 surface. All figures are shown in top view and side view. (a) $K_2Si_9H_{14}$ cluster; (b) $K_1Si_{17}H_{22}$ cluster; (c) $K_1Si_{23}H_{30}$ cluster; (d) $K_1Si_{31}H_{38}$ cluster; (e) $K_1Si_{29}H_{28}$ cluster. The hydrogen atoms are not shown in this figure.

respectively. In Section 1, we pointed out that the bond length of K–Si at the pedestal site varies with the coverage, and that in the higher (~ 1 monolayer) coverage the K–Si bond length is almost same for the cave and pedestal site. (The reason why the bond length changes with the coverage is still under investigation.) Thus, it is expected that the K–Si bond length when K atoms adsorb on the pedestal site at monolayer coverage will be close to the cave site value. Now that the energy difference between cave and pedestal is very small (~ 0.01 eV), either of the two sites cannot be ruled out as the possible chemisorption site for potassium atoms.

Table 4: The adsorption energy, ΔE, the Na–Si bond length d_{Na-Si} and the vertical distance between Na atom the silicon surface $d_\perp$. Energies are in eV unit, distances in Å.

	Cave $Na_1 Si_{23} H_{30}$	Pedestal $Na_1 Si_{29} H_{28}$
ΔE	-2.14	-2.40
d_{Na-Si}	2.93	3.07
$d_\perp$	1.28	2.11

The calculation including lattice relaxation on Na/Si(100) gives less controversial results (shown in Table 4). Obviously, the pedestal site has lower energy than that of the cave site; thus it can be concluded that the pedestal site is the preferred chemisorption site for sodium. This conclusion is supported by the most recent low-energy electron diffraction (LEED) analysis (Wei *et al.*) and by a pseudopotential calculation (Batra, 1990). The bond length at the pedestal site is 3.07 Å, but as stated above, this value is expected to be close to the cave site value (2.93 Å) at monolayer coverage. Recent photoemission extended X-ray absorption fine structure (EXAFS) studies gave a Na–Si bond length of 2.80 Å(Kim *et al.*), and the LEED analysis found the Na–Si bond length of 2.975 Å(Wei *et al.*). Our theoretical value lies in between the two experimental results.

Co-Adsorption of Oxygen and Potassium on the (2×1) Si(100) Surface

The oxidation (and nitridation) of semiconductor surfaces is interesting and important for both device applications and for understanding the fundamentals of surface physics. Most recently, experimental observations (Soukiassian *et al.*, 1988; Franciosi *et al.*, 1987) indicated that the existence of an alkali overlayer not only promotes the oxidation (or nitridation) of the Si(100) surface but also that the alkali serves as a true catalyst in that it can be desorbed easily after catalytic oxidation. As a first step to provide such potentially useful information, we present in this section results of the first extensive studies on the K enhanced oxidation of the $Si(100)2 \times 1$ surface (Ye *et al.*).

We first examined a single free O_2 molecule. The calculated bond length of 2.684 a.u. (1.4203 Å) is very close (within 0.15%) to the observed value (Landolt–Bornstein, 1987) of 1.4181 Å. Next we used the total energy approach to find appropriate sites for O_2 to be adsorbed on a clean $Si(100)$ surface. For this, we employed several smaller clusters to simulate the different adsorption sites on $Si(100)$ (Ye *et al.*, 1989). An $O_2Si_9H_{14}$ cluster (and Si_9H_{14}) was adopted for the cave site; an $O_2Si_7H_8$ cluster (and Si_7H_8) for the pedestal site and $O_2Si_9H_{12}$ (and Si_9H_{12}) for the bridge site.

The calculation showed that the O_2 molecule lying horizontal and parallel to the surface dimer at the bridge site is found to have the lowest energy (about 1.68 eV lower in energy than that for the pedestal site, and 2.36 eV lower than that for the cave site). Henceforth, we use bridge site chemisorption with the O_2 molecule lying horizontally above and parallel to the surface Si dimer. As shown in the last section, K atom adsorbed on cave and pedestal sites have almost the same adsorption energy; thus, here we chose the cave site.

The third step was to calculate separately K and O_2 adsorption on $Si(100)2 \times 1$, each with a larger cluster so that the results can be compared later on with calculations for $O_2/K/Si(100)2 \times 1$. An $O_2Si_{44}H_{46}$ cluster and its corresponding clean surface cluster, $Si_{44}H_{46}$, were studied (cf., Fig. 3 with the six K atoms removed). The following results were obtained: (i) there is a charge transfer of 0.16 e from the Si surface to each O atom; (ii) after chemisorption and charge transfer, the O_2 bond length is stretched by 11.2% from 2.684 a.u. to 2.984 a.u.; (iii) the Si–O bond length is found to be around 3.59 a.u. (1.90 Å); and (iv) the chemisorption energy calculated from Eq. (2) is found to be 4.53 eV/O_2.

The K/Si(100) system was calculated separately with a $K_6Si_{44}H_{46}$ cluster (cf., Fig. 3 with the O_2 molecule removed), having all six K atoms on the cave sites. A K–Si bond length of 3.23 Å is obtained which is very close to our previous result (Ye *et al.*, 1989) (3.22 Å) for a chain model, $K_6Si_{39}H_{44}$; charge transfer from each K atom to the Si surface is found to be 0.63 e in this model instead of 0.53 e for the chain model.

It is very interesting that the optimized height of O_2 on $Si(100)$ (3.10 a.u.) is almost the same as that of K on $Si(100)$ (3.05 a.u.) in the two separate calculations above, but with each sitting on its own appropriate site — the cave site for K and bridge site for O_2. Now, the separation between the two K chains is 14.518 a.u., which is wide enough for an oxygen molecule to fall in-between onto the bridge site.

For this reason, we studied the oxidation of $Si(100)2 \times 1$ in the presence of a K overlayer using the cluster $O_2K_6Si_{44}H_{46}$ shown in Fig. 3. The following results were obtained from total energy calculations:

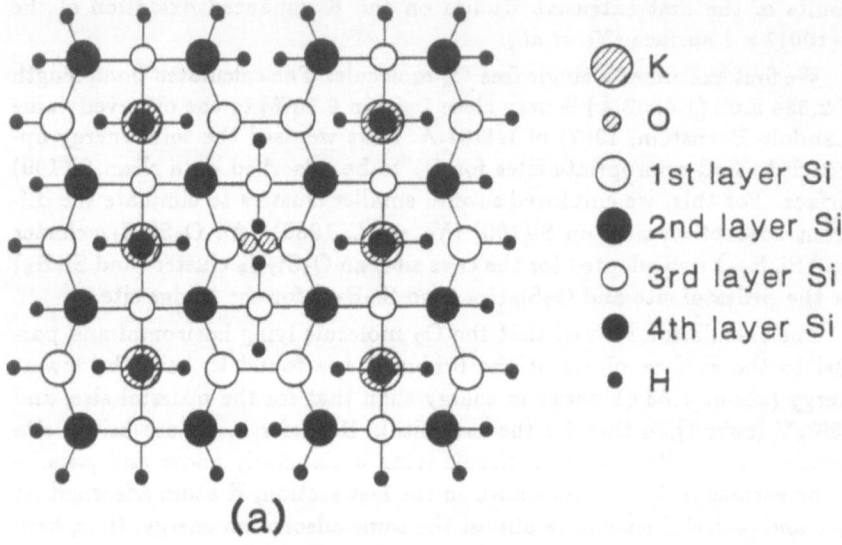

(a)

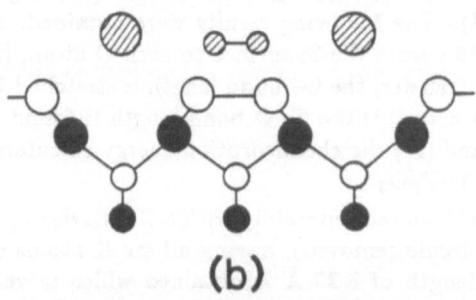

(b)

Figure 3: The $O_2K_6Si_{44}H_{46}$ cluster used in the calculation of $O_2/K/Si(100)$: (a) top view and (b) a simplified side view.

i. There is a change of height between K and O_2. The height of O_2 is lowered a little bit from 3.10 a.u. (without K) to 3.05 a.u., while that of K is raised to a substantially higher level; if all the six K atoms are raised simultaneously, a lower total energy is achieved at a height of 3.20 a.u.; when only the two K atoms next to O_2 are raised, with the other four fixed at a height of 3.0 a.u., the total energy is lowered even more, and reaches a maximum for a height of 3.45 a.u. for the two K atoms. It thus appears likely that, due to the insertion of an O_2 molecule, adjacent K atoms tend to be "squeezed" upward. In the process, the K–Si bond length is increased from 3.23 Å to 3.36 Å(by 4.0%), which implies a weaker binding of K to the Si surface. Thus, the experimental result (Soukiassian *et al.*, 1988; Franciosi *et al.*, 1987) that the alkali metal catalysts were removable from the surface after catalytic oxidation, can be easily understood on the basis of these results.

ii. The charge transferred to each O_2 molecule is increased from 0.32 e (for chemisorption on clean Si(100)) to 0.56 e (on K covered Si). This charge, transferred from the positively charged K atom to the negatively charged O_2, goes into the previously unoccupied antibonding orbitals and thereby promotes the tendency to dissociate.

iii. While no additional change in the bond length of O_2 is found when K is co-adsorbed, the vibrational frequency is reduced by $\sim 8\%$ — again indicating a tendency to dissociate.

iv. The other essential result of this calculation is that more binding energy is found for O_2 to be chemisorbed. Using Eq. (1), we determined the calculated chemisorption energies for O_2 on the K covered Si surface (~ 4.77 eV/O_2) and for O_2 on the clean Si surface (-4.54 eV/O_2) to yield, an increase of 0.23 eV/O_2, due to the existence of the K overlayer, while the K adsorbate becomes less bound.

Acknowledgment

We are grateful to our colleagues in the work cited for close collaboration. This work was supported by the Office of Naval Research (Grant No. N01004-89-J-1290).

References

Barone, V.; Del Re, G.; Le Lay, G.; Kern, R.; 1980, *Surf. Sci.* 99:223.

Batra, I.P.; 1990, private communication.

Bechstedt, F.; Reichardt, D.; 1988, *Surf. Sci.* **202**:83.

Chadi, D.J.; 1979, *Phys. Rev. Lett.* **43**:43.

Chambers, S.A.; Anderson, S.B.; Weaver, J.H.; 1985, *Phys. Rev.* **B32**:581.

Chou, S.-H.; Freeman, A.J.; Grigoras, S.; Gentle, T.M.; Delley, B.; Wimmer, E.; 1987, *J. Am. Chem. Soc.* **109**:1880.

Chou, S.-H.; Freeman, A.J.; Grigoras, S.; Gentle, T.M.; Delley, B.; Wimmer, E.; 1989, *J. Chem. Phys.* **89**:5177.

Delley, B.; 1986, *Chem. Phys.* **110**:329.

Delley, B.; 1990, *J. Chem. Phys.* **92**:508.

Delley B.; to be published.

Delley, B.; Ellis, D.E.; Freeman, A.J.; 1983a, *Phys. Rev. Lett.* **50**:488.

Delley, B.; Ellis, D.E.; Freeman, A.J.; Baerends, E.J.; Post, D.; 1983b, *Phys. Rev.* **B27**:2132.

Franciosi, A.; Soukiassian, P.; Philip, P.; Chang, S.; Wall, A.; Raisanen, A.; Troullier, N.; 1987, *Phys. Rev.* **B35**:910.

Gentle, T.M.; Owen, M.J.; unpublished results.

Haneman, D.; 1987, *Rep. Prog. Phys.* **50**:1045.

Holland, B.W.; Duke, C.B.; Paton, A.; 1984, *Surf. Sci.* **140**:L269.

Julg, A.; Allouche, A.; 1982, *Intern. J. Quantum Chem.* **22**:739.

Kendelewicz, T.; Soukiassian, P.; List, R.S.; Woicik, J.C.; Pianetta, P.; Lindau, I.; Spicer, W.E.; 1988, *Phys. Rev.* **B37**:7115.

Kim, S.T.; Kapoor, S.; Hurych, Z.; Barbier, L.; Soukiassian, P.; to be published.

Landolt–Bornstein, New Series II/7., 1976, *Structure Data of Free Polyatomic Molecules*, p. 179, Springer–Verlag, New York.

Le Lay, G.; 1983a, *J. Vac. Sci. Technol.* **B1**:354.

Le Lay, G.; 1983b, *Surf. Sci.* **132**:169.

Le Lay, G.; Manneville, M.; Kern, R.; 1978, *Surf. Sci.* **72**:405.

Levine, J.D.; 1973, *Surf. Sci.* **34**:90.

Lewis, K.M.; McLeod, D.; Kanner, B.; 1988, in: *Catalysis 1987*, (Ed. J.W. Ward), p. 415, Elsevier Science Publishers B.V., Amsterdam.

Ramirez, R.; 1989, *Phys. Rev.* **B40**:3962.

Redondo, A.; Goddard III, W.A.; McGill, T.C.; 1982, *J. Vac. Sci. Technol.* **21**:649.

Ringersen, F.; Derrien, J.; Dangy, E.; Layet, J.M.; Mathiez, P.; Salram, F.; 1983, *J. Vac. Sci. Technol.* **B1**:546.

Soukiassian, P.; Bakshi, M.H.; Starnberg, H.I.; Bommannavar, A.S.; Hurych, Z.; 1988, *Phys. Rev. B* **37**:6496 and references therein.

Stöhr, J.; Jaeger, R.; 1982, *J. Vac. Sci. Technol.* **21**:619.

Stöhr, J.; Jaeger, R.; Rossi, G.; Kendelewicz, T.; Lindau, I.; 1983, *Surf. Sci.* 134:813.

Tang, S.P.; Freeman, A.J.; Delley, B.; to be published.

van der Veen, J.F.; 1985, *Surf. Sci. Rep.* 5:199.

von Barth, U.; Hedin, L.; 1972, *J. Phys.* C5:1629.

Wei, C.M.; Huang, H.; Tong, S.Y.; Glander, G.S.; Webb, M.B.; to be published.

Ye, L.; Freeman, A.J.; Delley, B.; 1989, *Phys. Rev.* B39:10144.

Ye, L.; Freeman, A.J.; Delley, B.; *Surface Science* to appear.

Yin, M.T.; Cohen, M.L.; 1981, *Phys. Rev. B* 24:2303.

Yokotuka, T.; Kono, S.; Susuki, S.; Sagawa, T.; 1983, *Surf. Sci.* 127:35.

Zheng, Qing-qi; Zeng, Zhi; Han, Rushan; 1988, *Surf. Sci.* 195:L173.

6
Gaussian-based Density Functional Methodology, Software, and Applications

DENNIS R. SALAHUB, RENÉ FOURNIER, PIOTR MŁYNARSKI,
IMRE PAPAI, ALAIN ST-AMANT, AND JIRO USHIO

We review some of the progress made by the Université de
Montréal group over the last few years, and especially over recent
months, towards developing, testing, and applying Gaussian-
based density functional techniques of increased accuracy, speed
and functionality. These improvements include : the use of
(relativistic) model core potentials, basis set preparation, more
efficient integral evaluation (following Obara and Saika), the
calculation of energy derivatives, automatic geometry
optimization, calculation of vibrational frequencies, a "hybrid"
method for simultaneous geometry and electronic structure
optimization that incorporates elements of the Car-Parrinello
approach along with conventional potential energy surface
walkers, and the use of non-local functionals involving the
gradient of the density. These elements have been incorporated
into a software package called deMon that is being tested and
applied over a wide range of systems and interaction types
(organic and inorganic molecules, transition-metal complexes,
clusters, and surface models, van der Waals interactions, hydrogen
bonds).

Introduction

Although Density Functional Theory can trace its roots far back to the early days
of quantum theory, to the Thomas-Fermi and Thomas-Fermi Dirac models of the
electronic structure of atoms and even, arguably, to the pre-quantum mechanical
model of Drude for the conduction electron gas in metals, it is only recently that
it has shown strong signs of having an appreciable impact on "mainstream"
theoretical chemistry (as evidenced, for example, by the workshop that lead to
this book). The development of "Density Functional Quantum Chemistry" has
been far from linear. Indeed, the earliest "real" chemical calculations were
offshoots of techniques developped by the solid state physicists for calculating
the band structure of solids. In fact, these were going on well before the formal

basis of DFT became clear following the papers of Hohenberg and Kohn (1964) and Kohn and Sham (1965). John C. Slater, his school and his spirit reigned over this "$X\alpha$" era which provided a wealth of information on solids and shaped a sizable part of modern solid state theory.

The first inroads to chemistry were made with the $X\alpha$–scattered-Wave method suggested by Slater and implemented by Keith Johnson (Slater, 1974; Johnson, 1973). The method borrowed the muffin-tin potential and the multiple scattering formalism from the band structure world and combined them with molecular boundary conditions. The result was a rather fast method that could be applied to complex (mainly inorganic) systems. Much basic insight into the electronic structure and spectroscopy of transition metal complexes, clusters, chemisorption models and the like was obtained in this "era", despite the "ugliness" of the muffin-tin potential, with its approximations of spherical (near the atoms) or constant (between atoms) potentials. This period was also characterized by some genuinely fierce arguments between proponents and opponents of the $X\alpha$–SW methodology. Part of the "problem" (besides the natural tendencies for polemics to arise when controversial, promising discoveries are made) resided in the difficulty of making comparisons with more conventional techniques on a "fair" basis. The $X\alpha$-SW method was at its best for (fixed geometry) calculations of globular compact systems; metals, and primarily transition metals, were at the forefront and these are extremely difficult to treat, with an adequate description of electron correlation, by the usual *ab initio* techniques. So a "dialogue de sourds" set in. The $X\alpha$-ists pointed to many *bona fide* triumphs (and to the fact that Hartree-Fock was terrible for metals and "routine" correlated methods were not much better); the *ab initio* practitioneers rightly (righteously?) replied that a method which couldn't even provide optimized geometries shouldn't really aspire to too lofty a position. Almost no one among the chemists was very concerned about Density Functional Theory and the formal basis of the $X\alpha$ method, although there were some interesting discussions about whether $X\alpha$ should be viewed as incorporating some account of electron correlation or whether it was "simply" an approximate version of the Hartree-Fock method (Sabin et al., 1984). It took some time for the $X\alpha$ community as a whole to "realize" that it was involved with DFT calculations and that Kohn-Sham exchange ($\alpha = 2/3$) could be combined to advantage with an electron gas correlation potential.

A major step both towards the development of more powerful methodology and also towards removing some of the confusion about the comparison of *ab initio* and $X\alpha$ or other DFT methods came with the proposal of an LCAO-$X\alpha$ technique by Sambe and Felton (1975) and its extension and improvement by the Florida group (Dunlap et al., 1979). First, this eliminated the muffin-tin potential and opened the possibility of geometry optimizations, vibrational analysis, etc. But, equally importantly, the method involved numerical approximations (orbital and auxiliary basis sets, grid points for exchange-correlation fitting) that could, at least for small systems, be expanded to reasonably converged limits so that one could be confident that remaining errors were primarily those of the density functional itself and not of the particular numerical machinery used. This was the first step, then, toward the hard "quantum chemistry" comparisons that are becoming increasingly abundant (e.g. this and other chapters of this book).

In Montreal, we adopted the LCAO methodology in the early 80's and since then we (and several other laboratories) have built on the early work of Dunlap et al. (1979) and have extended it in several directions, in terms of the methodology and computer programs and also the areas of application.

Our "bread and butter" research interests have been in the area of transition metal clusters and their interactions and reactions with atoms and molecules. There we wish to push the methodology as far as possible into the realm of reaction dynamics. But, for want of space and so as not to cloud some other important messages, there will be little of those subjects here (see e.g. Salahub 1987; Rochefort et al., 1990; Papai et al., 1990 for entries into that literature). Rather, the rest of the paper will emphasize methodological improvements and comparisons with other computational approaches. It will be a "mini-review", our main purposes being to give the "newcomer" as faithful a representation as possible of the performance of the methodology (functionality, advantages, and limitations) for systems where hard comparisons can be made. We have striven to include several very challenging situations, cases where it is known that *ab initio* treatments, while feasible, are "non trivial".

We hope that the few details we can give here will at least be enough to provoke the reader's curiosity and lead him to the literature. We also hope that some will become interested enough to "join the game". The systematic work needed for true progress is only beginning and there is a great need, and opportunity, for new participants.

The LCGTO-MCP-DF method and the program deMon

Over the last two years, St-Amant and Salahub, (A. St-Amant, Ph.D. thesis, to be published) have developed deMon (density of Montreal) a linear combination of gaussian type orbitals-model core potential-density functional (LCGTO-LSD-DF) program package which allows users to perform single point SCF calculations as well as geometry optimizations employing the analytical energy gradient expression of Fournier et al. (1989). (The calculation of energy derivatives is reviewed in the next section.) Additionally, deMon may be used to study such properties of chemical interest as dipole moments, Mulliken populations, and harmonic frequencies. These calculations make use of the near LSD limit exchange-correlation (XC) potentials of Vosko et al. (1980) and Wilk et al. (1982) (VWN); Perdew's non-local corrections to the exchange (Perdew et al., 1986a) and correlation (Perdew, 1986b) potentials may in turn be added to the VWN potential.

In deMon, the one-electron orbitals $\{\Psi_i\}$, are given by an LCGTO expansion :

$$\Psi_i(r) = \sum_\mu C_{\mu i}\, \mu(r) \tag{1}$$

where $\{\mu\}$ forms the basis of atom-centered GTO's, and $\{C_{\mu i}\}$ is the set of expansion coefficients for the ith one-electron orbital. The gaussian basis sets used in deMon have typically been obtained by optimizing Huzinaga's (1984) basis sets for LSD calculations (Andzelm et al., 1985). Good results are still obtained for simple organic molecules if flexible Hartree-Fock basis sets are used without first optimizing for LSD calculations.

Following the work of Sambe and Felton (1975) and that of Dunlap et al. (1979), the charge density (CD), ρ, and XC potentials, v^σ_{xc} ($\sigma = \alpha$ or β), are fitted by an LCGTO expansion in an auxiliary basis, i.e.,

$$\tilde{\rho}(r) = \sum_i a_i\, f_i(r) \tag{2}$$

$$\tilde{v}^\sigma_{xc}(r) = \sum_i b_i^\sigma\, g_i(r) \tag{3}$$

are respectively fits (fitted quantities will be indicated by tildes) to ρ and v^σ_{xc}, where $\{f\}$ is the auxiliary basis and $\{a\}$ the set of coefficients for fitting ρ, and $\{g\}$ is the auxiliary basis and $\{b^\sigma\}$ the set of coefficients for fitting v^σ_{xc}. These fits reduce integral evaluation from an N^4 process to an N^2M process (where N is the number of GTO's in the orbital basis and M is the total number of GTO's in the auxiliary basis and is typically of comparable size to N). The two-electron integrals,

$$[\mu\nu \mid 1/r_{12} \mid f] \tag{4}$$

$$<\mu\nu \mid g> \tag{5}$$

as well as the one electron integrals (which are identical to those in the Hartree-Fock formalism) are evaluated with the efficient formulas of Obara and Saika (1986) (this part of deMon has yet to be vectorized).

The coefficients for the fit to ρ, $\{a\}$, are obtained by a least squares fitting (LSF) procedure which minimizes the error in the Coulomb repulsion energy (Dunlap et al., 1979)

$$[\rho - \tilde{\rho} \,|\, 1/r_{12} \,|\, \rho - \tilde{\rho}] \tag{6}$$

subject to the constraint,

$$< \tilde{\rho} > = N_e \tag{7}$$

which ensures that the fitted charge density is normalized to the total number of electrons, N.

The coefficients for the fits to the XC potentials, $\{b^\sigma\}$, cannot be obtained analytically as above for $\{a\}$ (though Dunlap (1981) has developed an analytical method in the specific case of $X\alpha$). The LDF procedure must therefore be performed over a set of grid points centered about each atom. The LSF procedure minimizes the error in the fitted potentials over the sum of the grid points,

$$\sum_I \left(v_{xc}^\sigma (I) - \tilde{v}_{xc}^\sigma (I) \right)^2 W(I) W'(I) \tag{8}$$

where $W(I)$ is the volume element associated with the Ith point and $W'(I)$ is an arbitrary weight which may be assigned to the Ith point. An arbitrary weight may be introduced since the purpose of the grid is to perform a fit to the XC potentials and not to perform a numerical integration. Dunlap et al. (1986a, 1986b) have proposed the arbitrary weighting function,

$$W'(r) = \rho(r) / \varepsilon_{xc}(r) \tag{9}$$

where ε_{xc} is the exchange correlation energy density. This significantly reduces noise in the total energy when the volume elements associated to the grid points are altered by simple rotations of the system (Dunlap et al., 1986a, 1986b).

Many types of grids for fitting the XC potentials have been proposed. Most grids for molecular calculations are based on superimposed atom-centred grids (Pederson and Jackson (1990) have recently introduced a promising algorithm for the generation of grid points which partitions space into atomic spheres, excluded cubic regions, and interstitial parallelepipeds). For these atom-centred grids, many radial grids have been proposed, notably Herman and Skillman (1963), Dunlap (1986b), and Becke (1988). In addition to the radial grid, a suitable angular grid must be chosen. Angular grids designed for accurate integration on the unit sphere have been developed by McLaren (1963), Stroud (1971), Sobolev (1962), and Lebedev (1975, 1976). The latter has introduced quadrature formulas with 50, 110, and 194 points which respectively integrate exactly all spherical harmonics of order 11, 17, and 23 or less. These three angular grids are those employed in deMon. The radial grid used in deMon was inspired by Becke (1988a), differing only in the fact that it uses the Gauss-Legendre quadrature scheme instead of Gauss-Chebyshev.

Having chosen a suitable radial and angular grid, the approach taken to treating the effect of overlapping atom-centred grids in the molecular system must be addressed. DeMon uses Becke's (1988a) approach to decomposing the molecular integration into a sum of one-centre, atom-like integrations. In this approach, a relative weight, $W'(I)$, is assigned to each grid point. The value of $W'(I)$, ranges from unity when it is near the nucleus about which the grid point is centred to zero when it is in the vicinity of any other nucleus and does so in a smooth fashion.

Previous experience (Fournier et al., 1989) indicated that modest XC fitting bases are inadequate when calculating the XC contribution to the LCGTO-MCP-DF energy gradient. The XC term of the energy gradient expression is therefore calculated numerically over an augmented set of grid points while the XC contributions to the Fock matrices, in the preceding SCF procedure are calculated analytically following the numerical LSF procedure using a modest XC auxiliary basis and a much smaller subset of grid points, as suggested originally by Dunlap et al. (1979). This "analytical energy - numerical gradient" discrepancy may be esthetically unappealing but seems to be necessary if one wishes to have both the CPU savings in the SCF step afforded by the XC LSF procedure and also an accurate energy gradient. The discrepancy does not seem to be a source of error since the agreement between the optimized geometries obtained by minimizing the norm of the energy gradient and the true LCGTO-MCP-DF optimized geometries (calculated by fitting a set of single point SCF calculations) is as good and often better than previously reported results (Papai et al.,1990 and certain results presented later in the present article) where larger XC fitting bases were employed and the XC term of the energy gradient expression, was calculated analytically and not numerically. Results of this new approach to calculating the XC contribution will be presented elsewhere (St-Amant and Salahub, unpublished).

The CD and XC auxiliary fitting bases obviously cannot, as was the case for the orbital bases, be taken from previous work within the Hartree-Fock formalism. Auxiliary bases may be optimized so as to reproduce numerical DF atomic calculations in a fashion similar to the optimization of the orbital basis sets (Andzelm et al., 1985a). A simpler method is the creation of atomic auxiliary bases with exponents (the auxiliary bases are uncontracted so contraction coefficients need not be considered) determined by the exponents of the corresponding atomic orbital basis. The justification behind such an approach is that if the orbital basis is suitable for describing a one-electron orbital, ψ_i, then a CD fitting basis consisting of doubled exponents should well describe $\rho = \Sigma \mid \psi_i \mid^2$. Further, if these exponents in the CD fitting basis are divided by three, an adequate XC fitting basis should be created since it is known that the XC potentials scale roughly as $\rho^{1/3}$. For the LCGTO-MCP-DF calculations to be feasible, not all exponents in the orbital basis should be chosen to create the auxiliary bases. The selection of the appropriate exponents is somewhat of an "art" and calculations on known systems the only test of their quality.

In addition to auxiliary bases, further computational savings may be afforded in the LCGTO-MCP-DF formalism by using model core potentials (Andzelm et al., 1985b). As in the Hartree-Fock formalism (Sakai et al., 1982; Huzinaga et al., 1984), MCP's are used to replace the effect of frozen, chemically uninteresting core electrons. The use of MCP's may, in addition to reducing CPU times, lead to improved results over "all-electron" calculations since the effect of basis set superposition error (BSSE) may be significantly reduced and relativistic effects in the system may be treated approximately by incorporating scalar relativistic effects into the MCP and treating the valence electrons with a non-relativistic Hamiltonian.

An MCP consists of two parts. The first, V_{MCP}, is a potential given by

$$V_{MCP}(r) = \sum_k A_k \exp\left\{-\alpha_k r^2\right\} / r \tag{10}$$

with the constraint

$$\sum_k A_k = Nc \tag{11}$$

where Nc is the number of core electrons. To ensure orthogonality of the valence orbitals to the frozen core orbitals, the MCP also includes a projection operator of the form

$$-\sum_c 2\varepsilon_c |\psi_c><\psi_c| \tag{12}$$

where c runs over all of the core orbitals and ε_c is the corresponding core orbital energy. To make the use of MCP's more efficient, V_{MCP} and the projection operator are averaged over both spins, a good approximation since any core effects from spin-polarization are swamped by a large Coulomb potential (Andzelm et al., 1985b).

The procedure taken to optimize MCP's is discussed in Andzelm et al. (1985b). Briefly, the core and valence orbitals are first fitted by an LCGTO expansion in a LSF procedure designed to minimize the error relative to a numerical atomic calculation. Orthogonality between all of the orbitals, core and valence, is maintained throughout. The potential of the core electrons is then fitted to the form of V_{MCP} in equation (10). Due to basis set incompleteness in the valence space, it is then desirable to slightly readjust the exponents and coefficients of V_{MCP} to reproduce valence orbital properties obtained in the numerical calculation.

Car-Parrinello   Techniques   and   deMon's "Hybrid" Geometry Optimization Scheme

Historically, deMon was written as an attempt to incorporate the ideas of Car and Parrinello (CP) (1985) into the LCGTO-MCP-DF formalism. The CP approach is a unified approach to molecular dynamics (MD) and quantum chemistry. Previous MFD simulations relied on empirical potentials (albeit sometimes from sophisticated quantum mechanical calculations) which provided no information on the evolution of the system's electronic structure. An MD simulation in the CP formalism is in principle exact within the approximations that :

a) the electronic structure as calculated in the DF formalism (or whatever formalism, HF, MP2, CI, etc.) is exact and

b) the electronic and nuclear coordinates can be decoupled (the Born-Oppenheimer approximation).

The CP algorithm is based on the introduction of a fictitious Lagrangean to define the motion of the nuclear coordinates, $\{R_I\}$, as well as the "motion" of the electronic parameters (in basis set techniques such as the LCGTO-MCP-DF method, these would be the LCGTO expansion coefficients, $\{c_{\mu i}\}$, and possibly even the exponents and contraction coefficients of the contracted GTO's themselves) used to define the occupied molecular orbitals, $\{\psi_i\}$. This Lagrangean is defined as

$$L = \sum_i 1/2\, m_i < \dot{\psi}_i | \dot{\psi}_i > + \sum_i 1/2\, M_I\, \dot{R}_I^2 - E\left(\{\psi_i\}, \{R_I\}\right)$$

(13)

where E is the true energy of the system, M_I is the mass assigned to the Ith nucleus (which may or may not be the actual mass), m_i is the mass assigned to the ith occupied molecular orbital (clearly an arbitrary and fictitious quantity), and a dot above a parameter indicates differentiation with respect to time.

From the above Lagrangean, equations of motion are derived for both the $\{R_I\}$ and the $\{\psi_i\}$. If appropriate masses are chosen for the electronic parameters and a suitable time step is employed for the numerical integration, an exact MD simulation within the DF formalism and the Born-Oppenheimer approximation may be performed. Many recent papers have been devoted to such MD simulations on amorphous systems (Car et al., 1988; Andreoni et al., 1987).

If at some point in the simulation, the system becomes "frozen", i.e., the $\{R_I\}$ and $\{\psi_i\}$ become stationary, and the forces on these parameters have vanished, then the above Lagrangean reduces to $E\left(\{\psi_i\}, \{R_I\}\right)$ and the system is in its ground state at a local minimum in its potential energy surface (PES). This simultaneous scheme for optimizing the nuclear and electronic structures is different from the traditional approach to investigating PES (Schlegel, 1987), i.e.,

i) optimize the electronic structure at the present geometry via an SCF procedure,

ii) calculate the forces on the nuclei in the system,

iii) displace the nuclei to create a new geometry and return to step i) if the forces calculated in step ii) are non-negligible (otherwise stop because the system is at a local extremum in its PES).

The CP approach has been used to elucidate the equilibrium structures of a wide range of molecules (e.g. Needels et al., 1988; Tarnow et al., 1989; Allan et al., 1987; Payne et al., 1988). However, this approach has been used with delocalized basis sets (floating s-type gaussian (Pederson et al., 1988)) and most usually plane wave bases which are inadequate when used to describe the structure of transition metal systems. An attempt was therefore made to incorporate the CP technique into the LCGTO-MCP-DF formalism. Somewhat analogous techniques have recently been examined for *ab initio* methods by Head-Gordon and Pople (1989). A straight implementation of the technique proved to be inefficient since CP simulations or geometry optimizations require many hundreds or thousands of time steps. With a plane wave basis, integral evaluation is rapid and only the Hellmann-Feyman force (Pulay, 1979) need be calculated to obtain the nuclear energy gradient. In our formalism, integral evaluation is not as rapid and Pulay-corrections ((Pulay, 1979) to the gradient are required since the basis sets used are "atom-centered" and any scheme requiring hundreds of intermediate geometries is presently unattractive computationally compared to established algorithms (Schlegel, 1987) which converge to local minima within roughly one geometry per internal degree of freedom.

A "hybrid" geometry optimization scheme (St-Amant et al, 1990) was therefore proposed for deMon which used both the fictious CP Lagrangean and the previously existing algorithms for geometry optimizations designed to keep the number of intermediate geometries to a minimum. Briefly, the "hybrid" scheme goes as follows :

i) the forces acting on the $\left\{C_{\mu i}\right\}$ and $\left\{R_I\right\}$ (the derivatives of the total energy with respect to these parameters) are calculated,

ii) the $\left\{C_{\mu i}\right\}$ and $\left\{R_I\right\}$ are assigned masses and are displaced simultaneously, the former by steepest descent and the latter by the conjugate-gradient method (Schlegel, 1987) (we are only interested currently with locating minima so we dispense with MD *per se*),

iii) the orthonormality constraints which the $\left\{C_{\mu i}\right\}$ must satisfy are reestablished by the SHAKE algorithm (Ryckaert et al., 1977).

iv) the displacement in the system having been too drastic for the $\left\{C_{\mu i}\right\}$ to remain on the Born-Oppenheimer surface, a "mini-SCF" comprising of roughly two to four iterations is performed to bring the system nearer to, but not necessarily back onto, the Born-Oppenheimer surface, and

v) steps i) to iv) are repeated until the $\left\{C_{\mu i}\right\}$ and $\left\{R_I\right\}$ become stationary.

The "hybrid" scheme therefore manages to keep the number of geometries under control while still making use of CP's fictitious Lagrangean to simultaneously optimize nuclear and electronic degrees of freedom. The number of geometries required by this scheme is greater than preexisting schemes but these schemes require SCF solutions at intermediate geometries while the "hybrid" scheme performs only "mini-SCF" processes which then yield SCF electronic structures as the system approaches a local minimum and the nuclear displacements become smaller. In the majority of cases tested (St-Amant et al., 1990), the CPU time required to optimize a molecular geometry within the "hybrid" scheme corresponded to less CPU time than that required to perform one SCF process per internal degree of freedom. Such promising results at an early stage of development of an algorithm is encouraging and further refinements to the procedure could surely make the "hybrid" scheme much more efficient.

ENERGY DERIVATIVES

The computation of analytic energy derivatives with increasingly sophisticated *ab initio* methods has been a major development in quantum chemistry in the past 20 years (Pulay, 1977; Pulay, 1987). Geometry optimization and the mapping of reaction coordinates (Schlegel, 1987) can now be done efficiently with routines using first energy derivatives, obtained from Hartree-Fock or correlated wavefunctions. Harmonic and cubic force constants for a number of small molecules have been obtained by calculating analytic second and third derivatives (Lee et al, 1989; Duran et al., 1989). Mixed derivatives with respect to nuclear coordinates and electric field components give infra-red and Raman band intensities (Pulay, 1987).

The so called analytic derivative methods are based on equations obtained by explicitly differentiating the energy expression within a given method. Two alternative ways of calculating E^{α} are usually much more time consuming ; i) numerical differentiation requires 2N energy calculations, N being the number of degrees of freedom; ii) using the Hellmann-Feynman (HF) theorem necessitates large bases to attain sufficient accuracy (Nakatsuji et al., 1982; Pulay, 1983a; Nakatsuji et al., 1983).

The first calculations of analytical gradients in DFT have apparently been made by Satoko less than 10 years ago (Satoko, 1981). Various authors later reported DFT gradients (Bendt et al., 1983; Satoko, 1984; Averill et al, 1985; Averill et al., 1986; Verluis et al., 1988; Fournier et al., 1989) but only recently has it been routinely used in geometry optimization and calculation of vibrational frequencies for polyatomics (Fan et al., 1988; Papai et al., 1990; St-Amant et al., 1990; see also various authors in this volume). No implementation of higher DFT energy derivatives have been reported but the equations for the LCGTO-LSD method have recently been derived (Fournier, 1990).

The derivation of formulas for derivatives in DFT follows closely that for the Hartree-Fock (Pople et al., 1979) and MCSCF (Pulay 1983b) methods.

However a number of differences arise because DFT methods use fits and/or numerical integration in evaluating the exchange-correlation (XC) -and sometimes other- contribution(s) to the energy and to the one-electron effective potential. The literature of DFT gradients can be confusing because expressions are often derived *assuming implicitly* perfect fits and/or perfect numerical integration and little discussion is made of the resulting inaccuracy of the calculated gradient. In the following we will discuss the differences between *ab initio* and DFT derivatives with emphasis on the LCGTO-LSD method (Sambe et al., 1975; Dunlap et al., 1979)

The HF theorem states that under certain conditions (e. g., if complete bases are used) E^α is :

$$E_{HF}^\alpha = \sum_i n_i < \phi_i / h^\alpha / \phi_i > + V_{NN}^\alpha \qquad (14)$$

where ϕ_i are spin-orbitals (SO), n_i are occupation numbers, h is the core hamiltonian operator and V_{NN} is the internuclear repulsion potential. It has been recognized long ago and emphasized by Pulay (Pulay, 1969; Pulay, 1987) that the HF gradient is not equal to the energy gradient in methods using incomplete atom-centered bases. Pulay advocated the evaluation of a term correcting the HF gradient for the incompleteness of the orbital basis set thus yielding the exact Hartree-Fock gradient. The expression including this "orbital basis correction" (OBC) can be written as :

$$E_{HF, OBC}^\alpha = E_{HF}^\alpha + 2 \sum_i n_i < \phi_i^{(\alpha)} | (F - \varepsilon_i) | \phi_i > = E_{HF}^\alpha + E_{OBC}^\alpha \qquad (15)$$

In (15-2) $\phi_i^{(\alpha)}$ denotes the derivative of the ith SO keeping the expansion coefficients constant and the ϕ_i's and ε_i's are from a fully converged SCF calculation. In every implementation of the KS equations there are other sources of error apart from OB incompleteness and lack of SCF convergence and, although most work on DFT gradients used Eq. (15), this expression *is only approximate*. However it has been shown (Fournier et al., 1989) that, in the LCGTO-LSD method, Eq (15) can be modified to include an explicit correction, E^α_{DBC}, for the incompleteness of the density fit basis. This is possible because in this method the energy is stationary with respect to variations (subject to a normalization constraint) of the fitting coefficients a_k (Dunlap et al., 1979).

$$\frac{\partial}{\partial a_k} \left(E - \lambda \int \rho(r) d^3 r \right) = 0 \text{ for all } k \qquad (16)$$

where :

$$\rho(r) = \sum_k a_k k(r)$$

Multiplying (16) by $a_k{}^\alpha$ and summing over k shows that all terms involving $a_k{}^\alpha$ in the gradient cancel *exactly* :

$$\sum_k \frac{\partial E}{\partial a_k} a_k{}^\alpha = 0 \qquad (17)$$

The gradient expression corrected for both orbital and density bases incompleteness is :

$$E^\alpha{}_{HF, OBC, DBC} = E^\alpha{}_{HF} + E^\alpha{}_{OBC} + \sum_k a_k \left[k^\alpha \mid r_{12}{}^{-1} \mid (\rho - \not{p}) \right] \quad (18)$$

Averill and Painter later obtained a similar result for the Harris energy expression (Averill et al., 1990). We stress that $E^\alpha{}_{DBC}$ is not small. Test calculations on diatomics (Fournier et al., 1989) showed that this term is larger in absolute value than $E^\alpha{}_{OBC}$ for commonly used bases. The numeral procedure of Verluis et al. (Verluis et al., 1988) however, which involves elimination of spurious one-center forces, succeeds in bringing this error to a near-zero value.

Eq. (18) is still not exact because of the incompleteness of the basis and grid used for evaluation of the XC energy. There is no easy way to evaluate the missing term $E^\alpha{}_{XCBC}$ because the XC energy evaluation is non-analytic and the fit of V_{XC} is non-variational. Fortunately, as discussed above, this term can be made very small by evaluating E_{XC} and related terms in the gradient by numerical integration while still retaining an analytical representation of the XC potential (See also Dunlap et al., 1990).

For computation of second derivatives (Fournier, 1990), one has to set up and solve the coupled perturbed Kohn-Sham (CPKS) equations to obtain the derivatives of the various coefficients. The first step in DFT methods that use fits is to relate the derivatives of the fitting coefficients to those of the SO coefficients. For the XC fit coefficients b_i of the LCGTO-LSD method this is done simply by differentiating the least-squares fit equation :

$$(b = Y^{-1} \cdot y)^\alpha = b^\alpha = -Y^{-1} \cdot Y^\alpha \cdot Y^{-1} \cdot y + Y^{-1} \cdot y^\alpha \qquad (19)$$

where Y is the overlap matrix of auxiliary functions and y the projection of the XC potential $v_{xc}{}^\alpha$ on the auxiliary basis, *both evaluated numerically*. Then, using the chain rule to rewrite $v_{xc}{}^\alpha$ as $v_{xc}{}^\alpha = (dv_{xc}/d\rho)\rho^\alpha$ and using this in Eq. (19) leads to a relation of the form :

$$b^{\alpha} = b_{o}^{\alpha} + L{:}P^{\alpha} \tag{20}$$

in which b_{o}^{α} and L can be readily evaluated and P^{α} is the derivative of the density matrix. A similar equation is found for density fit coefficients. From there on the derivation follows closely that of the Hartree-Fock method (Pople et al., 1979). We note that once the CPKS equations are solved, the *exact* gradient can be obtained as a by-product at a negligible extra computational expense.

To obtain a useful expression for the third derivatives one again makes use of the variational property of the density fit. The key step is to rewrite $v_{xc}^{\beta\gamma}$ using the chain rule and arrive at the following equality :

$$\int \left[\rho^{\alpha}(r) v_{xc}^{\beta\gamma}(r) - \rho^{\beta\gamma}(r) v_{xc}\alpha(r) \right] d^{3}r = \int \rho^{\alpha}(r) \rho^{\beta}(r) \rho^{\gamma}(r) \frac{d^{2}v_{xc}}{d\rho^{2}}(r) d^{3}r \tag{21}$$

In the implementation of $E^{\alpha\beta\gamma}$ an important aspect will be the efficient and accurate evaluation of the right hand side of Eq. (21) by numerical integration.

A full derivation of the equations for $E^{\alpha\beta}$ and $E^{\alpha\beta\gamma}$ can be found in (Fournier, 1990). We will conclude this section with a few remarks.

First, the equations for LCGTO-LSD energy derivatives are readily generalized to XC energy functionals depending on derivatives of the density to any order. Second, contrary to the equations derived for other DFT implementations, those for LCGTO-LSD derivatives include an exact correction for the incompleteness of the density fit basis. Inaccuracies in the gradient and third derivatives can result from the incompleteness of the basis and grid used in evaluating the XC energy. However, tests have shown that by evaluating certain terms numerically, the error on the gradient can be made virtually zero. We expect that a similar procedure will give very high accuracy on third derivatives as well. We believe that LCGTO-LSD analytic second derivatives will offer a moderate gain in efficiency compared to numerical differentiation but that the savings will be considerable for third derivatives. We are planning to implement second and third derivatives in the near future. While not an easy task, the programming work involved should not be much more difficult than that for the second and third derivatives of the Hartree-Fock energy. The availability of LCGTO-LSD higher derivatives would make possible the computation of force constants and vibrational band intensities for larger and more complex (e.g., containing transition metal atoms) molecules than currently possible with standard *ab initio* methods.

Selected Applications of Local Density Functional Theory

The LCGTO-MCP-LSD method has been applied for a wide range of systems in the last five years (see e.g. Salahub, 1987; Jones et al., 1989; Papai et al., 1990;

Rochefort et al., 1990, for entries into the literature). In spite of the encouraging results it has not been used to nearly the same degree as the HF method in geometry optimizations. This has been primarily due to the absence of analytical energy gradients, thus demanding numerous total energy calculations to construct potential energy surfaces. With the appearance of an analytical energy gradient expression (Fournier et al., 1989), complete geometry optimizations are computationally feasible within the LCGTO-MCP-LSD formalism. In addition, by numerical differentiation of analytical gradients (Pulay, 1969, 1977, 1983a, 1983b), vibrational frequencies can be evaluated providing normal coordinate analyses of polyatomic molecules.

We have recently shown (Papai et al., 1990) that the equilibrium geometry and the vibrational frequencies of polyatomic main group and organometallic molecules obtained by the LCGTO-MCP-LSD method are in good agreement with experiments, typically better than the HF method, and the results approach the quality of those of the MP2 method. As an illustration of this statement some examples are given below.

In table I. equilibrium bond distances and bond angles of four organic molecules (acetylene, ethylene, ethane and ethylene oxide) calculated by the LCGTO-MCP-LSD method are compared with the HF and MP2 results, as well as with experimental data. It is seen that the C-H bond lengths are systematically overestimated by about 0.02 Å relative to experiment. The same conclusion has been drawn in (Papai et al., 1990) for X-H bond distances (X = C, N, O, Si, P, and S). The error range is smaller ($\sim$0.01Å) for the C-C and C-O bond distances, but the discrepancy is not so systematic as for the X-H bonds. The bond angles agree typically within $1°$ of the experimental data. The X-H bond lengths provided by the HF and MP2 methods are in somewhat better agreement with experiment than the LCGTO-MCP-LSD results (see Table II of Papai et al., 1990). For the C-C and C-O distances, and for bond angles, LCGTO-MCP-LSD seems to give smaller errors than HF, and similar errors to the MP2 method.

The harmonic frequencies of ethylene oxide calculated by the LCGTO-MCP-LSD, HF and MP2 methods are compared with experimental data in Table II. It is seen that the LCGTO-MCP-LSD results fall very close to the experimental fundamentals ; the largest deviation is -79 cm^{-1} and the average relative error is 2.6%. The errors are not as systematic as they are for HF and MP2 harmonics for which in all but one case the frequencies are overestimated. The average relative errors for HF and MP2 are 11.6% and 4.9% respectively, but they would decrease, if we compared the theoretical frequencies with harmonized experimental values. Nevertheless, as it was concluded for main group molecules (Papai et al., 1990), the harmonics provided by the LCGTO-MCP-LSD method seem to be in better agreement with experiment than the HF results, and close to the quality of MP2.

Finally, we present results on the $^1A_1$ states of $Ni(C_2H_4)$ and $Pd(C_2H_4)$. The equilibrium geometries and vibrational frequencies are given in Table III. Four fundamentals have been assigned in the infrared spectra of both $Ni(C_2H_4)$ and $Pd(C_2H_4)$ obtained by matrix cocondensation techniques (Ozin et al., 1977). The calculated vibrational frequencies for both complexes are very close to the experimental values. The similarity in the IR spectra of $Ni(C_2H_4)$ and

Table I. Equilibrium geometries obtained by LCTGO-MCP-LSD, HF, MP2 and experiment (all distances in Å and angles in degrees).

Molecule	Parameter	LSD	HF[a]	MP2[a]	Exp[a]
C_2H_2	r_{CC}	1.212	1.185	1.218	1.203
	r_{CH}	1.088	1.057	1.066	1.061
C_2H_4	r_{CC}	1.333	1.317	1.336	1.339
	r_{CH}	1.107	1.076	1.085	1.085
	α_{HCH}	116.3	116.4	116.6	117.8
C_2H_6	r_{CC}	1.514	1.527	1.527	1.531
	r_{CH}	1.112	1.086	1.094	1.096
	α_{HCH}	107.3	107.7	107.7	107.8
C_2H_4O	r_{CO}	1.418	1.404[b]	1.441[b]	1.431[c]
	r_{CC}	1.468	1.459	1.474	1.466
	r_{CH}	1.108	1.078	1.086	1.085
	α_{COC}	62.4	62.6	61.5	61.6
	α_{HCH}	115.7	115.5	116.2	116.6
	α_{HCO}	115.4	115.3	115.0	114.4
	α_{HCC}	119.5	119.7	119.3	119.6

[a] As compiled in reference (Hehre et al., 1986); 6-31G* basis sets in HF and MP2

[b] Simandiras et al. (1988); double-ζ plus polarization basis set.

[c] Cunningham et al., 1951; Hirose, 1974.

Table II. Harmonic frequencies of ethylene oxide (C_2H_4O) obtained by LCGTO-MCP-LSD, HF, MP2 and experiment (all frequencies in cm^{-1})

Sym.	LSD	HF[a]	MP2[a]	Exp[b]
a_1	3002 (-22)[c]	3296 (272)	3202 (178)	3024
	1484 (-13)	1687 (190)	1588 (91)	1497
	1298 (28)	1419 (227)	1318 (48)	1270
	1093 (-27)	1294 (174)	1178 (58)	1120
	901 (24)	986 (109)	909 (32)	877
a_2	3078 (13)	3375 (310)	3303 (238)	3065
	1109	1247	1200	-
	1006 (-14)	1144 (124)	1055 (35)	1020
b_1	2993 (15)	3283 (305)	3193 (215)	2978
	1426 (-44)	1630 (160)	1548 (78)	1470
	1082 (-79)	1277 (118)	1141 (-18)	1159
	895 (73)	974 (152)	858 (36)	822
b_2	3086 (21)	3390 (325)	3317 (252)	3065
	1122 (-25)	1278 (131)	1176 (29)	1147
	794 (-14)	877 (69)	839 (31)	808

[a] Simandiras et al., 1988; double-ζ plus polarization basis sets.
[b] Nakanaga, 1980.
[c] Deviation from experimental values.

Table III. Equilibrium geometries and vibrational frequencies of Ni(C_2H_4) and Pd(C_2H_4) obtained by the LCGTO-MCP-LSD method (all distances in Å, angles in degrees, and frequencies in cm^{-1})

		Ni(C_2H_4)		Pd (C_2H_4)	
		LSD	CASSCF[a]	LSD	CIPSI[b]
	r_{MC}	1.862	1.972	2.053	2.26
	r_{CC}	1.425	1.454	1.414	1.36
	r_{CH}	1.111	-	1.109	-
	α_{HCH}	114.4	-	115.1	-
	tilt[c]	21.8	21.0	20.8	15.0
		calc.	exp.[d]	calc.	exp.[d]
a_1	v^1 (CH str)	2976	2961	3010	2952
	v^2 (CC str)	1445	1497	1473	1502
	v^3 (CH_2 def)	1176	1159	1204	1223
	v^4 (CH_2 wag)	894	901	902	913
	v^5 (MC str)	565	-	481	-
a_2	v^6 (CH str)	3058	-	3109	-
	v^7 (CH_2 rock)	760	-	792	-
	v^8 (CH_2 twist)	570	-	576	-
b_1	v^9 (CH str)	2968	-	3006	-
	v^{10} (CH_2)	1364	-	1374	-
	v^{11} (CH_2 wag)	920	-	933	-
	v^{12} (MC str)	500	-	433	-
b_2	v^{13} (CH str)	3042	-	3088	-
	v^{14} (CH_2 rock)	1140	-	1150	-
	v^{15} (CH_2 twist)	810	-	869	-

[a] Widmark et al., 1985.
[b] Novaro, 1989.
[c] Tilt is the angle between the bisector of the HCH bond angle and the C-C bond.
[d] Ozin et al., 1977.

$Pd(C_2H_4)$ suggests that the structure of the coordinated ethylene is likely similar in the two complexes (for example, only a 5 cm^{-1} difference is observed in the C-C stretching modes). LCGTO-MCP-LSD gives 28 cm^{-1} difference for the C-C stretching frequencies, and the bond lengths are within 0.011 Å of each other. There is a significant discrepancy between the LCGTO-MCP-LSD equilibrium geometries and those obtained by previous *ab initio* calculations (Widmark et al., 1985; Novaro, 1989). However, the calculated equilibrium geometries of the coordinated ethylene are quite different in $Ni(C_2H_4)$ (Widmark et al., 1985) and $Pd(C_2H_4)$ (Novaro, 1989) indicating the possibility of a large error in one or the other of the *ab initio* techniques. Our results agree more closely with those of CASSCF calculations (Widmark et al., 1985).

Non-local, Density Gradient Corrected Functionals

Although the successes of the LDF approach are many, and well documented, its limitations are being revealed more clearly as the user community becomes wider and more demanding. Difficulties in attaining a quantitative account of atomic exchange and correlation energies and of molecular dissociation energies are well known (typically LDF overestimates bond energies). In addition, little is known about the behavior of LDF far away from potential minima, near transition states, or for cases of weak or intermediate interaction strength. In this section we present some (still fragmentary) very recent results using non-local functionals that involve the gradient of the density as well as the density itself.

The concept of the exchange-correlation hole in DFT (Slater, 1974; Gunnarsson et al., 1976) provides a starting point for various methods that go beyond the LSD approximation. The seminal work of Langreth and Mehl (1983), using a wave vector analysis to provide cut-offs for singularities of the gradient expansion, established an avenue for introducing useful gradient corrections to the exchange and correlation energy functionals and potentials. Since then, work by Perdew (1985, 1986), Hu and Langreth (1985), Becke (1986, 1988b, 1988c), Lee et al. (1988), De Pristo and Kress (1987), and Perdew and Wang (1986) has provided a number of realistic functionals.

We have recently implemented some of these, both in the older Hermite-Gaussian programs (derivatives of the Dunlap-Connolly-Sabin code) (Mlynarski et al., 1991) and in deMon (A. St-Amant and D. R. Salahub, unpublished). the functional derivatives of the energy have been taken, to derive the potentials and the implementations are fully self consistent. We have found (Mlynarski et al., 1991) that the generalized gradient approximation (CGA) to non-local exchange (Perdew, 1985; Perdew et al., 1986) in concert with the non-local correlation energy functional of Perdew (1986) gives an effective one-body non-local (NL) potential which behaves correctly over a wide range of electron densities and has correct asymptotic properties in the large gradient limit.

We have applied this to a number of cases where LDF has distinct quantitative difficulties (these are also difficult cases for *ab initio* techniques). The applications range over a wide variety of bonding situations : the chemical bonds in O_2, and CH_2 (Mlynarski et al., 1991), weak, essentially van der Waals

Table IV. Binding energies (eV) for O_2 and Mg_2

	LDA	NL	Exp[a]
O_2	7.46	6.22[b]	5.21
		5.39[c]	
Mg_2	0.171	0.034	0.05[d]

(a) : experimental values taken from Schaeffer, 1971.
(b) : spherical density has been used for atom.
(c) : nonspherical density.
(d) : experimental values taken from Purvis et al. (1978).

binding in Mg_2 (Mlynarski et al., 1991) and (using deMon) a tough case of hydrogen bonding in malonaldehyde (I. Papai, A. St-Amant and D. R. Salahub, unpublished).

The LDF binding energy of the first row diatomics is worst for the oxygen molecule (Table IV). The gradient correction leads to a distinct improvement, as does the use of a non-spherical atomic density (Kutzler et al., 1989). The equilibrium distance remains within 0.02 Å of the experimental value and the vibrational frequency is within 1%.

For Mg_2, the LDF overbinding is dramatically reduced and improvement is also observed for the equilibrium distance (6.50 a.u (LDF), 6.95 (NL), 7.35 (exp.)) and the vibrational frequency (114 cm^{-1} (LDF), 77(NL), 51 (exp.)).

Methylene, CH_2, was a long history of experimental and quantum chemical study, much of which has focused on the energy separation between the ground, $^3B_1$, state and the excited, $^1A_1$, state. In conventional *ab initio* approaches, the latter requires at least two configurations for a reasonable first order description (Hartree-Fock puts the singlet much too high, more than 25 kcal/ mole above the triplet (vs the experimental separation of 9 kcal/mol.). We think it is particularly interesting to see how the inherently monoconfigurational Kohn-Sham approach fares for this type of problem when various local and non-local functionals are employed. The exact KS functional, of course, would perform the same work as the multiconfigurational wave-function approach and we are hopeful that comparisons could yield suggestions in the search for new, more accurate functionals. As can be seen from Table V, the NL functional leads to a marked improvement of the singlet-triplet separation (the exact agreement is coincidental) while retaining good accuracy for the geometry.

Finally, in Table VI, we show results for the equilibrium O-H bondlengths involved in the internal hydrogen bond of malonaldehyde, the most stable isomer of β-hydroxyacrolein (Fig. 1). LSD gives unacceptable O-H and O...H distances, the optimized geometry is close to C_{2v} symmetry. The NL results are greatly improved, somewhat better than HF, but still significantly worse than MP2.

Table V. Geometries and energy difference (ΔE) between singlet and triplet states in methylene. Angles are in deg, bond lengths in Å, and ΔE in kcal/mole.

Method	$^3B_1$		$^1A_1$		ΔE
	H-C-H	R_{C-H}	H-C-H	R_{C-H}	
LDA	137	1.08	102	1.12	13.2
NL	130	1.08	99	1.12	9.1
exp.	134	1.07	102	1.11	9.1

Table VI. Equilibrium O...H and O-H distances in malonaldehyde obtained by LCGTO-MCP-LSD, non-local corrections (NL), HF, MP2 and experiment (all distances in Å).

	LSD	NL	HF[a]	MP2[a]	Exp[b]
O...H	1.228	1.567	1.880	1.694	1.680
	(-0.452)[c]	(-0.113)	(+0.200)	(+0.014)	
O-H	1.196	1.042	0.956	0.994	0.969
	(+0.227)	(+0.073)	(-0.013)	(+0.025)	

[a]Binkley et al., 1986.; 6-31G** basis set.
[b]Baughcum et al., 1981.
[c]Deviation from experimental bond distances in parentheses.

These few results, along with others in the literature, indicate that, while not perfect, the currently available non-local density gradient-corrected functionals can provide distinct improvements over LDF. More testing is clearly needed but we believe these NL functionals should become the "default" level of theory over the next few years.

Concluding remarks

We hope that the above overview has shown the current possibilities of the Gaussian-based density functional methodology. Many of the technical advances are to a large extent grafts onto the DFT methods of features developped in traditonal *ab initio* circles over the last two decades or so. Their success has depended on cross-fertilization and it is a very healthy sign that there is steady growth in the number of *ab initio* quantum chemists who are not only interested in DFT peripherally but, indeed, are making key contributions to the development of DFT methodology. It is still too early to know the ultimate role that DFT will play in mainstream quantum chemistry. There is certainly a

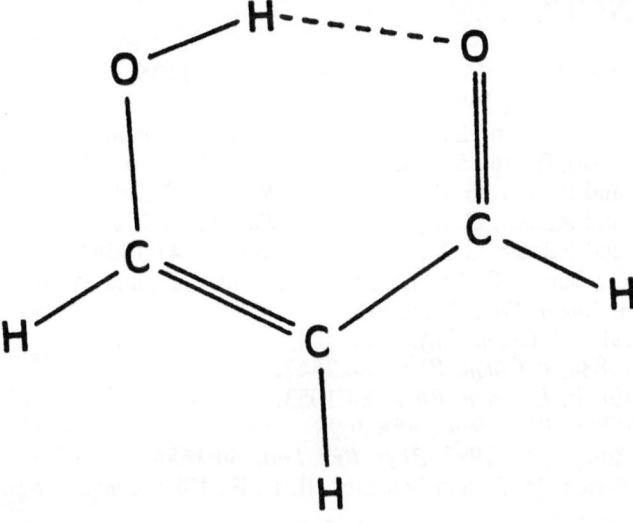

Fig. 1. Malonaldehyde

wide class of problems where it is the methodology of choice because of its ability to treat some types of electron correlation at a reasonable level and because it can offer computational economics. We believe its role will only increase in importance and that Density Functional concepts and methods will help define the mainstream theoretical chemistry of tomorrow.

Acknowledgements

The research summarized here has been made possible by the generous support of NSERC (Canada) FCAR (Québec), l'Institut Français du Pétrole, Hitachi, Kodak, and the Canada Council-Killam Research Fellowship program. Acknowledgement is made to the donors of the Petroleum Research Fund, administered by the ACS, for partial support of this research. We thank NSERC, Cray Canada and the Services Informatiques de l'Université de Montréal for computing resources, and Peter Fillmore (Cray Canada) for his interest and support.

Jan Andzelm's continued interest, collaboration and enthusiasm are highly appreciated.

REFERENCES

Allan, D. C., Teter, M. P., 1987, *Phys. Rev. Lett.*, **59**:1136.

Andreoni, W., Ballone, P., 1987, *Physica Scripta*, T**19**:289.

Andzelm, J., Radzio, E., and Salahub, D. R., 1985a, *J. Comput, Chem.*, **6**:520.

Andzelm, J., Radzio, E., and Salahub, D. R., 1985b, *J. Chem. Phys.*, **83**:4573.

Averill, F. W., and Painter, G. S., 1985, *Phys. Rev. B*, **32**:2141.

Averill, F. W., and Painter, G. S., 1986, *Phys. Rev. B*, **34**:2088.

Averill, F. W., and Painter, G. S., 1990, *Phys. Rev. B*, **41**:10344.

Baughcum, S. L., Duerst, R. W., Rowe, W. F., Smith, Z., and Wilson, E. B., 1981, *J. Am. Chem. Soc.*, **103**:6296.

Becke, A. D., 1986, *J. Chem. Phys.*, **85**:7184.

Becke, A. D., 1988a, *J. Chem. Phys.*, **88**:2547.

Becke, A. D., 1988b, *J. Chem. Phys.*, **88**:1053.

Becke, A. D., 1988c, *Phys. Rev.*, A**38**:3098.

Bendt P., and Zunger, A., 1983, *Phys. Rev. Lett.*, **50**:1684.

Binkley, J. S., Frisch, M. J., and Schaefer III, H. F., 1986, *Chem. Phys. Lett.*, **126**:1.

Car, R., and Parrinello, M., 1985, *Phys. Rev. Lett.*, **55**:2471.

Car, R., and Parrinello, M., 1988, *Phys. Rev. Lett.*, **60**:204.

Cunningham, G. L., Boyd, A. W., Myers, R. J., Gwinn, W. D., and Le Van, W. I. J., 1951, *J. Chem. Phys.*, **19**:676.

DePristo, A. E., Kress, J. D., 1987, *J. Chem,. Phys.*, **86**:1425.

Dunlap, B. I., Connolly, J. W. D., and Sabin, J. R., 1979, *J. Chem. Phys.*, **71**:3396, 4993.

Dunlap, B. I., 1986a, *J. Phys. Chem.*, **90**:5524.

Dunlap, B. I., and Cook, M., 1986b, *Int. J. Quantum Chem.*, **29**:767.

Duran M., Yamaguchi, Y., Remington, R. B., Osamura, Y., and Schaefer III, H. F., 1989, *J. Chem. Phys.*, **90**:334.

Fan, L., Verluis, L., Ziegler, T., Baerends, E. J., and Ravenek, W., 1988, *Int. J. Quantum Chem. Symp.*, **22**:173.

Fournier, R., Andzelm, J., and Salahub, D. R., 1989, *J. Chem. Phys.*, **90**:6371.

Fournier, R., 1990, *J. Chem. Phys.*, **92**:5422.

Gunnarsson, O., Lundqvist, B. I., 1976, *Phys. Rev.*, B**13**:4274.

Head-Gordon, M., Pople, J. A., and Frisch, M. J., 1989, *Int. J. Quantum Chem.*, S**23**:291.

Hehre, W., Radom, R., Schleyer, P. V. R., and Pople, J. A., 1986, *Ab initio Molecular Orbital Theory*, Wiley, New York.

Herman, F., and Skillman, S., 1963, *Atomic Structure Calculations*, Prentice-Hall, Englewood Cliffs.

Hirose, C., 1974, *Bull. Chem. Soc. Jpn.*, **47**:1311.

Hohenberg, P., and Kohn, W., 1964, *Phys. Rev.*, **136**:B864.

Hu, C. D., Langreth, D. C., 1985, *Physica Scripta*, **32**:391.

Huzinagas, S., Ed., 1984, *Gaussian Basis Sets for Molecular Calculations*, Elsevier, Amsterdam.

Huzinaga, S., Klobukowski, M., and Sakai, Y., 1984, *J. Phys. Chem.*, **88**:4880.

Johnson, K. H., 1973, *Adv. Quantum Chem.*, 7:143.

Jones, R. O., Gunnarsson, O., 1989, *Rev. Mod. Phys.*, 61:689.

Kohn, W., and Sham, L. J., 1965, *Phys., Rev.*, 140:A1133.

Kutzler, F. W., and Painter, G. S., 1987, *Phys. Rev. Lett.*, 59:1285.

Langreth, D. C., Melhl, M. J., 1983, *Phys. Rev.*, B28:1809.

Lebedev, V. I., 1975, *Zh. Vychisl. Mat. Mat. Fiz.*, 15:48; *ibid*, 1976, 16:293 (as cited in Becke, 1988a).

Lee, C., Yang, W., and Parr, R. G., 1988, *Phys., Rev.*, B37:785.

Lee, T. J., Willets, A., Gaw, J. F., and Handy, N. C., 1989, *J. Chem. Phys.*, 90:4330.

McLaren, A. D., 1963, *Math Comput.*, 17:361.

Mlynarski, P., Salahub D. R., 1991, *Phys. Rev. B*, in press.

Nakanaga, T., 1980, *J. Chem. Phys.*, 73:5451.

Nakatsuji, H., Kanda, K., Hada, M., and Yonezawa, T., 1982, J. Chem. Phys., 77:3109.

Nakatsuji, H., Kanda, K., Hada, M., and Yonezawa, T., 1983, *J. Chem. Phys.*, 79:2494.

Needels, M., Payne, M. C., and Joannopoulos, J. D., 1988, *Phys. Rev. B*, 38:5543.

Novaro, O. A., 1989, in *The Challenge of d and f Electrons*, Salahub, D. R., and Zerner, M. C. Eds., *ACS Symp. Ser.*, 394:228.

Obara, S., and Saika, A., 1986, *J. Chem. phys.*, 84:3963.

Ozin, G. A., and Power, W. J., 1977, *Inorg. Chem.*, 16:212.

Papai, I., St-Amant, A., Ushio, J., and Salahub, D. R., 1990, *Int. J. Quantum Chem.* (Symposium Series), to be published.

Payne, M. C., Bristowe, P. D., and Joannopoulos, J. D., 1988, *Phys. Rev. Lett.*, 58:1348.

Pederson, M. R., Klein, B. M., and Broughton, J. Q., 1988, *Phys. Rev. B*, 38:3825.

Pederson, M. R., and Jackson, K. A., 1990, *Phys. Rev. B*, 41:7453.

Perdew, J. P., 1985, *Phys. Rev. Lett.*, 55:1665.

Perdew, J. P., and Wang, Y., 1986, *Phys. Rev. B*, 33:8800.

Perdew, J. P., 1986, Phys. Rev. B, 33:8822; erratum in *Phys. Rev. B*, 34:7406.

Pople, J. A., Krishnan, R., Schlegel, H. B., and Binkley, J. S., 1979, *Int. J. Quantum Chem. Symp.*, 13:225.

Pulay, P., 1969, *Mol. Phys.*, 17:197.

Pulay, P., 1977, in *Applications of Electronic Structure Theory*, ed. H. F. Schaefer, Plenum, New York.

Pulay, P., 1983a, *J. Chem. Phys.*, 79:2491.

Pulay, P., 1983b, *J. Chem. Phys.*, 78:5043.

Pulay, P., 1987, *Adv. Chem. Phys.*, 69:241.

Purvis, G. D., and Bartlett, R. J., 1978, *J. Chem. Phys.*, 68:2114.

Rochefort, A., Andzelm, J., Russo, N., and Salahub, D. R., 1990, *J. Amer. Chem. Soc.*, in press.

Ryckaert, J. P., Ciccotti, G., and Berendsen, H. J. C., 1977, *J. Comput. Chem.*, 23:327.

Sabin, J. R., and Trickey, S. B., 1984, in *Local Density Approximations in Quantum Chemistry and Solid State Physics*, (edited by Dahl, J. P., and Avery, J.), Plenum, New York, p. 333.

Sakai, Y., and Huzinaga, S., 1982, *J. Chem. Phys.*, **76**:2537, 2552.

Salahub, D. R., 1987, *Adv. Chem. Phys.*, **69**:747.

Sambe, H., and Felton, R. H., 1975, *J. Chem. Phys.*, **62**:1122.

Satoko, C., 1981, *Chem. Phys. Lett.*, **83**:111.

Satoko, C., 1984, *Phys. Rev. B*, **30**:1754.

Schaefer III, H. F., 1971, *J. Chem. Phys.*, **54**:2207. and references therein.

Schlegel, H. B., 1987, *Adv. Chem. Phys.*, **67**:249.

Simandiras, E. D., Amos, R. D., Handy, N. C., Lee, T. J., Rice, J. E., Remington, R. B., and Schaefer III, H. F., 1988, *J. Am. Chem. Soc.*, **110**:1388.

Slater, J. C., 1974, "The Self-Consistent Field for Molecules and Solids", vol. 4, McGraw-Hill, New York.

Sobolev, S. L., 1962, *Sibirsk. Mat. Zh.*, 3:769 (as cited in Becke, 1988a).

St-Amant, A., and Salahub, D. R., 1990, *Chem. Phys. Lett.*, **169**:387.

Stroud, A. H., 1971, *Approximate Calculation of Multiple Integrals*, Prentice-Hall, Englewood Cliffs.

Tarnov, E., Joannopoulos, J. D., and Payne, M. C., 1989, *Phys. Rev. B*, **39**:6017.

Verluis, L., and Ziegler, T., 1988, *J. Chem. Phys.*, **88**:322.

Vosko, S. H., Wilk, L., and Nusair, M., 1980, *Can. J. Phys.*, **58**:1200.

Widmark, P. O., Roos, B. O., and Siegbahn, P. E. M., 1985, *J. Phys. Chem.*, **89**:2180.

Wilk, L., and Vosko, S. H., 1982, *J. Phys. C*, **15**:2139.

7
DMol Methodology and Applications

B. DELLEY

Abstract

An overview of the DMol method, an implementation of the local density functional (LDF) approach for molecules, is presented. A brief review of the key features of the method, namely orbital expansions with accurate numerical basis sets (readily available for the whole periodic system), charge density and potential representation and automatic three dimensional integration is given. The now available analytical energy derivatives are discussed. Accurate calculated molecular geometries from the Hedin-Lundqvist local exchange correlation model for some transition metal compounds are compared to experiment and standard ab initio results. The geometry of planar Fe(II)porphine is investigated and compared to the assumed geometry for the molecules in the gas phase. The ground state is calculated to have ^3E symmetry and the first excited state $^3A_{2g}$ symmetry. Further calculations of low lying states of Fe(II)porphine show that both singlet and quintet spin states are higher in energy.

Method

Density functional theory in its local density functional (LDF) approximation is becoming a complementary tool to ab initio quantum chemical methods for studying the electronic structure of larger molecules and clusters. LDF theory has been put on a solid theoretical footing by the work of Kohn and Sham (1965) and the density functional approach is now regarded as a complete theory (Parr 1989). Definitely, today's LDF implementations avoid problems intrinsic to the HF approximation such as the logarithmic divergence in infinite systems with Coulomb interaction. LDF also does not suffer from the excessively ionic contributions to the HF approximate wavefunction showing up in underestimates of bonding, but also sometimes as underestimates of bond length because of dissociation into high lying ionic states. LDF may be thought of as an effective molecular orbital (MO) theory which lends to a qualitative MO interpretation of structure and bonding. The MO picture is an approximate theory and clearly LDF is not the exact density functional so some sacrifice in the detailed description of the electronic structure is implicit in this method as it is in any finite calculation for a large system. For systems with only a few electrons, having been spared the calculation of an n-electron wavefunction may not count as much as the loss of not having an approximation to this function beyond the single determinant at hand. In principle, however, there is no problem since the density functional is known (Levy 1979) and in that spirit it is possible to reconstruct an n-electron wavefunction from the n-representable density. For larger molecules and clusters, however, some loss of detail is unavoidable or even desired and LDF can be expected to yield quantum mechanical results to an excellent degree of approximation, without using semiempirical parameters.

The DMol method solves the LDF equations for polyatomic molecules and clusters variationally. Its merits rely on four related key features described below:

First the method uses numerically defined basis sets, which are easy to handle and which guarantee exact solutions in limiting cases. Exact atomic LDF functions, internally handled as a spline tabulation, form part of a standard basis set. This will recover an exact solution in the separated atom limit. In the standard basis set, valence functions of the corresponding positive ions help to describe radial contractions and expansions on bonding. The ionic charges are optimised to maximise bonding energies for strong bonds. Since it is an extremum, the exact value of the charge is not critical and yields nearly maximal binding energy in other cases. These basis sets are referred to as double numerical basis sets. An excellent representation of spin densities and reasonably good representation of bonding can be expected. Larger basis sets use polarization functions, generated also from spherical ion calculations. Routinely, the additional basis functions are orthogonalised against the atomic functions in order to better condition the overlap matrix arising in solving the single particle equations. It is found that increasing the angular cutoff of the basis set by one lets the binding energy approach the limit by approximately an order of magnitude, provided sufficient flexibility is maintained in the the lower angular momentum components.

Second, a three dimensional fully automatic integration is used to calculate the matrix elements of the algebraic eigenvalue problem. The density (and spin-density) have to be reconstructed on a real space integration mesh for the calculation of the local exchange correlation potential. It may seem that using three dimensional integrations can pose a bottleneck in the calculation. Actually, in any density functional method, the density has to be reconstructed in real space in order to evaluate the nonlinear exchange and correlation functions. The density and the effective potential have qualitatively the same properties. The matrix elements have a similar sharply varying behaviour near the nucleus and an exponential decay further away, which implies that the calculation of integrals is of comparable difficulty. On these grounds, the density reconstruction becomes an important if not dominating step in a well designed LDF method. The overhead of calculating also matrix elements numerically in the present method pays off in higher accuracy for a given basis set size, since the most adequate basis sets can be used. The very effective integration scheme of DMol uses a set of partition functions to divide up functions over all space into nearly atomic contributions, which can then be integrated in a fast converging way using radial and angular integrations.

Third, the partitioning method is also used to divide up the density (and spin density) into atomic contributions in order to get an auxiliary density which can be used to generate the static potential in an efficient way. Projection methods allow then fully representing the atomic contribution as an angular expansion. Contrary to least squares fitting methods, a perfectly flexible representation of the static and exchange-correlation potential is obtained in an N^2 algorithm. This makes higly accurate potentials available in routine calculations by just using the default cutoff for angular representation and maximal radial flexibility. A more detailed account of these first three points is given in (Delley 1990a).

Finally, the algorithm now incorporates analytical energy derivatives of the total energy function with respect to nuclear coordinates. Since the method is designed to work with localized orbitals, clearly a Pulay (1969) like orbital truncation term must be calculated. Numerical integration allows one to calculate these terms for all derivatives with the total operation count being comparable to three extra SCF iterations or fewer. Further derivative terms described in (Delley 1990b) account for the truncation of the auxiliary density in way that is consistent with its second order effect on the total energy. At the same time, this term also makes the energy derivative second order in a density difference between current density in an iteration and converged density. Energy derivatives are also calculated with N^3 scaling when using the frozen core approximation with explicit Schmitt orthogonalization of valence functions against core functions. A basic geometry optimizer working currently in Cartesian and in symmetry coordinates was used for some of the examples decribed below. Vibrational frequency calculations are available through finite differencing of analytic energy derivatives.

In its implementation, DMol is set up to make full use of symmetry for real Hamiltonian matrices. The orthogonal transformation matrix from atomic like orbitals to symmetry adapted orbitals has to be supplied for higher groups. A symmetry program for some Abelian groups is presently included in DMol. The algorithm is set up to fully vectorize in its N^3 and N^2 parts on machines with vector capability. The program is structured allowing use of parallel (vector) processors at a higher internal level. The program is written in rather conservative Fortran 77 for excellent portability. It contains as a single machine dependent parameter the maximum worksize depending on machine and probably preferred job class. It optimizes the vector length given the amount of workspace. A consideration of cache-size might be built in any time.

Before embarking into the applications it seems worthwile to discuss briefly the scaling of computing times or number of numeric operations when studying larger and larger systems in general. We assume that the molecule or system of molecules and clusters or irreducible unit cell, in short the system, is characterized by N the number of atoms. We further assume that the average composition of the system and the quality of the calculation remain unchanged. In this case the computing time is generally characterized by a polynomial plus $log(N)$ times another polynomial as long as computations that grow exponentially with system size are not considered. Of course, the coefficients depend on the algorithm and the machine used, but no real machine, be it vector and/or parallel, can make a leading term in these polynomials disappear. While other algorithms may have another order of polynomial, it appears however that any code solving quantum mechanical equations is at least of leading order N^3. But, for calculations done today, the algorithm may be dominated by a lower term which leads to quotations of $NMlog(M)$ scaling (where M may denote the number of basis functions) and scaling laws with fractional exponent from fitting timing data. It is one of the reasons of favoring the local density functional methods that they easily get leading order N^3 by using an auxiliary density for the calculation of the static potential, while in HF it takes making use of locality of basis sets to make the N^4 order from exchange disappear. If locality is used in local basis LDF methods, the N^3 coefficient reduces to the one of the particular generalized algebraic eigenvalue algorithm. A further upshot of these considerations is that quoted fractional exponents of scaling with system size are valid only for the size range included in the fitting. For larger systems the effective exponent deteriorates approaching the leading exponent of the polynomial.

Applications

Geometries of some transition metal complexes

An important area of application of the LDF methods has always been transition metal complexes. In Tab.1 optimized structures are shown for some transition metal complexes and organometallics. LDF calculations were done with a double numerical basis with p polarization functions on hydrogen and double d-polarization functions on C,N,O. No f-polarization functions were used on the transition metal. Comparison with experimental data quoted from Hehre (1986) shows that the metal bonds tend to be too short by up to 0.05Å for the carbonyl complexes and probably less for organometallics. A similar underestimation of bond length has also been found in other LDF calculations (Versluis 1988). C-C and C-O bonds are in excellent agreement with the experimental data. Obviously the HF geometries (Hehre 1986) were obtained with a rather modest basis set which very likely obscures which deficiencies have to be assigned to the HF approximation and which are is due to basis set incompleteness. As the HF data stand, they show deviations exceeding ±0.1Å. Obviously such HF calculations are already taxing today's computational resources.

Table 1: Calculated and experimental geometries for some transition metal complexes. ^{a} means assumed geometry for planar Fe(II)porphine as inferred from X-ray diffraction measurements on Fe(II)porphine and derivates. HF and other exp data cited from (Hehre 1986). Units are in Å and deg.

molecule	parameter	LDF	exp	HF(STO3G)
$Cr(CO)_6$	$r(CrC)$	1.874	1.92	1.789
O_h	$r(CO\ )$	1.146	1.16	1.167
$Cr(C_6H_6)_2$	$r(CrC)$	2.109	2.150	2.095
D_{6h}	$r(CC)$	1.411	1.423	1.426
	$\angle(HC\text{-plane})$	175.9	175.3	178.6
$Fe(CO)_5$	$r(FeC_{as})$	1.778	1.824	2.016
D_{3h}	$r(FeC_{eq})$	1.776	1.824	1.643
	$r(C_{as}O_{as})$	1.145	1.145	1.147
	$r(C_{eq}O_{eq})$	1.148	1.145	1.171
CuF	$r(CuF)$	1.746	1.745	1.592
Fe-porphine	$r(FeN)$	1.960	1.974^a	
D_{4h}	$r(NC_1)$	1.374	1.388	
	$r(C_{1py}C_{2py})$	1.427	1.445	
	$r(C_{1py}C_4)$	1.373	1.390	
	$r(C_{2py}C_{3py})$	1.359	1.351	
	$r(C_{3py}H_1)$	1.092	1.080	
	$r(C_4H_2)$	1.091	1.081	

The geometry of planar unsubstituted Fe(II)porphine in the gas phase has been inferred on the basis of various X-ray crystallographic measurements for nonplanar crystalline porphine (Collman 1975) and from more planar substituted porphine derivatives (Sontum 1983). Since the size of the porphyrin hole in particular is of crucial importance to the ground state symmetry and spin state (Sontum 1983), it is interesting to calculate the geometry within LDF. The same basis set as for the other transition metal compounds was used with the same truncation threshold for the test integrals of $10.^{-4}$. It is interesting to see that agreement between the optimized structure and the assumed geometry is very much in line with the previous data, with a slightly short metal-ligand bond distance.

Low Lying states of Fe(II)Porphine

In the assumed geometry quoted above, low lying states were calculated by specifying the states of different symmetry by their configuration. A theoretical justification for this simple procedure has been given by Gunnarsson (1976). The present calculation clearly predicts an intermediate spin ground state with $^3E(A)$ symmetry. This state is not only the ground state on the basis of total energy comparisons, it also satisfies the Fermi occupation rule. The Jahn-Teller effect should further lower the ground state slightly, but this has not been calculated so far. The first excited state, $^3A_{2g}$, is only 0.25eV higher (see Tab.2). This leaves the possibility open that the real ground state actually may have a significant $^3A_{2g}$ admixture because of spin-orbit coupling or vibrational coupling. In fact, interpretations of experiments are contradictory in favoring either the $^3E(A)$ or the $^3A_{2g}$ state. The X-ray deformation density (Coppens 1984) gives the most clean indication on the ground state symmetry: it clearly suggests a ^3E ground state.

Table 2: Low Lying states of Fe(II)Porphine. Energies in eV with $^3E(A)$ as reference. HF_O from (Obara 1982), HF_R and CI_R from (Rawlings 1986) and CI_B INDO-CI from (Edwards 1986).

state	configuration					LDF	HF_O	HF_R	CI_R	CI_B
	x^2-y^2	z^2	xy	yz	xz					
$^5B_{2g}$	1	1	2	1	1	1.74	-1.30	-1.19	-0.83	0.39
$^5A_{1g}$	1	2	1	1	1	1.44	-1.53	-1.40	-0.10	0.28
^5E_g	1	1	1	1.5	1.5	1.29	-1.18	-1.05	0.09	0.63
$^3E_g(B)$	0	2	1	1.5	1.5	0.88	1.43	1.42	1.12	1.03
$^3B_{2g}$	0	1	1	2	2	0.53	0.56	0.51	0.20	0.39
$^3A_{2g}$	0	2	2	1	1	0.25	-0.32	-0.29	0.47	-0.03
$^3E_g(A)$	0	1	2	1.5	1.5	0.	0.	0.	0.	0.
$^1A_{1g}$	0	2	0	2	2	3.67				3.50
$^1A_{1g}$	0	0	2	2	2	1.20	1.39	1.36	1.06	0.98

It is also interesting to compare high and low spin states of Fe(II)porphine with HF and CI calculations. LDF puts both low and high spin states more than one eV above the ground state. HF puts the quintet spin states lowest in energy. This can be understood on the ground that HF exaggerates the ionic character of the wavefunction and thus underestimates bonding between metal and ligand. This leads in the present case to a high spin ground state that follows Hund's rule. It is well known that correlation effects tend to diminish exchange effects. This effect of raising the quintet states relatively can be seen from the limited configuration interaction calculations quoted in Tab.2. In fact it is believed that the ground state should actually be a triplet state.

Finally, we should note that many more applications of Dmol have been presented at the workshop, the reader is referred to the corresponding articles in this volume.

Acknowledgements

The author is indebted to George Fitzgerald formerly at Biosym for making the commercially available code.

References

Collman J.P., Hoard J.L., Kim N. Lang G., Reed C.A., 1975, *J. Am. Chem. Soc.* 97:2676.

Coppens P., Li L., 1984, *J. Chem. Phys.* 81:1983.

Edwards W.D., Weiner B., Zerner M.C., 1986, *J. Am. Chem. Soc.* 108:2196

Delley B., 1990a, *J. Chem. Phys.* 92:508

Delley B., 1990b, in preparation

Gunnarsson O., Lundqvist B.I., 1976, *Phys. Rev. B* 13:4274

Hedin L., Lundqvist B.I.,1971, *J. Phys. C*4:2064.

Hehre W.J., Radom L., Schleyer P.V.R., Pople J.A., 1986 *Ab Initio Molecular Orbital Theory*, Jown Willey & Sons, New York.

Kohn W., Sham L.J.,1965, *Phys. Rev.* 140:A1133.

Levy M., 1979, *Proc. Natl.Acad. Sci.(USA)*, 76:6062.

Obara S.and Kashiwagi H, 1982, *J. Chem. Phys.* 77:3155.

Parr R.G., Yang Weitao, 1989, *Density-Functional Theory of Atoms and Molecules* Oxford University Press, New York.

Pulay P., 1969, *Mol.Phys.* 17:197

Rawlings D.C., Gouterman M., Davidson E.R., Feller D, 1985, *Int. J. Quant. Chem.* 18:773

Sontum S.F., Case D.A., Karplus M., 1983, *J.Chem.Phys.* 79:2881

Versluis L., Ziegler T., 1988, *J.Chem.Phys.* 88:322

8
Local Density Functional Approaches to Spin Coupling in Transition Metal Clusters

Louis Noodleman, David A. Case,
and Evert Jan Baerends

Introduction

Since the earliest days of Xα theory, there has been widespread interest in applying density functional theories (as they are now called) to practical problems in transition metal systems (Slater, 1974). In earlier times this interest was driven in large part by the computational difficulties involved in more conventional calculations for clusters containing hundreds of electrons. While such concerns are still in force, there is growing point of view that treats density functional theory as a complementary first-principles approach which may be developed into something more than an inexpensive substitute for *ab initio* wavefunctions calculations (Parr and Yang, 1989).

One notable difficulty for applications of density functional theories that follow the Kohn-Sham ideas is that many interesting clusters have a large number of unpaired electrons, arranged in such a way that no (reasonable) single determinant can be constructed that has the same symmetry properties as the ground-state wavefunction. This can lead to interesting difficulties in constructing and interpreting the Kohn-Sham orbitals (Dunlap, 1987). In this paper, we illustrate a somewhat pragmatic approach to this problem, which begins with *spin unrestricted, broken symmetry* solutions to the orbital equations of density functional theory (Noodleman, 1981; Noodleman and Davidson, 1986). The energies and properties of these solutions are then related to those of the "true" wavefunctions (which have proper space and spin symmetry) through an *ansatz* that involves a spin Hamiltonian description of both symmetrized and broken symmetry states. This paper summarizes the fundamental ideas of this approach to electronic structure calculations.

Our examples will be drawn from work on iron-sulfur clusters that form the active sites of a variety of metalloproteins. These systems are of considerable interest in both inorganic chemistry and biochemistry, and spin couplings have important consequences for their redox potentials and electron transfer kinetics. In addition, these systems can exhibit both delocalized and localized mixed-valent states; electron delocalization and spin coupling interact in interesting ways. Figure 1 illustrates some of the active sites and their associated ground state spins that are encountered in metalloproteins.

A large number of density functional calculations on iron-sulfur clusters have used the multiple scattering method to determine approximate orbital energies and wavefunctions (Norman and Jackels, 1975; Yang *et al.*, 1975; Norman *et al.*, 1978, 1980; Aizman and Case, 1982; Noodleman *et al.*; 1985,

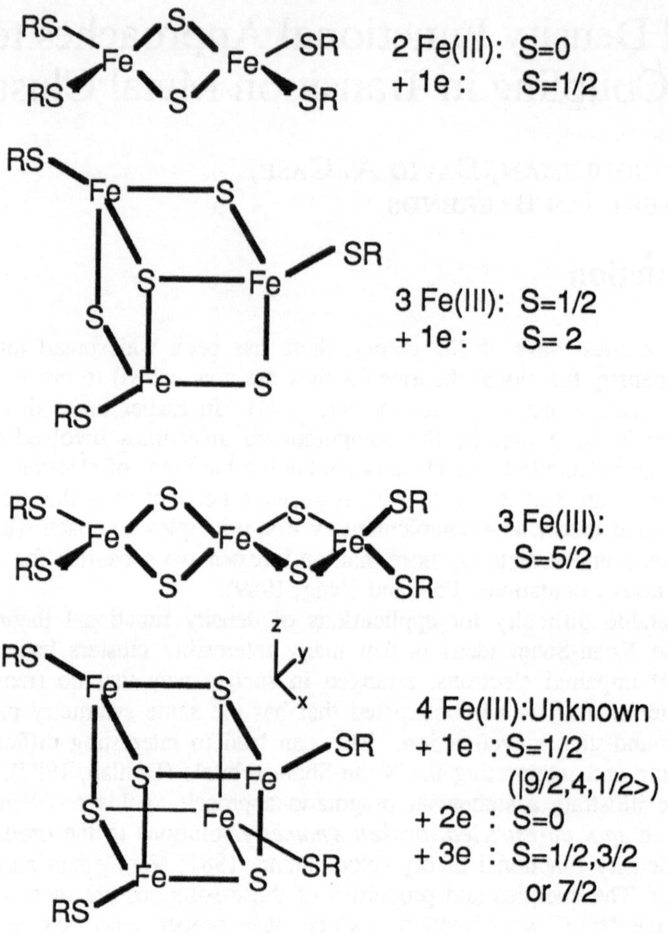

Figure 1. *Geometries and ground spin states for some iron-sulfur clusters*

1988, 1989; Sontum *et al.*, 1989). More recently, it has been possible to carry out calculations using a fairly large basis set of Slater orbitals as an expansion set (Geurts *et al.*, 1980; Noodleman and Baerends, 1984; present work). Some of the early work utilized only spin restricted calculations (Yang *et al.*, 1975, Geurts *et al.*, 1980, Norman *et al.*, 1978). These calculations are inherently incapable of describing high spin transition metal centers or their spin coupling. In the presence of the weak tetrahedral ligand field of the inorganic sulfur and thiolate ligands, both Fe(II) and Fe(III) sites are high spin, yet determinants made of doubly-occupied molecular orbitals describe only low-spin configurations. For a single iron site (as in rubredoxin) it is sufficient to perform spin unrestricted calculations (Norman *et al.*, 1975). For two or more spin-coupled iron sites, the energy of the high spin configuration (having parallel high spin Fe sites) can be compared to broken symmetry configurations where two or more sites are aligned in an antiparallel fashion. In this paper we illustrate the ways in which such a broken symmetry analysis can be carried out, with examples taken from iron-sulfur clusters that form active sites of metalloproteins.

Survey of iron-sulfur clusters

Figure 1 illustrates some common cluster geometries for iron-sulfur clusters and representative ground spin states. There is a rich variety of behavior, and many qualitative features that require explanation. For example, the reduced 2Fe2S ferredoxins exhibit trapped Fe^{2+}–Fe^{3+} valence sites in the mixed valence state whereas the 3Fe and 4Fe clusters at comparable oxidation levels have delocalized valence electron distributions over specific pairs of sites, with an effective oxidation state of 2.5 (Sands and Dunham, 1975; Middleton *et al.*, 1978; Münck *et al.*, 1988).

As we outline below, this switch is the result of competition between Heisenberg exchange and resonance delocalization (also called double exchange). Resonance delocalization is energetically favorable over a pair of sites with parallel or nearly parallel spin vectors (S_i, S_j) so that the corresponding intermediate spin quantum number S_{ij} is large. For a dimer, resonance delocalization must compete directly with the Heisenberg term, which is usually antiferromagnetic. In trimers or tetramers, the Heisenberg pairwise couplings are also typically antiferromagnetic. However, in the presence of a number of coupled centers, there is less cost in energy to make some spin pairs parallel since others may remain antiparallel. The resonance delocalization interaction further favors this situation in the spin ground state (Münck *et al.*, 1988; Noodleman, 1988).

Other aspects of the spin states shown in Figure 1 may appear puzzling. For example, the oxidized form of the "cubane" 3Fe cluster has a ground spin state S=1/2, whereas the "linear" cluster below it has S=5/2 (Girerd *et al.*, 1984; Münck *et al.*, 1988). In the "high potential" form of 4Fe clusters [with three Fe(III) and one Fe(II)], the ferrous-ferric pair couple to an intermediate spin of $S_{34} = 9/2$ (maximal), but the ferric-ferric pair to an intermediate spin of $S_{12} = 4$ (less than maximal) to give a total system spin S=1/2 (Noodleman, 1988). (Recent experimental measurements (Jordanov *et al.*, 1990) indicate that $|S_{34}S_{12}S\rangle = |7/2\ 3\ 1/2\rangle$ is nearly degenerate with the ground spin state quoted above.) Finally, the "reduced" forms of these clusters [with one Fe(III) and three Fe(II)] have ground states with S=1/2, 3/2, 7/2 or statistical mixtures of these in different proteins or synthetic analogues, with excited spin states appearing much closer in energy than in the more oxidized clusters (Carney *et al.*, 1988; Papaefthymiou *et al.*, 1982). Although we do not have space to discuss all of these features, they arise in a natural and convincing way from the theory we shall outline.

Broken symmetry analysis

We can illustrate the basic ideas of the broken symmetry approach with reference to the "cubane" 3Fe clusters in the oxidized state, where the three iron sites are geometrically equivalent and are all formally in the +3 oxidation state. [A theory for inequivalent sites is outlined in Noodleman *et al.* (1988).] The crux of our computational approach arises from the recognition that broken symmetry wavefunctions (in which otherwise equivalent metal sites have different spin populations) are relatively easy to compute and interpret. By contrast, the "correct" wavefunctions (which are eigenfunctions of $\hat{S}^2$) are generally multiconfiguration states that are considerably more difficult to approximate and understand. Hence, we choose to fit an (assumed) spin Hamiltonian to energies computed from broken symmetry wavefunctions, and use the resulting parameters to estimate the locations of the pure spin states, including the pure-spin ground state.

We assume that the true electrostatic interactions that couple iron spins together can be replaced by an interaction of the Heisenberg type:

$$\hat{H} = J (\hat{S}_1 \cdot \hat{S}_2 + \hat{S}_1 \cdot \hat{S}_3 + \hat{S}_2 \cdot \hat{S}_3) \tag{1}$$

Griffith (1972) has worked out the eigenstates of this Hamiltonian, and related methods may also be used to estimate the energies of broken symmetry states in which each iron atom is either spin-up (with $S = M_S = 5/2$) or spin-down (with $S = 5/2$, $M_S = -5/2$). If for convenience we denote these monomer states as $| \alpha \rangle$ and $| \beta \rangle$, respectively, then broken symmetry kets will have forms such as

$$| \alpha \alpha \beta \rangle \equiv | \alpha_1 \rangle | \alpha_2 \rangle | \beta_3 \rangle \tag{2}$$

which represents a state with $M_S = 5/2$ which can be approximately identified with a broken symmetry molecular orbital wavefunction that places five unpaired spin-up d-electrons on centers 1 and 2, and five spin-down d-electrons on center 3. These broken symmetry kets are not intended to approximate eigenstates of the Hamiltonian; rather, they represent mixed states whose energies can be computed by both spin-unrestricted molecular orbital and by spin Hamiltonian methods, so that the two may be compared.

The energies of kets like that in Eq. (2) are relatively straightforward to evaluate since each term in the Hamiltonian couples only two centers at a time. Hence, with $S' \equiv S_1 + S_2$,

$$\langle \alpha \alpha \beta | \hat{S}_1 \cdot \hat{S}_2 | \alpha \alpha \beta \rangle = \langle \beta | \beta \rangle \langle \alpha \alpha | \hat{S}_1 \cdot \hat{S}_2 | \alpha \alpha \rangle$$

$$= (1/2)[S'(S'+1) - S_1(S_1+1) - S_2(S_2+1)] = (25/4) \tag{3}$$

since $| \alpha \alpha \rangle$ is a pure spin state with $S' = 5$. The mixed-spin states are slightly more complicated to evaluate, but also yield simple results:

$$\langle \alpha \alpha \beta | \hat{S}_2 \cdot \hat{S}_3 | \alpha \alpha \beta \rangle = \langle \alpha \alpha \beta | \hat{S}_1 \cdot \hat{S}_3 | \alpha \alpha \beta \rangle = -(25/4) \tag{4}$$

For a three-iron system we can define a high-spin ket $| \alpha \alpha \alpha \rangle$ (which is a pure spin state with $S = S_{max} = 15/2$,) and three equivalent broken symmetry states,

$|\alpha\alpha\beta\rangle$, $|\beta\alpha\alpha\rangle$, and $|\alpha\beta\alpha\rangle$, all with $M_S = 5/2$. Their energies are $E(S_{max}) = 75J/4$ and $E(B) = -25J/4$. J can thus be estimated by comparing the energy differences arising from these formulas with those computed from a broken symmetry molecular orbital approach, and estimates of the pure spin state energies are then made from the resulting parametrized spin Hamiltonian. By comparing the energies of the high spin and broken symmetry states, we can generate the entire Heisenberg spin state ladder. This method of "spin state interpolation" depends on the adequacy of the postulated spin Hamiltonian, but successive refinements of this are possible; for example, there can be different J's for different pairs in place of the single J of Eq. (1). Vibronic or exchange striction effects may also be important; we have considered some of these possibilities elsewhere (Noodleman *et al.*, 1988). Nevertheless, recent magnetization measurements for the oxidized 3Fe protein in *D. gigas* ferredoxin II yield a lower limit for $J > 200$ cm^{-1} (Day *et al.*, 1988), in acceptable agreement with our theoretical value of $J = 369$ cm^{-1}.

The basic idea of this approach is illustrated in Fig. 2, which shows the expected energy level diagram for a three-iron cluster where all of the J's are equal, as in our cubane model. In this case the pure-spin states have energies that are a simple function of the total spin S (Griffith, 1972):

$$E(S) = (J/2)[S(S+1)-S_1(S_1+1)-S_2(S_2+1)-S_3(S_3+1)] \quad (5)$$

For three Fe(III) sites, $S_1 = S_2 = S_3 = 5/2$. Experimental estimates of J (*e.g.* from magnetic susceptibility measurements) essentially use the difference of the $S=1/2$ and $S=3/2$ states, as shown on the left. Our computational method, on the other hand, determines J from the difference of energy between the $S=15/2$ state and a broken symmetry state with $M_S = 5/2$, as shown on the right.

It is worth noting that the ground state of this cluster would be very difficult to describe in "pure" density functional terms. As we mentioned above, a determinant of doubly-occupied Kohn-Sham orbitals describes a system with low-spin irons, and in any event has a much higher computed total energy than our broken symmetry state. Our method has some similarity to the use of Slater sum rules to deduce the energies of (multiconfigurational) multiplets in atoms from computed information on single determinant configurations (Ziegler *et al.*, 1977). It would be valuable to explore extensions of density functional theory which allow a detailed description of such multiconfigurational states.

Adding resonance delocalization effects

For Fe(III) sites in these clusters, the five parallel-spin d-electrons have fairly low orbital energies, and there is little delocalization among metal sites, although there is still substantial iron-sulfur covalency. When these sites are reduced, however, the "extra" electrons can be distributed over more than one site, and this delocalization can interact strongly with spin coupling effects. To account for this behavior we must extend the usual Heisenberg description, so that for a dimer (for example,) the dimer energies are not given by

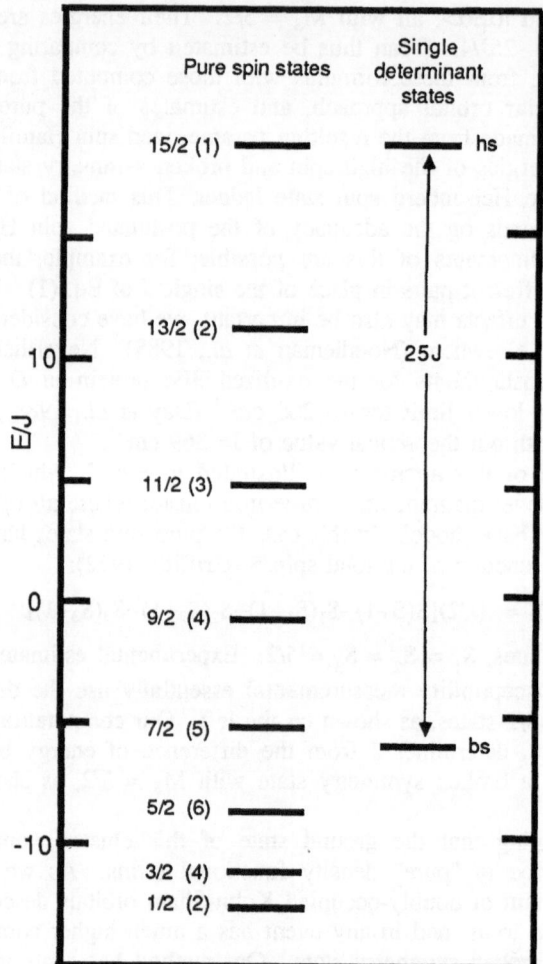

Figure 2. Energy levels for an oxidized 3Fe cluster. Total spin quantum numbers and degeneracies are indicated at the left.

just the Heisenberg formula, $E = (J/2)S'(S'+1)$, but are more closely approximated as

$$E = (J/2) S'(S'+1) \pm B(S'+\tfrac{1}{2}) \tag{6}$$

Here B is a resonance parameter, and the proportionality factor $(S'+\tfrac{1}{2})$ can be derived in a variety of ways, perhaps most transparently from the "double exchange" model of Anderson and Hasagawa (1955). Their work has been developed further by various groups (Girerd, 1983; Borsch *et al.*, 1985,1989; Belinskii, 1989; and others cited below). This provides a simple description of why low-spin Fe(II)--Fe(III) dimers appears as distinct valences, whereas high-spin analogues (which form parts of larger clusters) are delocalized in an effective +2.5 oxidation state. For the antiferromagnetic dimer ($S'=\frac{1}{2}$), the resonance splitting is quite small, so that minor static or dynamic distortions that break the strict equivalence of the iron sites will lead to trapped valence states. In the ferromagnetic ($S'=9/2$) case, however, the resonance interaction predicted by Eq. (2) is much larger, and delocalized states are obtained. In a simple language, it is easier for the final d-electron to "hop" between iron sites when the spins are parallel than when they are opposed, since no net change in exchange interactions is involved in the former shift. A detailed analysis along these lines for Fe_2S_2 dimers has been given by Noodleman and Baerends (1984). For these clusters J is positive, favoring low S', while the resonance interaction stabilizes states of high S', so that the final spin state often represents a balance of opposing forces.

The extension of this theory to polynuclear clusters has been pioneered by Münck, Girerd and their co-workers (Papaefthymiou *et al.*, 1987; Münck *et al.*, 1988). The initial motivation came from analysis of the reduced three iron cluster in *D. gigas* ferredoxin, which showed clearly that one component of this cluster was a delocalized Fe(II)/Fe(III) dimer with $S'=9/2$. Later studies have confirmed that a spin Hamiltonian that leads to energies analogous to those given in Eq. (6) can explain many of the properties in a variety of clusters.

We outline here how these ideas can be incorporated into our broken symmetry analysis. Consider the same 3Fe cluster we discussed above, but now with one additional electron. In the high spin configurations, the first five d-electrons on each site are aligned in a parallel fashion, say spin-up. We form three basis configurations by allowing the final d-electron (which must be spin-down) to reside in turn on each of the three sites. The Heisenberg matrix elements are computed as we outlined above; for the delocalization terms we assume that a single parameter B' characterizes resonance interactions between each pair of sites. Hence, the spin Hamiltonian matrix becomes:

$$\mathbf{H_{hs}} = \begin{bmatrix} (65/4)J & 5B' & 5B' \\ 5B' & (65/4)J & 5B' \\ 5B' & 5B' & (65/4)J \end{bmatrix} \qquad (7)$$

Here and below, the diagonal elements represent the system energy in the absence of delocalization, and the off-diagonal elements give the specific resonance delocalization effects, recognizing that $(S'+\frac{1}{2}) = 5$ for parallel spin Fe(II)/Fe(III) dimers. The eigenvalues are $E_1 = (65/4)J + 10B'$, and $E_{2,3} = (65/4)J - 5B'$ (doubly degenerate.) For these clusters we find B'<0, and hence E_1 lies lowest.

For the broken symmetry state, the first five d-electrons of one of the iron atoms (which we call "a") is of opposite spin to that of an equivalent pair "b". There are still three basis configurations, corresponding to the three possible locations of the last d-electron. We will adopt the simplest delocalization hypothesis, that resonance interaction is important only between the two irons of the same spin, pair "b". The spin Hamiltonian matrix becomes:

$$\mathbf{H}_{bs} = \begin{bmatrix} -(25/4)J & 0 & 5B \\ 0 & -(15/4)J & 0 \\ 5B & 0 & -(25/4)J \end{bmatrix} \quad (8)$$

Here we have allowed the delocalization parameter in the broken symmetry state, B, to differ from that in the high spin state. Eigenvalues for the broken symmetry case are $E_{1,2} = -(25/4)J \pm 5B$ and $E_3 = -(15/4)J$. The splitting between E_1 and E_2 (which is 10B in this model) reflects the difference between bonding and antibonding orbitals delocalized over the "b" sites, as we discuss below.

The J's and B's can thus be estimated by comparing the energy differences arising from these formulas with those computed from a broken symmetry molecular orbital approach, and estimates of the pure spin state energies (including the ground state energy and its spin value) are then made from the resulting parametrized spin Hamiltonian. For the simple model used here, the eigenstates for various values of B and J are given by Papaefthymiou et al. (1987), and numerical estimates are reported by Sontum et al. (1989).

Comparison of LCAO and Scattered Wave Results

Most of our early calculations on iron-sulfur clusters had used the scattered-wave model, but improvements in computing speed and cost, and in program development, have made LCAO calculations increasingly feasible (Boerrigter et al., 1988; Bickelhaupt et al., 1990; Ravenek, 1990). Here we offer some comparisons between the methods for $[Fe_4S_4(SCH_3)_4]^{2-}$, which has an S = 0 ground state as depicted in Figure 1. We also consider the influence of various exchange-correlation potentials on estimates of the spin-coupling parameters J and B.

For this cluster in C_{2v} symmetry, our Xα-LCAO calculations employed a double ζ Fe(4s), triple ζ Fe(3d), single ζ Fe(4p) basis set on Fe and double ζ quality on all other atoms. In all, we have 28 atoms, 110 valence and 270 total electrons, and 376 Cartesian Slater-type basis functions. The set-up time for integrals and other quantities required about 10 minutes CPU time on a CRAY XMP/SE computer. Each iteration of the SCF procedure requires about 1 minute, with about 190 iterations to converge the initial SCF calculation. Subsequent SCF calculations require much less overall time. For example, changing the spin alignment from $M_S = 9$ to $M_S = 0$ requires about 60 iterations. Bond energy analysis takes an additional 6 minutes for each completed SCF calculation. Given the improved quality and accuracy of LCAO vs the SW method, the amount of computer time required compares very favorably with the simpler Xα–SW method. For the latter, each iteration requires 0.3 min. with about 100 to 200 cycles required for convergence of the initial SCF calculation. On the other hand, in this implementation the LCAO codes require about 0.5 Gbyte of temporary disk storage, whereas the scattered wave calculations need less than 1 Mbyte.

It is likely that the number of iterations quoted above can be substantially reduced with more careful attention to convergence parameters. Our more recent calculations on tetrahedral manganese clusters (in C_{3v} symmetry) have required about half the number of iterations we needed for the iron-sulfur cluster. Even so, the entire LCAO analysis for the 4Fe cluster used less than 5 hours of cpu time.

Figures 3 and 4 show the minority spin ligand field levels for the two calculations. As described earlier by Aizman and Case (1982), there is a "mirror" symmetry, such that spin-β orbitals at the top of the cluster are equivalent to the spin-α orbitals at the bottom. The formal oxidation state for this cluster is two Fe(III) and two Fe(II) sites, and the two highest occupied orbitals (indicated by arrows in the figures) correspond to the two sixth d-electrons. The remaining orbitals shown are unoccupied. Although the B parameters can be determined from configuration energy differences by the methods outlined above, they are also quite close to the differences in one-

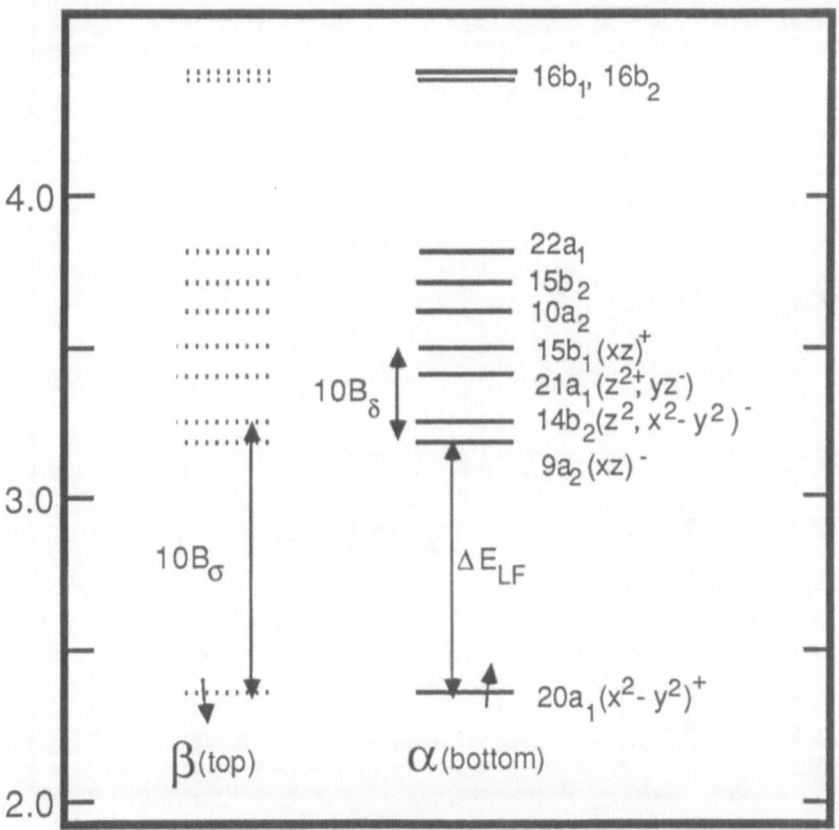

Figure 3. Orbital energies (eV) for ligand field levels from the LCAO model.

electron energies shown in the figure, *i.e.*, the B_σ parameter is determined by the difference between the σ bonding ("+") and antibonding ("-") combinations of the x^2-y^2 orbitals. (These orbitals bond Fe pairs across the top and bottom face diagonals of the cube; see Fig. 1.) Figures 3 and 4 also show how two other parameters, a δ resonance and a ligand field splitting, are estimated.

Overall, the ligand field descriptions of the two methods are fairly consistent. In each case, the lowest three orbitals are $20a_1$, $9a_2$ and $14b_2$, with similar orbital character. The largest difference is in the relative position of the $21a_1$ orbital, but this has little effect on the predicted properties. There is also a large shift in the absolute one-electron energies. This arises from the use of a Watson sphere in the scattered wave calculations (about 5 eV stabilization) and from the special effect that the intersphere potential in that model; such rigid shifts also have little or no contribution to predicted properties. More extensive comparisons between LCAO and scattered wave orbital

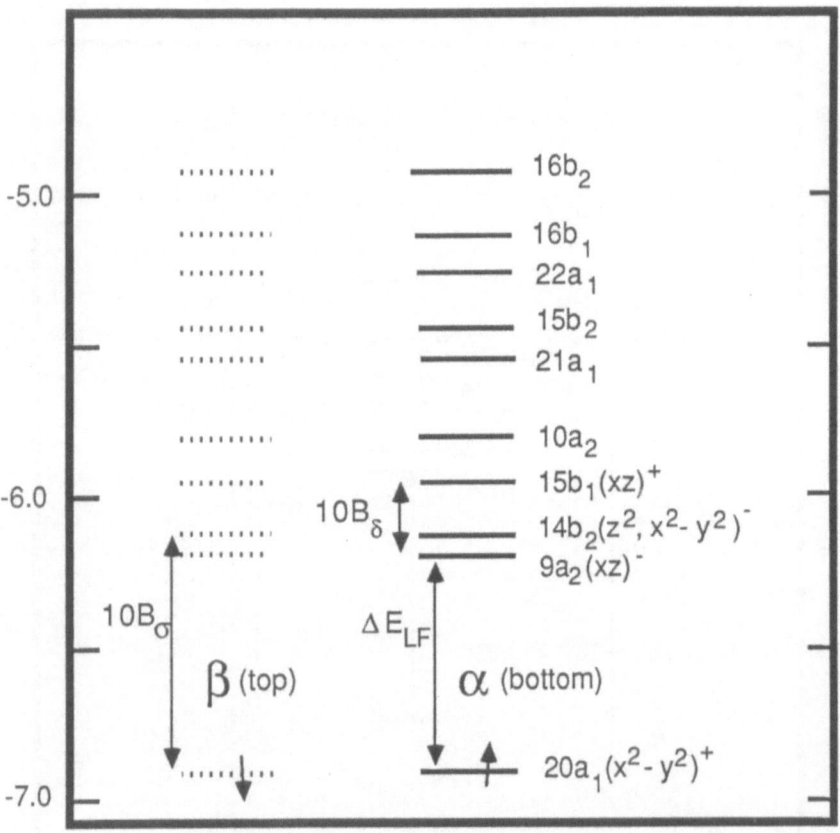

Figure 4. Orbital energies (eV) for ligand field levels from the scattered wave model.

energies for these complexes have been reported earlier (Aizman and Case, 1982; Noodleman and Baerends, 1984), and support the notion that, for these relatively spherical clusters, there are not large differences.

In Table 1, we compare spin coupling parameters calculated with Xα-LCAO to those previously obtained by Xα-SW. It is seen that the results are generally comparable, but that J parameters, the B parameter, and ligand field excitation energies are somewhat larger with the LCAO method.

It is also of interest to see how estimated spin coupling parameters depend upon the exchange-correlation potential. We have looked at the 4Fe clusters using various alternative density functional potentials, specifically the Xα potential with α=0.7334, the Vosko-Wilk-Nusair potential (Ceperly and Alder, 1980; Vosko et al., 1980; Painter, 1981), this plus the Stoll correction potential (Stoll et al., 1980), and both of these including the Becke (1986) energy correction for exchange. Results are collected in Table 2. These results show that the qualitative features of our results are not sensitive to the details

Table 1. Spectroscopic and Spin Coupling Parameters for $[Fe_4S_4(SCH_3)_4]^{2-}$

Parameter	Xα–LCAO	Xα–SW
J	454	401
B_σ	-740	-585
B'(interlayer)	-282	-212
B_δ	248	239
$\Delta E_{LF}(20a_1 \rightarrow 9a_2)$	6751	5243
$\Delta E(\sigma \rightarrow \delta)$	4291	3517

Notes:

Experimental results: For the J parameter, Holm's group has obtained $J = 464 cm^{-1}$ and $500 cm^{-1}$ for an analogous cluster (one and three parameter fits respectively) using a Heisenberg Hamiltonian (Papaefthymiou et al., 1982). However, the resonance delocalization parameter B_σ above is part of a combined Resonance + Heisenberg Hamiltonian. Fitting to this will change both the B and J parameters substantially.

Experimental values of the other spectroscopic parameters have not yet been measured; direct measurements should be possible with infrared spectroscopy. The last two optical parameters calculated, $\Delta E_{LF}(20a_1 \rightarrow 9a_2)$ and $\Delta E(\sigma \rightarrow \delta)$ indicate a rather large ligand field splitting of the e type orbital (in the idealized Fe site symmetry T_d). This splitting is further increased by the resonance term so that $\Delta E_{LF}(20a_1 \rightarrow 9a_2)$ exceeds $\Delta E(\sigma \rightarrow \delta)$ by $(|B_\sigma|-|B_\delta|)(S_{34} + 1/2)$, where S_{34} is the subdimer spin quantum number of the mixed valence pair $S_{34} = 9/2$.

of the exchange-correlation potential; we are planning further studies (such as on 2Fe clusters) to gain a better understanding of the quantitative aspects of these calculations.

An additional advantage of the LCAO codes lies in their ability to analyze bond and total energies as a function of spin state and geometry. As an initial test we have calculated the mean Fe-S bond energy (average over the 16Fe-S bonds of the cluster) for the cluster $[Fe_4S_4(SCH_3)_4]^{2-}$. With the Xα potential, the mean Fe-S bond energy is 64 kcal/mol, while the VWNB result is less, 50 kcal/mol and the VSB result is 44 kcal/mol. (The lower bond energy with VSB is the expected effect from the accumulated experience with density functional methods. Where experimental results are known, VSB usually gives better binding energies in both simple molecules and organometallic systems.) These results are consistent with previous Xα-LCAO calculations on the $[Fe_2S_2]^{2+,1+}$ (Noodleman and Baerends, 1984). In the oxidized form, the mean calculated Fe-S bond strength in the 2Fe system is 67 kcal/mol, while for the reduced form, this is lowered to 49 kcal/mol. The trend that the Fe-S bond strength is lowered as the Fe oxidation state is lowered from $2Fe^{3+}$ to $Fe^{2+}-Fe^{3+}$ is expected because Fe-S antibonding orbitals are occupied on reduction. The higher calculated bond energy for the $[Fe_4S_4]^{2+}$ cluster than for the $[Fe_2S_2]^{1+}$ cluster with the same mean Fe oxidation state (comparing $2Fe^{2.5+}$ for the delocalized 4Fe with $Fe^{2+}-Fe^{3+}$ for the 2Fe system) is certainly a consequence of the resonance delocalization which contributes Fe-Fe bonding density in the 4Fe cluster.

The predicted bond energies above compare well with what little is known of experimental bond energies. The gas phase ion dissociation for $Fe^{+}-S$ was measured as $D^0(Fe^{+}-S) = 61±6$ kcal/mol (MacMahon et al., 1989), while the literature value for the mean Fe-Cl bond energy in $FeCl_3$ is 82 kcal/mol (Huheey, 1979). Further, it is valuable to compare the mean Fe-S bond energy with the energy of stabilization due to spin coupling alone. Considering the interlayer interaction of the two $[Fe_2S_2(SR)_2]^{1-}$ subdimers to constitute 4Fe-S interactions (alternatively 4Fe-Fe interactions) we find that the associated spin stabilization of the S=0 state with respect to the center of gravity

Table 2. Estimates of J for $[Fe_4S_4(SCH_3)_4]^{2-}$

Potential		LCAO	SW
Xα		454	401
Vosko-Wilk-Nusair	(VWN)	662	574
+ Stoll	(VS)	740	
+ Becke	(VWNB)	598	
+ Stoll & Becke	(VSB)	654	

(CG) of the spin ladder gives 8 kcal/mol stabilization for each Fe-S interlayer interaction from Xα-LCAO (vs. 10.7 kcal/mol from VWNB and 11.7 kcal/mol for VSB). As a fraction of the mean Fe-S bond energy, spin coupling accounts for approximately 13% (Xα), 21% (VWNB) or 27% (VSB) of the total interlayer binding energy. This is a substantial proportion, and is consistent with the qualitative notion of the plasticity of the FeS core.

Conclusions

In the preceding sections, we have considered the spin coupling and valence delocalization properties of some representative FeS clusters, showing how broken symmetry density functional calculations can be related to spin Hamiltonian methods. In this final section, we want to mention some forefront areas currently under investigation.

Iron-sulfur clusters in their oxidized form exhibit fairly strong antiferromagnetic coupling parameters J of the order of 300 to 500 cm^{-1}. In the oxidized 2Fe2S and 3Fe4S clusters, all irons are Fe(III), while the high potential (superoxidized) and the oxidized forms of the 4Fe4S, are formally 3Fe(III)/1Fe(II) and 2Fe(III)/2Fe(II), respectively. In the reduced form of 4Fe4S clusters with formal oxidation level 3Fe(II)/1Fe(III), there is a drastic reduction in the Heisenberg constant, and high spin states with S>1/2 may appear at low energy; these may even be degenerate with or below the expected S=1/2 ground state. This reduction in the Heisenberg J parameters is also a feature of the doubly reduced Zn-3Fe4S cluster with formal oxidation state Zn(II)/2Fe(II)/Fe(III) according to our theoretical work (Noodleman et al., 1989). The reduction in J affects both the mixed valence ferrous-ferric (J_1) interaction and the diferrous interaction (J_2); both of these are less than $100cm^{-1}$ in our model, and J_2 can even be negative (ferromagnetic). Based on the Zn3Fe4S work, we have developed a phenomenological model for analyzing spin crossover phenomena and spin dependent properties of reduced 4Fe4S and 4Fe4Se clusters using Racah algebra, and a Heisenberg plus resonance delocalization Hamiltonian (Noodleman, 1990). Spin crossovers are seen as a function of J_2/J_1 and B. Quantitative density functional work on reduced 4Fe4S systems in various geometries will be used to understand and characterize these highly complex spin state equilibria, and to integrate the broken symmetry calculations with the phenomenological model.

Although we have emphasized spin coupling in this paper, there are also significant charge and spin density changes upon oxidation or reduction; these are more subtle than the changes in formal Fe oxidation states (which attribute all charge and spin population changes to the Fe atoms alone) because of significant Fe-S covalency. We have examined such spin and charge density shifts on oxidation/reduction for both 2Fe and 4Fe clusters (Noodleman and Baerends, 1984; Noodleman et al., 1985). In recent work, Moulis et al. (1988) have shown that the bond strengthening on oxidation of high potential iron-sulfur protein is asymmetric, and fits well with the expected orbital electron removed ($20a_1$), and with relaxation effects due to charge flow from the surrounding cysteine sulfurs to the oxidized iron sites.

Since most of the iron sulfur proteins are electron transfer agents, it is important to understand the detailed electron transport pathways into and from the cluster (electron transfer kinetics) as well as electronic contributions to the redox potential. There can be strong interactions of the cluster with the protein environment through hydrogen bonding, electrostatic interactions, and solvation effects. Spin coupling and valence delocalization are also expected to affect electron transfer pathways and redox potentials (Bertrand and Gayda, 1982), but this is not well understood. Close collaborative work between experimental and theoretical groups should lead to advances in our understanding of this important class of proteins with benefits also in understanding the properties of related chemical and solid state systems.

This work was supported by NIH grant GM39914. We thank Eckard Münck for many valuable discussions, and Pieter Vernooijs for computational assistance.

References

Aizman A. and Case, D.A., 1982, *J. Am. Chem. Soc.* **104**:3269-3279.

Anderson, P.W. and Hasagawa, H., 1955, *Phys. Rev.* **100**:675-681.

Becke, A.D., 1986, *J. Chem. Phys.* **84**:4524-4529.

Becke, A.D., 1989, *The Challenge of d and f Electrons*, (Salahub, D.R., Zerner, M.C., Eds.) pp. 165-179, American Chemical Society, Washington, D.C.

Belinskii, M.I., 1987, *Mol. Phys.* **60**:793-819.

Bertrand, P. and Gayda, J.-P., 1982, *Biochim. Biophys. Acta* **680**:331-225.

Bickelhaupt, F.M., Baerends, E.J., Ravenek, W., 1990, *Inorg. Chem.* **29**:350-354.

Boerrigter, P.M., Te Velde, G., Baerends, E.J., 1988, *Int. J. Quantum Chem.* **33**:87-113.

Borshch, S.A., Kotov, I.N., Bersuker, I.B., 1985, *Sov. J. Chem. Phys.* **3**:1009-1016.

Borshch, S.A., Chibotaru, L.F., 1989, *Chem. Phys.* **135**:375-380.

Carney, M.J., Papaefthymiou, G.C., Spartalian, K., Frankel, R.B., Holm, R.H., 1988, *J. Am. Chem. Soc.* **110**:6084-6095.

Ceperly, D.M., Alder, B.J., 1980, *Phys. Rev. Lett.* **45**:566-569.

Day, E.P., Peterson, J., Bonvoisin, J.J., Moura, I. and Moura, J.J.G., 1988, *J. Biol. Chem.* **263**:3684-3689.

Dunlap, B.I., 1987, *Ab Initio Methods in Quantum Chemistry--II* (Edited by Lawley, K.P.) pp. 287-318, John Wiley, New York.

Girerd, J.J., 1983, *J. Chem. Phys.* **79**:1766-1775.

Girerd, J.J., Papaefthymiou, G.C., Watson, A.D., Gamp, E., Hagen, K.S., Edelstein, N., Frankel, R.B., Holm, R.H., 1984, *J. Am. Chem. Soc.* **106**:5941-5947.

Geurts, P.J.M., Gosselink, J.W., van der Avoird, A., Baerends, E.J., and Snijders, J.G., 1980, *Chem. Phys.* **46**:133.

Huheey, J.E.,1979, *Inorganic Chemistry, 2nd ed.*, p.839-851. Harper and Row International Edition, New York.

Jordanov, J., Roth, E.K.H., Fries, P.H., Noodleman, L. 1990,*Inorg. Chem.*, in press.

MacMahon, T.J., Jackson, T.C., Freiser, B.S., 1989, *J. Am. Chem. Soc.***111**:421-428.

Middleton, P., Dickson, D.P.E., Johnson, C.E., Rush, J.D. 1978,*Eur. J. Biochem.***88**:135-141.

Moulis, J.-M., Lutz, J., Gaillard, J. and Noodleman, L., 1988, *Biochemistry* **27**:8712-8719.

Münck, E., Papaefthymiou, V., Surerus, K.K., Girerd, J.J., 1988, *Metal Clusters in Proteins, ACS Symposium Series* 372, (Edited by Que, L., Jr.), Chapter 15, American Chemical Society, Washington, D.C.

Noodleman, L., 1981, *J. Chem. Phys.* **74**:5737-5743.

Noodleman, L. and Baerends, E.J., 1984, *J. Am. Chem. Soc.* **106**:2316-2327.

Noodleman, L., Norman, J.G. Jr., Osborne, J.H., Aizman, A. and Case, D.A., 1985, *J. Am. Chem. Soc.* **107**:3418-3426.

Noodleman, L. and Davidson, E.R., 1986, *Chem. Phys.* **109**:131-143.

Noodleman, L., 1988, *Inorg. Chem.* **27**:3677-3679.

Noodleman, L., Aizman, A. and Case, D.A., 1988, *J. Am. Chem. Soc.* **110**:1001-1005.

Noodleman, L., Case, D.A. and Sontum, S.F., 1989, *J. Chim. Phys.* **86**:743-755.

Noodleman, 1990, *Inorg. Chem.* (in press).

Norman, J.G. Jr., and Jackels, S.C., 1975, *J. Am. Chem. Soc.* **97**:3833.

Norman, J.G. Jr., Kalbacher, B.J., Jackels, S.C., 1978, *J. Chem. Soc. Chem. Commun.* 1027.

Norman, J.G. Jr., Ryan, P.B. and Noodleman, L., 1980, *J. Am. Chem. Soc.* **102**:4279.

Papaefthymiou, G.C., Laskowski, E.J., Frota-Pessoa, S., Frankel, R.B., Holm, R.H. 1982, *Inorg. Chem.* **21**:1723-1728.

Papaefthymiou, V., Girerd, J.J., Moura, I., Moura, J.J.G., Münck, E., 1987, *J. Am. Chem. Soc.* **109**:4703-4710.

Painter, G.S., 1981,*Phys. Rev. B* **24**:4264-4270.

Parr, R.G. and Yang, W., 1989, *Density-Functional Theory of Atoms and Molecules*, Oxford University Press, New York.

Ravenek, W., 1989, *Scientific Computing on Supercomputers* (Edited by J.T. Devreese and P.E. Van Camp), pp. 201-218, Plenum, New York.

Sands, R.H., Dunham, W.R., 1975, Quart. Rev. Biophys. **7**:443-504.

Slater, J.C., 1974, *Quantum Theory of Molecules and Solids*, McGraw-Hill, New York.

Sontum, S.F., Noodleman, L. and Case, D.A., 1989, *The Challenge of d and f Electrons* (Edited by D. Salahub and M.C. Zerner), pp. 366-377, American Chemical Society, Washington, D.C.

Stoll, H., Pavlidou, C.M.E., Preuss, H., 1978, *Theoret. Chim. Acta* **149**:143-149.

Stoll, H., Golka, E., Preuss, H., 1980, *Theoret. Chim. Acta* **55**:29-41.

Vosko, S.H., Wilk, L., Nusair, 1980, *Can. J. Phys.* **58**:1200-1211.

Yang, C.Y., Johnson, K, Holm, R.H., Norman, J.G. Jr., 1975, *J. Am. Chem. Soc.* **97**:6596.

Ziegler, T., Rauk, A. and Baerends, E.J., 1977, *Theor. Chim. Acta* **43**:261.

9
Local-Density Functional Electronic Structure of Helical Chain Polymers

J.W. MINTMIRE

Introduction

Polymers represent an area of chemistry intermediate between molecular chemistry and solid-state physics. The one-dimensional periodicity leads to delocalization of electron states analogous to that in crystals, but the finiteness in other dimensions requires the use of localized wavefunctions as in molecular approaches. Although *ab initio* methods have been used to calculate both total energies and electronic structures of chain polymers for several decades (Del Re, et al., 1967; André, 1969), the ability to calculate accurate total energies for polymers (or even large molecular clusters) within a local-density functional (LDF) approach has come about principally within the last decade. Our group at the Naval Research Laboratory has had great success in recent years in applying local-density functional methods to this area of chemistry, in particular to polyacetylene (Mintmire and White, 1983abcd, 1987ab) and polysilane systems (Mintmire, 1989ab), using methods developed for polymer chains with translational symmetry, as well as the calculation of electronic properties on molecular species (Kutzler, et al., 1986; White, et al., 1986; Mintmire, et al., 1987). The techniques for polymers are based on one-electron wavefunctions constructed from linear combinations of Gaussian-type orbitals (LCGTO), using algorithms equivalent to an infinite chain limit of the molecular scheme introduced by Dunlap, et al. (1979ab).

In addition to pure translational symmetry, many chain polymers exist in helical conformations. Model calculations on such helical systems performed taking only translational symmetry into account can require excessively large unit cells, increasing the needed computational resources. For example, in recent local-density functional calculations on polysilane helical conformations, (Mintmire, 1989a) calculations on the 7/3 helical structure of polysilane, $-(SiH_2)-$, required a unit cell with 7 silicon atoms and 14 hydrogen atoms, although the minimal unit cell using helical symmetry is the SiH_2 unit. Using the helical symmetry to reduce the computational size of the electronic structure calculation was first proposed for semiempirical methods (Imamura, 1970; Fujita and Imamura, 1970), with an *ab initio* scheme later reported by Blumen and Merkel (1977). More recently the *ab initio* approach using Gaussian basis sets has been applied to a range of realistic chain polymer systems (Karpfen and Beyer, 1984; André, et al., 1984; Teramae and Takeda, 1989).

We have recently extended the LCGTO local-density functional approach originally developed for molecular systems (Dunlap, et al., 1979ab), and extended to two-dimensionally periodic systems (Mintmire, et al., 1982) and chain polymers (Mintmire and White, 1983ac), to calculate the total energies and electronic structures of helical chain polymers. We present below a brief overview of this approach, with special attention paid to those aspects of this scheme that are specific to helical polymer systems. In particular we focus on how the fitting schemes developed for the original molecular approach (Dunlap, et al., 1979ab; Mintmire and Dunlap, 1982) are adapted to an infinite system. Finally, we present some preliminary total energy results for polyethylene, which we compare with previous *ab initio* (Karpfen and Beyer, 1984) and local-density functional linear muffin-tin orbital (Springborg and Lev, 1989) results.

Formalism

Let us begin by defining the systems that we shall consider. Borrowing concepts from solid-state band-structure theory, we shall define a helical chain polymer as a nuclear lattice constructed from a finite basis of atoms and a screw operation $S(a, \phi)$. For mathematical convenience we define the screw operation in terms of a translation a units down the z axis in conjunction with a right-handed rotation ϕ about the z axis. That is,

$$S(a, \phi)\mathbf{r} \equiv \begin{pmatrix} x \cos \phi - y \sin \phi \\ x \sin \phi + y \cos \phi \\ z + a \end{pmatrix}. \tag{1}$$

with the screw operation henceforth denoted as S, with arguments a and ϕ implicitly understood. Because the symmetry group generated by the screw operation S is isomorphic with the one-dimensional translation group, Bloch's theorem can be generalized so that the one-electron wavefunctions will transform under S according to

$$S^m \psi_i(\mathbf{r}; \kappa) = e^{i\kappa m} \psi_i(\mathbf{r}; \kappa). \tag{2}$$

The quantity κ is a dimensionless quantity which is conventionally restricted to a range of $-\pi < \kappa \leq \pi$, a central Brillouin zone. For the case $\phi = 0$ (i.e., S a pure translation), κ corresponds to a normalized quasimomentum; i.e., $\kappa \equiv ka$, where k is the traditional wavevector from Bloch's theorem in solid-state band-structure theory.

The one-electron wavefunctions ψ_i are constructed from a linear combination of Bloch functions φ_j, which are in turn constructed from a linear combination of nuclear-centered Gaussian-type orbitals $\chi_j(\mathbf{r})$ (in this case, products of Gaussians and the real solid spherical harmonics),

$$\psi_i(\mathbf{r}; \kappa) = \sum_j c_{ji}(\kappa) \varphi_j(\mathbf{r}; \kappa) \tag{3}$$

$$\varphi_j(\mathbf{r}; \kappa) = \sum_m e^{-i\kappa m} S^m \chi_j(\mathbf{r}). \tag{4}$$

The one-electron density matrix is then given by

$$\rho(\mathbf{r}; \mathbf{r}') = \sum_i \frac{1}{2\pi} \int_{-\pi}^{\pi} d\kappa \, n_i(\kappa) \, \psi_i^*(\mathbf{r}'; \kappa) \, \psi_i(\mathbf{r}; k)$$

$$= \sum_{jj'} \sum_m P_{j'j}^m \sum_{m'} \chi_{j'}^{m+m'}(\mathbf{r}') \, \chi_j^{m'}(\mathbf{r}), \tag{5}$$

where $n_i(\kappa)$ are the occupation numbers of the one-electron states, χ_j^m denotes $S^m \chi_j(\mathbf{r})$, and P_{ij}^m are the coefficients of the real lattice expansion of the density matrix given by

$$P_{j'j}^m = \sum_i \frac{1}{2\pi} \int_{-\pi}^{\pi} d\kappa \, n_i(\kappa) \, c_{j'i}^*(\kappa) \, c_{ji}(\kappa) \, e^{i\kappa m} \tag{6}$$

The total energy for the polymer system is given by

$$E = \sum_{ij} \sum_m P_{ij}^m \left\{ -\frac{1}{2} \langle \chi_i^m | \nabla^2 | \chi_j^0 \rangle + \langle \chi_i^m | \epsilon_{xc}[\rho(\mathbf{r})] | \chi_j^0 \rangle \right\}$$

$$+ \frac{1}{2} \sum_{m''} \left\{ \left(\sum_n' \sum_{n'} \frac{Z_n Z_{n'}}{|\mathbf{R}_n^0 - \mathbf{R}_{n'}^{m''}|} \right) \right.$$

$$+ \sum_{ij} \sum_m P_{ij}^m \left(\sum_{i'j'} \sum_{m'} P_{i'j'}^{m'} \left[\chi_i^m \chi_j^0 | \chi_{i'}^{m'+m''} \chi_{j'}^{m''} \right] \right.$$

$$\left. \left. -2 \sum_n \langle \chi_i^m | \frac{Z_n}{|\mathbf{r} - \mathbf{R}_n^{m''}|} | \chi_j^0 \rangle \right) \right\} \tag{7}$$

where Z_n and $\mathbf{R}_n$ denote the nuclear charges and coordinates within a single unit cell, $\mathbf{R}_n^m$ denotes the nuclear coordinates in unit cell m ($\mathbf{R}_n^m \equiv S^m \mathbf{R}_n$), and $[\rho_1 | \rho_2]$ denotes an electrostatic interaction integral,

$$[\rho_1 | \rho_2] \equiv \int d^3 r_1 \int d^3 r_2 \, \frac{\rho_1(\mathbf{r}_1) \rho_2(\mathbf{r}_2)}{r_{12}} \tag{8}$$

Rather than solve for the total energy directly as expressed in Eq. 7, we follow the suggestion of earlier workers (Sambe and Felton, 1974, 1975) to fit the exchange-correlation potential and the charge density (in the Coulomb potential) to a linear combination of Gaussian-type functions. The methods used are essentially a straightforward extension of fitting methods developed for molecular calculations (Dunlap et al., 1979ab; Mintmire and Dunlap, 1982); the primary innovations necessary for treating an extending system occur for fitting the charge density and will be discussed in detail in this work.

Exchange-Correlation Potential Fitting

We shall use the non-spin-polarized Xα (Connolly, 1976; Slater, 1974) local-density functional in the following description of our techniques, although extension to other local-density functionals and to the spin-polarized case is straightforward. Given a set of nuclear-centered (this restriction can be trivially relaxed if desired) Gaussian-type functions $G_i(r)$, we fit the exchange-correlation potential $\epsilon_{xc}(r)$ with a linear combination of periodic fitting functions,

$$\epsilon_{xc}(r) \approx \sum_i g_i \mathcal{G}_i(r) \tag{9}$$

$$\mathcal{G}_i(r) \equiv \sum_m S^m G_i(r). \tag{10}$$

The coefficients g_i are chosen by requiring a least-squares sum over a weighted set of points to yield the best-fit condition,

$$\sum_j g_j \sum_\ell w_\ell \mathcal{G}_j(r_\ell) \mathcal{G}_i(r_\ell) = \sum_\ell w_\ell \mathcal{G}_i(r_\ell) \epsilon_{xc}(r_\ell). \tag{11}$$

Dunlap and Cook (1986) have found that an optimum choice of weighting function w_ℓ results from multiplying the normal quadrature weight v_ℓ (appropriate for approximating an integral over all space) with a factor proportional to the charge density divided by the exchange-correlation energy density. In $X\alpha$ this reduces to

$$w_\ell = v_\ell \, \rho^{2/3}(r_\ell). \tag{12}$$

This fitting procedure is computationally implemented by calculating the values of $\mathcal{G}_i(r_\ell)$ and $\chi_i^m(r_\ell)$ before the self-consistent field (SCF) cycle begins, with $\rho(r_\ell)$ calculated in each iteration of the SCF cycle from Eq. 5. Values of χ_i^m are typically calculated for m less than 10. Our integration grids are chosen using multiple sets of nuclear-centered grids, with the quadrature weights for the grids reduced smoothly as the grid points leave their respective Voronoi polyhedra. The radial grid is chosen using the first Herman-Skillman grid point for the first step size (Herman and Skillman, 1963), and increasing the step sizes in a geometric progression (Dunlap, 1986). We currently use 26 points in each spherical shell using the procedure described in Jones, et al. (1988), although we are investigating the use of new techniques for smoothly combining nuclear-centered grids developed by Becke (1988).

Charge Density Fitting

Sambe and Felton (1974, 1975) suggested fitting $\rho(r)$ in the Coulomb interaction terms of the total energy expression for molecules as a means of reducing computational effort; i.e., fitting $\rho(r)$ markedly reduces the number of Coulomb matrix elements that must be evaluated. For the helical polymer, we wish to approximate the charge density with a linear combination of nuclear-centered Gaussian-type functions in a fashion analogous to that in Eqs. 9 and 10, although typically using a basis for the functions $F_i(\mathbf{r})$ different from those used for $G_i(\mathbf{r})$,

$$\rho(\mathbf{r}) \approx \sum_i f_i \, \mathcal{F}_i(\mathbf{r}) \equiv \tilde{\rho}(\mathbf{r}), \tag{13}$$

$$\mathcal{F}_i(\mathbf{r}) \equiv \sum_m \mathcal{S}^m F_i(\mathbf{r}). \tag{14}$$

Dunlap et al. (1979ab) showed that an improved fitting technique, relative to earlier (Sambe and Felton, 1974, 1975) least-squares fitting techniques, could be obtained by noting that minimising the self-interaction of the error in the charge density, $\Delta\rho(\mathbf{r}) \equiv \rho(\mathbf{r}) - \tilde{\rho}(\mathbf{r})$, leads to a variationally optimised estimate of the electron-electron Coulomb repulsion energy. The exact expression for the Coulomb energy E_c,

$$E_c \equiv \frac{1}{2}[\rho \,|\, \rho] = [\rho \,|\, \tilde{\rho}] - \frac{1}{2}[\tilde{\rho} \,|\, \tilde{\rho}] + \frac{1}{2}[\Delta\rho \,|\, \Delta\rho] \tag{15}$$

leads to the estimate for E_c given by $\tilde{E}_c$ given by

$$\tilde{E}_c \equiv [\rho \,|\, \tilde{\rho}] - \frac{1}{2}[\tilde{\rho} \,|\, \tilde{\rho}], \tag{16}$$

which is a lower bound to the true Coulomb interaction, E_c. Minimization of this difference with respect to the fitting coefficients f_i (in the molecular case these are, of course, coefficients for local Gaussian-type functions rather than periodic sums as in Eq. 14) in turn leads to the condition for best fit,

$$[\tilde{\rho} - \rho \,|\, F_i] = 0. \tag{17}$$

We write Eq. 17 in the above form because this expression remains well-defined as a molecular cluster increases in size to the limit of an infinite polymer chain. For finite molecules, we note that Eq. 17 does not require that the total integrated charge of $\tilde{\rho}$ equal the total charge of ρ, although this condition can be added as a constaint on the fitting procedure (Dunlap et al., 1979ab; Mintmire and Dunlap, 1982). For the expression in Eq. 17 to remain finite however, the total charge per unit cell in both charge distributions $\tilde{\rho}$ and ρ must be equal, as demonstrated earlier for two-dimensionally periodic systems (Mintmire et al., 1982).

Although the molecular fitting scheme for the exchange-correlation potential can be extended directly to the polymer chain, this is not possible for the charge density fit because of the long-range nature of the Coulomb potential. With the understanding that the notation $[\rho_1 \,|\, \rho_2]$ for periodic charge distributions denotes a Coulomb interaction per unit cell,

$$[\rho_1 \,|\, \rho_2] \equiv \int_\Omega d^3r_2 \int d^3r_2 \, \frac{\rho_1(\mathbf{r}_1)\rho_2(\mathbf{r}_2)}{r_{12}}, \tag{18}$$

we note that such an integral will only be finite for either ρ_1 or ρ_2 having zero total charge. Thus matrix elements such as those between fitting functions, $[\mathcal{F}_i \mid \mathcal{F}_j]$, will not in general be finite for polymers, although the analogous terms in the molecular case, $[F_i \mid F_j]$, are finite. We further note that a similar lack of convergence will occur for the electron-nuclear attraction and the nuclear-nuclear repulsion terms in Eq. 7. This leads us to extend Eq. 17 by including the nuclear point charges in our fitting scheme. Defining ρ_N as the charge distribution resulting from the set of nuclear point charges (with self-interaction terms between point charges omitted in Coulomb matrix elements), and

$$\eta_i \equiv \left\{ \int d^3r \, F_i(\mathbf{r}) \right\} / \sum_n Z_n, \tag{19}$$

we find,

$$[\tilde{\rho}-\rho_N \mid \mathcal{F}_i - \eta_i\rho_N] - [\rho-\rho_N \mid \mathcal{F}_i - \eta_i\rho_N] = \eta_i [\rho_N \mid \tilde{\rho}-\rho]. \tag{20}$$

Treating the electrostatic term on the right-hand side as an undetermined multiplier (i.e., $\lambda \equiv [\rho_N \mid \tilde{\rho}-\rho]$), we obtain a computationally practical condition for best fit algebraically similar to that of a least-squares fit with a constraint.

$$\sum_j [\mathcal{F}_i - \eta_i\rho_N \mid \mathcal{F}_j - \eta_j\rho_N] f_j = [\mathcal{F}_i - \eta_i\rho_N \mid \rho - \rho_N] + \lambda\eta_i, \tag{21}$$

$$\sum_i f_i\eta_i = 1. \tag{22}$$

Formally the approximate total Coulomb energy, $\tilde{E}_c$, includes the approximate electron-electron interaction and the exact electron-nuclear attraction and nuclear-nuclear repulsion terms; i.e.,

$$\tilde{E}_c \equiv [\rho - \rho_N \mid \tilde{\rho} - \rho_N] - \frac{1}{2}[\tilde{\rho} - \rho_N \mid \tilde{\rho} - \rho_N] \tag{23}$$

$$= [\rho \mid \tilde{\rho}] - \frac{1}{2}[\tilde{\rho} \mid \tilde{\rho}] - [\rho \mid \rho_N] + \frac{1}{2}[\rho_N \mid \rho_N]. \tag{24}$$

The value of λ is obtained by solving Eqs. 21 and 22; $\tilde{E}_c$ is then obtained from

$$\tilde{E}_c = \frac{1}{2} \left\{ \left(\sum_i f_i [\mathcal{F}_i - \eta_i\rho_N \mid \rho - \rho_N] \right) - \lambda \right\}. \tag{25}$$

Multipole Expansion Techniques

Most of the terms in the total energy expression of Eq. 7 (and the terms in the corresponding secular equation) can be expressed as a lattice summation with exponential convergence, and summed directly in our computational procedure to reasonable relative accuracies (we typically sum to a relative accuracy of $\sim 10^{-14}$). In the charge fitting procedure, however, the long-range nature of the Coulomb potential yields sums of terms which decrease as an inverse power of distance rather than exponentially, with the worst case decreasing as $1/r^3$. Consider the matrix element involving the two charge fitting functions $\mathcal{F}_i$ and $\mathcal{F}_j$,

$$[\mathcal{F}_i - \eta_i \rho_N \mid \mathcal{F}_j - \eta_j \rho_N] = \sum_m [F_i^0 - \eta_i \rho_N^0 \mid F_j^m - \eta_j \rho_N^m], \qquad (26)$$

where F_i^m denotes the local Gaussian-type function $F_i(\mathbf{r})$ translated to unit cell m (i.e., $F_i^m(\mathbf{r}) = S^m F_i(\mathbf{r})$), and ρ_N^m denotes the set of nuclear charges contained in unit cell m. Piela and Delhalle (1978) found that such sums could be treated more efficiently in *ab initio* calculations on polymers (i.e., with more rapid convergence and better precision for given length expansion) by examining the long-range terms of such expressions in a multiple expansion, and approximating the long-range behavior of such sums to infinity with a finite-order multipole expansion. Later work (Piela, et al., 1980; Delhalle, et al., 1980; André, et al., 1984) developed these techniques further and extended them to helical polymers within an *ab initio* framework, and similar techniques were introduced for two-dimensionally periodic systems within a local-density functional approach (Mintmire, et al., 1982). We present below a description of these multipole expansion techniques as we implement them for helical chain polymers within the above-mentioned charge fitting procedure.

Using a bipolar expansion (Steinborn and Ruedenberg, 1973; Weniger and Steinborn, 1985) of the Coulomb potential, $1/r_{12}$, we can express the long-range (i.e., at sufficient distance to neglect any overlap between the charge distributions) interaction between two charge distributions ρ_1^n and ρ_2^m localized in the origin unit cell and unit cell m (i.e., $\rho_i^n(\mathbf{r}) \equiv S^n \rho_i(\mathbf{r})$) by

$$[\rho_1^n \mid \rho_2^0] = \sum_{L=0}^{\infty} \hat{Z}_L(na) \sum_{\ell=0}^{L} (-1)^{L-\ell} \sum_{|m| \leq min[\ell, L-\ell]} M_1^n(\ell, m) \, M_2^0(L-\ell, -m). \qquad (27)$$

Here $M_i^n(\ell, m)$ is the overlap of charge distribution ρ_i^n with the modified regular solid spherical harmonic $\hat{\mathcal{Y}}_\ell^m(\mathbf{r})$ translated a distance na down the z axis,

$$M_i^n(\ell, m) \equiv \int d^3r \, \rho_i^n(\mathbf{r}) \, \hat{\mathcal{Y}}_\ell^m(\mathbf{r} - na\,\mathbf{e}_z) \qquad (28)$$

taking advantage of our use of the z axis as the helical chain axis. $\hat{Z}_L(d)$ is the value of the modified irregular solid spherical harmonic $\hat{Z}_L^0$ evaluated a distance d along the z axis, with the two modified solid spherical harmonics defined (Steinborn and Ruedenberg, 1973),

$$\hat{\mathcal{Y}}_\ell^m(\mathbf{r}) \equiv r^\ell Y_{\ell m}(\theta, \phi) \left[\frac{2\ell+1}{4\pi}(\ell-m)!\,(\ell+m)!\right]^{-1/2}, \tag{29}$$

$$\hat{\mathcal{Z}}_\ell^m(\mathbf{r}) \equiv r^{-(\ell+1)} Y_{\ell m}(\theta, \phi) \left[\frac{4\pi}{2\ell+1}(\ell-m)!\,(\ell+m)!\right]^{1/2}. \tag{30}$$

All of the summations for which we use the multipole expansion contain contributions from both the n and $-n$ unit cells. Taking advantage of the symmetries of the spherical harmonics, we find

$$[\rho_1^n + \rho_1^{-n} \mid \rho_2^0] = 2 \sum_{L=0}^{\infty} \sum_{m=0}^{[L/2]} \hat{\mathcal{Z}}_L(na)\, g_L(mna)\, Q_{Lm}, \tag{31}$$

where

$$g_L(\theta) = \begin{cases} \cos(\theta) & L \text{ even}, \\ \sin(\theta) & L \text{ odd}, \end{cases} \tag{32}$$

and

$$Q_{Lm} = \begin{cases} \displaystyle\sum_{\ell=0}^{L}(-1)^{L-\ell} M_1(\ell,0)\, M_2(L-\ell,0) & m=0,\ \text{else} \\[2ex] \displaystyle 2\sum_{\ell=m}^{L-m}(-1)^{\ell-m}\, \Re[M_1(\ell,m)\, M_2^*(L-\ell,m)] & L \text{ even}, \\[2ex] \displaystyle 2\sum_{\ell=m}^{L-m}(-1)^{\ell-m}\, \Im[M_1(\ell,m)\, M_2^*(L-\ell,m)] & L \text{ odd}, \end{cases} \tag{33}$$

where $M_i(\ell,m) \equiv M_i^0(\ell,m)$. In our computational algorithm for calculating summations such as those in Eq. 26, we use a truncated expansion up to $L=8$ in Eq. 31. For a given summation, we first calculate the required values of $M_i(\ell,m)$ and generate the values of Q_{Lm}. The sum in Eq. 26, for example, is then evaluated by starting with the exact expression in Eq. 26 for $m=0$, then summing the differences between the exact expression for a given value of m and the value that is evaluated from the truncated multipole expansion in Eq. 31. The truncation of this series is then determined by the convergence of these differences. The sum to infinity of the multipole interaction terms is then added to extrapolate the series in Eq. 26 for finite order in L. Evaluation of this term,

$$\sum_{n=1}^{\infty} Z_L(na)\, g_L(mna) = L!\, \zeta_{L+1}(m\phi)/a^{(L+1)} \tag{34}$$

requires the evaluation of the quantity $\zeta_L(\theta)$, which has the general form

$$\zeta_\ell(\theta) = \begin{cases} \sum\limits_{n=1}^{\infty} \cos(n\theta)/n^\ell & \ell \text{ odd,} \\[2mm] \sum\limits_{n=1}^{\infty} \sin(n\theta)/n^\ell & \ell \text{ even.} \end{cases} \tag{35}$$

As noted by other workers (Piela, et al. 1980), closed analytic expressions are known for particular values of the arguments; in particular $\zeta_L(0)$ is the Riemann zeta function, $\zeta(L)$, (Abramowits and Stegun, 1972) and values are also tabulated in this reference for $\theta = \pi/2$ and π. For computational convenience and precision we have chosen to use the identity (Gradshteyn and Ryzhik, 1965)

$$\zeta_1(\theta) = \sum_{n=1}^{\infty} \frac{\cos n\theta}{n}$$
$$= -\frac{1}{2}\ln[2(1-\cos\theta)]. \tag{36}$$

We used the *Mathematica* software package (Wolfram, 1988) to evaluate the coefficients for the series expansion to Eq. 36,

$$\zeta_1(\theta) = -\ln\theta + \frac{\theta^2}{24} + \frac{\theta^4}{2880} + \cdots, \tag{37}$$

Using the relation $\zeta_\ell'(\theta) = (-1)^\ell \, \zeta_{\ell-1}(\theta)$, we find that using the series expansion to 40th order and explicitly integrating each term to obtain ζ_ℓ for $\ell > 1$ yields a relative precision of $\sim 10^{-14}$ for $0 \le \theta \le \pi$.

Polyethylene Results

We have carried out a series of total energy calculations for polyethylene, $-(CH_2)-$, as a function of the backbone CCCC dihedral angle. No optimization of the geometry was performed at this step; the internal coordinates were chosen by using a standard CC bond length of 1.54 Å, a CH bond length of 1.08 Å, and tetrahedral bond angles. A 7s3p/3s uncontracted orbital basis set from van Duijneveldt (1974) was used. The primary purpose of these preliminary calculations was to test the stability of the numerical algorithms with respect to rotation around bond angles. Dunlap and Cook (1986) investigated the change in total energy of ethane as a function of bond rotation, and noticed that changes in the numerical grid for fitting the exchange potential could lead to effects on the total energy of the order of the rotational barrier of ~ 3 kcal/mole. Later work by Jones, et al. (1988) indicated that the calculated exchange-correlation energy could be stabilized with respect to fitting parameters (e.g., the choice of grid points) by several steps including random rotation of spherical shells in the numerical grids. Because of the typical small magnitude of changes in energy with respect to bond rotation, we are in the process of examining our numerical results to verify that no artifacts of the fitting procedure are producing systematic errors in our results.

Table I presents our calculated total energies for the polyethylene chain relative to the planar zigzag conformation (i.e., $\phi = 180°$). We compare our results with accurate *ab initio* Hartree-Fock calculations (Karpfen and Beyer, 1984) using helical symmetry for polyethylene with a 7s3p/3s contracted basis set. Because the HF calculations were performed over a regular set of helix twist angles (as opposed to our calculations over a set of dihedral angles) we have converted the helical twist angles from the HF results into the equivalent dihedral angles, and presented the closest comparisons to our results.

We find that the rotational barrier height at a dihedral angle of 120° (expected there because of the eclipsed conformation) is 4.09 kcal/mole, in good agreement with the HF rigid-rotor calculations. The primary qualitative difference between our LDF results and the HF results is in the height of the secondary *gauche* minimum at a dihedral angle of roughly 60°. We find using the LDF method a *gauche* conformation only 0.2 kcal/mole above the planar conformation. Our LDF result is thus somewhat lower than the ~ 1 kcal/mole difference calculated within the HF formalism (Karpfen and Beyer, 1984)—with the HF results in agreement with current experimental estimates of this quantity in gas-phase butane and other small alkanes.

We believe that this low value of the relative energy of the *gauche* conformation may result from residual effects of numerical grids used. For example, using our current numerical grids to integrate the charge density over the unit cell (which should yield 8 for polyethylene) yields 7.731 for the planar conformation and 7.797 for the *gauche*. The exchange-correlation energy (calculated using analytic matrix elements between orbital products, $\chi_i\chi_j$, and the exchange fitting functions $\mathcal{G}_m$) is -5.2733 hartrees for the planar conformation compared to -5.2812 hartrees for the *gauche*. A naive approach would be to expect a systematic error in the exchange-correlation energy proportional to the change in the integrated charge density; i.e., if a change in numerical grid enhanced the relative weight of the high-density regions of space, we would expect a corresponding increase in the magnitude of the calculated exchange-correlation energy. In this simplistic view we could expect the exchange-correlation energy to yield a result for the *gauche* conformation ~ 30 kcal/mole less than the planar conformation. The fitting procedure, of course, eliminates systematic errors of this magnitude (as we see by the actual change in exchange-correlation energy of only ~ 5 kcal/mole), but smaller errors on the order of 1 kcal/mole could still occur. We are currently evaluating the use of improved numerical grids (Becke, 1988) that we have found in a molecular scheme (Mintmire, 1990) improves the relative error in the integrated charge density from 10^{-2} (our

Table I. Comparison of rigid-rotor potential curves for polyethylene using LCGTO LDF (this work), *ab initio* Hartree-Fock (Karpfen and Beyer, 1984), and LMTO LDF (Springborg and Lev, 1989) methods. Dihedral angles are given in degrees for the CCCC backbone dihedral angle. because the HF calculations were evaluated for regular intervals in the helical twist angle, the exact dihedral angles used are given in the fourth column of the table.

Dihedral angle (LDF) (degrees)	LDF (kcal/mole)	LDF (LMTO) (kcal/mole)	Dihedral angle (HF) (degrees)	HF (kcal/mole)
180	0.0	0.0	180.0	0.0
150	+1.24	+46.1	155.9	+1.08
143	+2.09		143.7	+2.42
120	+4.09		118.9	+4.29
90	+0.76		92.8	+1.96
60	+0.20		63.6	+1.29
45	+7.56		45.9	+7.72

current grid) to $< 10^{-4}$.

We have also compared our results with recent results for polyethylene using a local-density functional scheme with LMTO basis sets (Springborg and Lev, 1989). As we see in Table I, the variation in energy on rotating the carbon backbone from planar to a dihedral angle of 150° in the LMTO approach is more than an order of magnitude greater than either our LCGTO results or the *ab initio* results, and such a large change in energy is inconsistent with the general trend of rotational barriers in alkanes to be on the order of 1–10 kcal/mole.

Summary

We have presented an overview of a new LDF approach for calculating the total energies and electronic structures of helical chain polymers using Gaussian basis sets. This approach represents a formal extension of earlier techniques developed for molecular systems (Dunlap, et al., 1979ab) using fitting techniques to treat the exchange-correlation potential and the Coulomb potential; we have discussed how these schemes must be modified to treat a one-dimensionally periodic chain polymer. Preliminary calculations were presented as a test of the stability of the numerical algorithms involved. A possible systematic error of the order of up to 1 kcal/mole was noted, potentially arising from the use of a finite grid whose effects vary as the conformation is altered. We are currently investigating minimizing even these errors by the introduction of improved grid techniques.

Acknowledgements

I would like to thank C. T. White, B. I. Dunlap, E. J. Weniger, and Z. H. Levine for many helpful conversations. This work was supported in part by the U. S. Department of Energy, Office of Basic Energy Sciences. Computational support for this project was provided in part by a grant from the Ohio Supercomputer Center.

References

Abramowitz, M., and Stegun, I. A., 1972, *Handbook of Mathematical Functions*, pp. 804–819. Dover, New York..

André, J. M., 1969, *J. Chem. Phys.* 50:1536–1542.

André, J. M., Vercauteren, D. P., Bodart, J. P., and Fripiat, J. G., 1984, *J. Comp. Chem.* 5:536–547.

Becke, A. D., 1988, *J. Chem. Phys.* 88:2547–2553.

Blumen, A., and Merkel, C., 1977, *Phys. Stat. Sol. B* 83:425–431.

Connolly, J. W. D., 1976, *Modern Theoretical Chemistry*, (Edited by Segal, G. A.) pp. 105–132. Plenum Press, New York..

Delhalle, J., Piela, L., Brédas, J.-L., and André, J. M., 1980, *Phys. Rev. B* 22:6254–6266.

Del Re, G., Ladik, J., and Bicsó, G., 1967, *Phys. Rev.* 155:997–1003.

Dunlap, B. I., Connolly, J. W. D, and Sabin, J. R, 1979a, *J. Chem. Phys.* 71:3396–3402.

Dunlap, B. I., Connolly, J. W. D, and Sabin, J. R, 1979b, *J. Chem. Phys.* 71:4993–4999.

Dunlap, B. I., 1986, *J. Phys. Chem.* 90:5524–5529.

Dunlap, B. I., and Cook, M., 1986, *Int. J. Quantum Chem.* 29:767–777.

Fujita, H., and Imamura, A., 1970, *J. Chem. Phys.* 53:4555–4566.

Gradshteyn, I. S., and Ryshik, I. M., 1965, *Tables of Integrals, Series, and Products*, (Translation Edited by Jeffrey, A.), p. 38. Academic Press, New York..

Herman, F., and Skillman, S., 1973, *Atomic Structure Calculations*, Prentice-Hall, Englewood Cliffs, NJ..

Imamura, A., 1970, *J. Chem. Phys.* 52:3168–3175.

Jones, R. S., Mintmire, J. W., and Dunlap, B. I., 1988, *Int. J. Quantum Chem. Symp.* 22:77–84.

Karpfen, A., and Beyer, A., 1984, *J. Comp. Chem.* 5:11–18.

Kutzler, F. W., White, C. T., and Mintmire, J. W., 1986, *Int. J. Quantum Chem.* 29:793–797.

Mintmire, J. W., and Dunlap, B. I., 1982, *Phys. Rev. A* 25:88–95.

Mintmire, J. W., Sabin, J. R., and Trickey, S. B., 1982, *Phys. Rev. B* 26:1743–1753.

Mintmire, J. W., and White, C. T., 1983a, *Phys. Rev. Lett.* 50:101–105.

Mintmire, J. W., and White, C. T., 1983b, *Phys. Rev. B* 27:1447–1449.

Mintmire, J. W., and White, C. T., 1983c, *Phys. Rev. B* 28:3283–3290.

Mintmire, J. W., and White, C. T., 1983d, *Int. J. Quantum Chem. Symp.* 17:609–612.

Mintmire, J. W., and White, C. T., 1987a, *Phys. Rev. B* 35:4180–4183.

Mintmire, J. W., and White, C. T., 1987b, *Int. J. Quantum Chem Symp.* 21:131–136.

Mintmire, J. W., Kutzler, F. W., and White, C. T., 1987, *Phys. Rev. B* 36:3312–3318.

Mintmire, J. W., 1989a, *Phys. Rev. B* 39:13350–13357.

Mintmire, J. W., 1989b, *Mat. Res. Soc. Symp. Proc.* 141:235–239.

Mintmire, J. W., 1990, *Int. J. Quantum Chem. Symp.* 24:in press.

Piela, L., and Delhalle, J., 1978, *Int. J. Quantum Chem.* 13:605–617.

Piela, L., André, J. M., Brédas, J.-L., and Delhalle, J., 1980, *Int. J. Quantum Chem. Symp.* 14:405–418.

Sambe, H., and Felton, R., 1974, *J. Chem. Phys.* 61:3862–3863.

Sambe, H., and Felton, R., 1975, *J. Chem. Phys.* 62:1122–1126.

Slater, J. C., 1974, *Quantum Theory of Molecules and Solids*, Vol. 4. McGraw-Hill, New York..

Springborg, M., and Lev, M., 1989, *Phys. Rev. B* 40:3333–3339.

Steinborn, E. O., and Ruedenberg, K., 1973, *Adv. Quantum Chem.* 7:1–80.

Teramae, H., and Takeda, K., 1989, *J. Am. Chem. Soc.* 111:1281–1285.

van Duijneveldt, F. B., 1974, *IBM Report RJ945*.

Weniger, E. J., and Steinborn, E. O., 1985, *J. Math. Phys.* 26:664–670.

White, C. T., Kutzler, F. W., and Cook., M., 1986, *Phys. Rev. Lett.* 56:252–255.

Wolfram, S., 1988, *Mathematica*, Addison-Wesley, Redwood City, CA..

10
Density Functional Theory as a Practical Tool in Organometallic Energetics and Dynamics

TOM ZIEGLER AND VINCENZO TSCHINKE

Introduction

The dearth of reliable experimental data on bond dissociation energies is felt throughout the field of organometallic chemistry. Accurate theoretical studies should afford a much needed supplement to the sparse available experimental data on metal-ligand bond energies, necessary for a rational approach to the synthesis of new transition metal complexes.

Recently, Density Functional investigations of molecular bond energies have gained novel impetus due to the introduction by Becke (1) of a gradient correction to the Hartree-Fock-Slater local exchange expression,

$$E_X^{LSD/NL} = E_X^{HFS} - \sum_\gamma \beta_B \int \frac{|\vec{\nabla}_1 \rho_1^\gamma(\vec{r}_1)|^2}{[\rho_1^\gamma(\vec{r}_1)]^{7/3}} \left\{ 1 + \gamma_B \frac{|\vec{\nabla}_1 \rho_1^\gamma(\vec{r}_1)|^2}{[\rho_1^\gamma(\vec{r}_1)]^{8/3}} \right\}^{-1} \delta\vec{r}_1 \ (1),$$

where ρ_1^γ is a spin density and β_B and γ_B are parameters. In conjunction with appropriate approximations for antiparallel spin correlations, the expression of Eq.(1) provides near-quantitative estimates (1) of bond energies in main-group compounds.

Computational Details

In this contribution we shall present several applications of the new method, which we shall refer to as LSD/NL, to the calculation of bond energies of transition metal complexes (2). We shall focus on trends along a transition period and/or down a transition triad. The following subjects will be discussed: a) metal-metal bonds in dimers of the group 6 transition metals; b) metal-ligand bonds in early and late transition metal complexes; c) the relative strength of metal-hydrogen and metal-methyl bond in transition metal complexes; d) the metal-carbonyl bond in hexa- penta- and tetra-carbonyl complexes.

In the present set of calculations we have used the functional proposed by Becke (1), which adopts a non-local correction to the HFS exchange, and treats correlation between electrons of different spins at the local density functional level. All calculations presented here were based on the LCAO-HFS program

Table 1   Calculated bond energies [D(M-M)] (eV) and metal-metal bond distances (R_{M-M}) (Å) for Cr_2, Mo_2 and W_2.

	D(M-M)		RM-M	
	Calc.	Exp.[11]	Calc..	Exp.[11]
Cr2	1.75	1.56±0.2	1.65	1.69
Mo2	4.03	4.18±0.2	1.95	1.93
W2	4.41(3.54)[a]	–	2.03(2.07)[a]	–

[a] Non-relativistic results.

system due to Baerends *et al.* (*3*) or its relativistic extension due to Snijders *et al.*(*4*), with minor modifications to allow for Becke's non-local exchange correction as well as the correlation between electrons of different spins in the formulation by Stoll *et al.* (*5*) based on Vosko's parametrization (*6*) from homogeneous electron gas data.   Bond energies were evaluated by the Generalized Transition State method (*7*), or its relativistic extensions (*7b*).

A double ζ-STO basis (*8*) was employed for the ns and np shells of the main group elements augmented with a single 3d STO function, except for Hydrogen where a 2p STO was used as polarization.   The ns, np, nd, $(n + 1)s$ and $(n + 1)p$ shells of the transition metals were represented by a triple ζ-STO basis (*3*). Electrons in shells of lower energy were considered as core and treated according to the procedure due to Baerends *et al.* (*2*).   The total molecular electron density was fitted in each SCF-iteration by an auxiliary basis (*9*) of s, p, d, f and g STOs, centred on the different atoms, in order to represent the Coulomb and exchange potentials accurately.

Metal-Metal Bond Strength of the Dimers Cr_2, Mo_2 and W_2 (*10*)

The bond energies (*11*) of the metal dimers Cr_2 and Mo_2 are accurately known experimentally and we note that several theoretical accounts of the bonding in these systems have already appeared, based on *ab initio* (*11a*) and Density Functional Theory (*12*). However, no calculation has been reported for W_2, nor are there any experimental data available.

The bond energies in Table 1 were calculated by evaluating the energy difference

$$\Delta E = 2E(^7S) - E(M_2) \tag{2}$$

between two metal atoms in the ^7S state corresponding to the $nd^5(n + 1)s^1$ configuration, and M_2.   For Cr_2 and Mo_2, ΔE represents the bond energies D(Cr-Cr) and D(Mo-Mo), respectively, since Cr and Mo have a spherical ^7S ground-state.   However, the W-atom has a ^5D ground-state with the configuration $5d^4 6s^2$, thus we have subtracted for W_2 the experimental energy difference (.37 eV) (*13*) between the ^5D and the ^7S states twice to arrive at D(W-W) of Table 1.

The calculated bond energies and equilibrium bond distances R_{M-M} for Cr_2 and Mo_2 are in good accord with experimental values, as can be seen from Table 1. In contrast to other calculations based on DFT, we have employed in the present work $(n + 1)f$ polarization functions. Their contribution to the bond energies are modest, 0.2 - 0.4 eV. On the other hand, the contributions to $D(M-M)$ from the non-local correction to the exchange are -1.8 and -2.4 eV for Mo_2 and Cr_2, respectively, and are thus important in determining the agreement with experiment.

We predict that W_2, after the inclusion of relativistic effects, should have a stronger metal-metal bond than Mo_2. Even in the non-relativistic case, the bonding interaction is stronger in W_2 than in Mo_2 if the two metal atoms are referred to the same ^7S reference state.

Metal-ligand Bond Strengths in the Early Transition Metal Systems Cl_3ML and Late Transition Metal Systems $LCo(CO)_4$ *(14)*

The way in which metal-ligand bond energies of early transition metals and f-block elements differ from those of middle to late transition metals, or metal-ligand bond energies of 3d and 4f elements differ from those of their heavier congeners, has been the subject of many experimental *(15)* as well as a few theoretical studies *(16)* over the past decade.

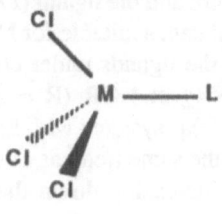

1

We shall present here calculations on the $D(M-L)$ bond strength in the Cl_3ML (**1**) model systems of the early transition metals M = Ti, Zr and Hf, as well as the $LM(CO)_4$ model system with the late transition metal M = Co, for a number of rudimentary ligands, L = H, CH_3, SiH_3, OH, SH, OCH_3, NH_2, PH_2 and CN. The calculated bond energies $D(M-L)$ are displayed in Table 2. The ligands L = OH, OCH_3, with the coordinating atoms of the highest electronegativity and the most polar Cl_3M-L bond have the largest $D(M-L)$ bond energies. The ligand L = SiH_3, with the coordinating atom of the lowest electronegativity and the least polar M-L bond, has a modest $D(M-L)$ bond energy. For the series of ligands NH_2, SH, CH_3 and H one finds in Table 2 that $D(M-L)$ follows the order of the electronegativity of the coordinating atom, with $NH_2 > SH > CH_3 > H$.

It is clear from Table 2 that zirconium, and even more so hafnium, form

Table 2   Calculated [D(M-L)] bond energies (kJ mol^{-1}) and optimized (R_{M-L}) bond distances (Å) in Cl$_3$ML.

L	D(M-L)			R_{M-L}[a]		
	M			M		
	Ti	Zr	Hf	Ti	Zr	Hf
H	250.7	297.2	313.5	1.70	1.82	1.80
CH$_3$	267.5	309.5	326.6	2.13	2.26	2.25
SiH$_3$	210.9	239.5	272.2	2.63	2.78	2.79
OH	453.2	527.2	535.9	1.83	1.95	1.97
OCH$_3$	426.9	484.5	506.6	1.86	1.99	2.01
SH	293.3	347.9	360.1	2.28	2.47	2.47
NH$_2$	364.7	420.6	439.1	1.87	2.01	2.04
PH$_2$	190.6	225.6	233.9	2.24	2.48	2.47
CN	410.4	457.6	477.9	2.06	2.21	2.23

[a] Optimized from a quadratic fit through three energy points corresponding to three different M-L distances.

stronger M-L bonds to the ligands under investigation than titanium in the Cl$_3$M-L systems. We have found that the calculated increase in D(M-L) down the triad is primarily caused by a corresponding increased overlap between the singly occupied 1a$_1$-orbital of Cl$_3$M and the singly occupied orbital on the ligand L, see Figure 1, which is in turn responsible for an increased σ-bonding interaction between the metal centre and the ligand (17).

There are few thermochemical data available for M-L bond involving group 4 metals. To our knowledge, of the ligands under consideration data (15a) are only available for L = CR$_3$, NR$_2$ and OR (R = alkyl), for the homoleptic M(CR$_3$)$_4$, M(NR$_2$)$_4$, and M(OR)$_4$ systems with M = Ti, Zr and Hf. The M-L bonds in these systems follow the same trend as observed here for the Cl$_3$M-L systems, with the bond energy increasing down the triad as well as with the increasing electronegativity of the corresponding atom on the ligand (O > N > C). Group 4 metals are known (18) to form several complexes involving M-L bonds with L = SiR$_3$, PR$_2$, and SR ligands. However, the corresponding D(M-L) bond energies have not been determined experimentally. Perhaps not surprisingly, we find that the M-L bonds of L = SiH$_3$, PH$_2$, and SH are weaker and less polar than the M-L bonds of the homologous ligands L = CH$_3$, NH$_2$, and OH (Table 2).

Comparative experimental data on M-H and M-CH$_3$ bond energies of early transition metals are not available. However, it has been asserted (19) that the M-H and M-CH$_3$ bond strengths of early transition metal complexes are quite similar. Indeed, we find D(M-H) and D(M-CH$_3$) to be quite similar in the Cl$_3$M-H and Cl$_3$M-CH$_3$ systems, respectively. We shall dedicate the next section to the relative strengths of the M-H and M-CH$_3$ bonds in systems ranging from early transition metal and f-block element complexes to middle and late transition metal complexes.

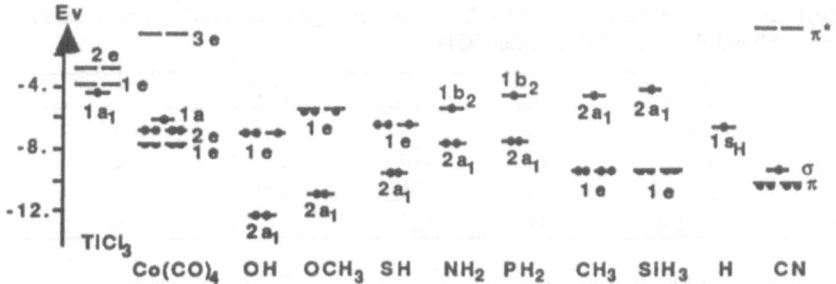

Fig. 1 Energies for the frontier orbitals of the two metal fragments TiCl3 and Co(CO)4 as well as the ligands L = OH, OCH3, SH, NH2, PH2, CH3, SiH3, CN and H.

We shall now consider the M-L bond energies in the late transition metal complexes LCo(CO)4 (2). The σ-bond in LCo(CO)4 is considerably less polar than in Cl3ML. This is primarily so because cobalt is more electronegative than the group 4 metals Ti, Zr and Hf, and as a consequence the $1a_1$ metal based frontier orbital, involved in the σ-bond, is of lower energy than the frontier orbital $1a_1$ of Cl3M (see Figure 1).

OC —— Co —— CO
        L
        θ
        C
        O

2

For ligands other than H we had in the Cl3ML systems favourable donor-acceptor interactions from occupied ligand orbitals to the two empty e-sets, see Figure 1.   In Co(CO)4 the metal based d-orbitals of e-symmetry are fully occupied and the corresponding interactions between occupied ligand orbitals and either 1e or 2e are as a consequence repulsive.   As a consequence, we find in accord with available experimental evidence[20] that all ligand except hydrogen form stronger bonds to the early transition metal Ti than to the late transition metal Co, see Table 2 and 3.

The Relative Strengths of the Metal-Hydrogen and the Metal-Methyl Bonds in Transition Metal Complexes (14,21,22)

The breaking or formation of metal-hydrogen and metal-alkyl bonds is an integral part of most elementary reaction steps in organometallic chemistry.   As a consequence, considerable efforts have been directed toward the determination of

Table 3   Calculated [D(Co-L)] bond energies (kJ mol^{-1}) and optimized (R_{Co-L}) bond distances (Å) in LCo(CO)$_4$.

L	D(Co-L)	R_{Co-L}	L	D(Co-L)	R_{Co-L}
H	230.	1.55	SH	168.9	2.49
CH$_3$	160.	2.11	NH$_2$	145.5	2.09
SiH$_3$	211.6	2.73	PH$_2$	145.5	2.43
OH	232.4	2.09	CN	304.3	2.04

M-H (15b) and M-alkyl bond strength (23) as a prerequisite for a full characterization of the reaction enthalpies of elementary steps in organometallic chemistry.

As already mentioned, the strengths of M-H and M-Alkyl bonds are comparable for early transition metals (19). According to sparse experimental data, the same trend is observed in actinide complexes (24). By contrast, data for alkyl (20a,15e,25,26) and hydride (20e,26,27) complexes of middle to late transition metals indicate that the M-H bond is stronger than the M-Alkyl bond by some 40-80 kJ mol^{-1}. This difference in strength has implications for the relative ease by which ligands can insert into the M-H and M-Alkyl bonds (28). Also, it is one of the thermodynamic factors, along with the relative order of the bond energies H$_2$ < H–Alky < Alkyl–Alkyl, which favour the oxidative addition (19,29) to metal centres of H$_2$ compared to H–Alkyl and Alkyl–Alkyl bonds.

In the preceding section we have presented results on the relative bond strengths of the M-H and M-CH$_3$ bonds of model complexes Cl$_3$M-R (R = H, CH$_3$) involving the early transition metals Ti, Zr and Hf, as well as the late transition metal complex R-Co(CO)$_4$. Here, we shall present additional results on actinide metal complexes as well as on middle and late transition metal complexes, in an attempt to supplement the rather sparse experimental data available.

We have conducted calculations on the bond energy D(M-R) (R = H, CH$_3$) of the model actinide complexes Cl$_3$M-R (M = Th and U) as well as the middle transition metal complexes R-M(CO)$_5$ (M = Mn, Tc and Re) and the late transition metal complexes R-M(CO)$_4$ (M = Co, Rh and Ir). These results are depicted in Figure 2, along with the corresponding results presented in the preceeding section.

The results depicted in Figure 2 seem to indicate that the trend which assigns comparable M-H and M-CH$_3$ bond strengths in early transition metal and actinide complexes but a stronger M-H bond in middle and late transition metal complexes, is of general validity. It also appear that both the M-H and the M-CH$_3$ bond strengths increase down a triad for transition metal complexes.

The reduced strength of the M-CH$_3$ bonds in middle and late transition metals can be readily explained in terms of destabilizing three- and four-electron two-orbital interactions which occur between the fully occupied 1σ (**3a**) and 1π (**3b**) methyl orbitals and the fully or singly occupied d-orbitals of matching symmetries present on the metal centres. By contrast, in early transition metal

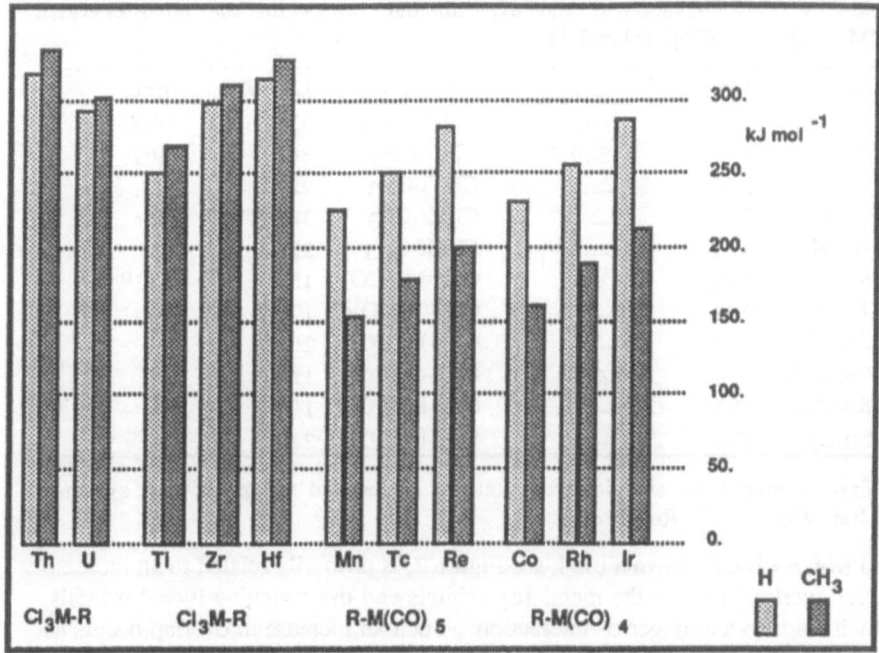

Fig. 2 Calculated M-H and M-CH3 bond energies for several actinide and transition metal complexes.

and actinide complexes, the metal orbitals of π-symmetry are vacant and are involved in stabilizing interactions with **3b**.   A destabilizing three-electron two-orbital interaction between **3a** and the single-occupied metal orbital $1a_1$ (Figure 1) is still present.   However, the M-CH3 bond is more polar in early transition metal complexes than in middle and late transition metal complexes and as a consequence electronic charge is transferred from the metal orbital $1a_1$ to the methyl ligand, thereby relieving the destabilizing interaction between $1a_1$ itself and **3a**.

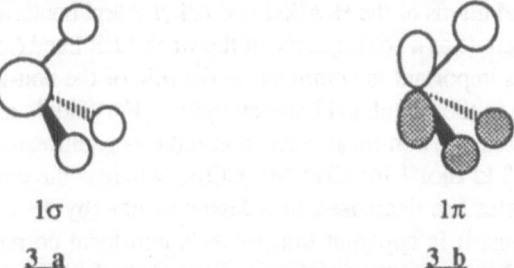

The comparable strengths of the M-H bonds in the complexes studied is perhaps not too surprising, since H is a simple one-orbital ligand without additional occupied orbitals involved in four-electron two-orbital interactions or π-donor-acceptor interactions.   Finally, the increase in strength of both the M-H

Table 4   Calculated and experimental values for the bond energies D(M-R) (R = H, CH₃) (kJ mol⁻¹).

M-H	LSD/NL	Exp.	M-CH$_3$	LSD/NL	Exp.
Cl$_3$Th-H	318.0	~335.[a]	Cl$_3$Th-CH$_3$	333.9	~335.
Cl$_3$U-H	293.3	319.7	Cl$_3$U-CH$_3$	302.1	302.
Cl$_3$Ti-H	250.7	—	Cl$_3$Ti-CH$_3$	267.5	—
Cl$_3$Zr-H	297.2	—	Cl$_3$Zr-CH$_3$	309.5	—
Cl$_3$Hf-H	313.5	—	Cl$_3$Hf-CH$_3$	326.6	—
H-Mn(CO)$_5$	225	213[b]	CH$_3$-Mn(CO)$_5$	153	153[b]
H-Tc(CO)$_5$	252	—	CH$_3$-Tc(CO)$_5$	178	—
H-Re(CO)$_5$	282	—	CH$_3$-Re(CO)$_5$	200	—
H-Co(CO)$_4$	230	238[c]	CH$_3$-Co(CO)$_4$	160	—
H-Rh(CO)$_4$	255	—	CH$_3$-Rh(CO)$_4$	190	—
H-Ir(CO)$_4$	286	—	CH$_3$-Ir(CO)$_4$	212	—

[a] Experimental bond energies from Ref. 24 correspond to Cp$_2$ MCl-R systems.
[b] Ref. 20c.   [c] Ref. 27e.

and M-CH₃ bonds down a triad, see Figure 2, is primarily related to an increase in the overlap between the metal 1a₁ orbitals and the matching ligand orbitals, which leads to a stronger σ−interaction.   Such an increase in overlap occurs as the metal d-orbitals become more diffuse down the triad.

A comparison between our calculated results for D(M-R) (R = H, CH₃) and the few available experimental data is presented in Table 4.   We find in general a good agreement with the experimental bond energies.   Also, the stability order D(M-L) > D(M-CH₃) in middle and late transition metal complexes supported by our theoretical study is consistent with data on organometallic reactions in which M-L and M-CH₃ bonds are formed or broken.   Thus, CO will readily insert into a M-CH₃ bond whereas the corresponding insertions into M-H bonds are virtually unknown (28), and methyl has likewise a larger migratory aptitude toward most other ligands than hydride.   The H₂ molecule is known to add oxidatively and exothermically to several metal fragments where the corresponding oxidative additions of the H-Alkyl and Alkyl-Alkyl bonds are unknown and probably endothermic as a consequence of the weak M-R bond (19).

At this stage, it is important to comment on the role of the non-local correction to the exchange in the calculated bond energies.   For middle and late transition metal complexes, the non-local correction reduces significantly the D(M-CH₃) values (by 105 kJ mol⁻¹ for CH₃-Mn(CO)₅) whereas the corresponding D(M-H) bond energies are decreased to a lesser extent (by 13 kJ mol⁻¹ for H-Mn(CO)₅).   Thus, it is apparent that Becke's non-local correction to the exchange is essential to assure the good agreement of the LSD/NL results with experiment, whereas the HFS and the LSD methods not only tend to give too large bond energies, but in some cases predict the wrong order for the M-H and M-CH₃ bond strengths.

Thermal Stability and Kinetic Lability of the Metal-Carbonyl Bond (30)

The extensive use of coordinatively saturated mono-nuclear carbonyls as starting materials in organometallic chemistry, along with their volatility and high molecular symmetry, has prompted numerous experimental (15a,31,32,33) and theoretical (34,35) studies on their structure and reactivity.   Special attention has been given to the degree of σ-donation and π-back-donation (34b-g,35a,35e) in the synergic (34k) M-CO bond.

However, in spite of many experimental (32) investigations, there is still a lack of basic data on the thermal stability and kinetic lability of the M-CO bond in essential metal carbonyls such as $M(CO)_6$ (M = Cr, Mo, W), $M(CO)_5$ (M = Fe, Ru, Os) and $M(CO)_4$ (M = Ni, Pd, Pt), particularly with respect to the carbonyls of the second- and third-row metals.

Theoretical methods have begun to play a role in determining the energetics of organometallics (35g) and ab initio type methods have recently been applied to calculation on the M-CO bond strength of $Cr(CO)_6$ (35d-e), $Fe(CO)_5$ (35a-c,f), and $Ni(CO)_4$ (35a,f), but not yet to M-CO bond strength of their second- and third-row homologues.   Here, we shall present LSD/NL calculations on the intrinsic mean bond energy D(M-CO) and first CO dissociation energy ΔH of $Cr(CO)_6$, $Fe(CO)_5$, and $Ni(CO)_4$ as well as their second- and third-row homologues.

4 a   **4 b**   **4 c**

We shall here be concerned with periodic trends in the strength of the M-CO bonding interaction within the triads M = Cr, Mo, W; M = Fe, Ru, Os; and M = Ni, Pd, Pt.   As measures for the M-CO bonding interaction in the hexacarbonyls (**4a**), pentacarbonyls (**4b**) and tetracarbonyls (**4c**), we will consider the intrinsic mean bond energy D(M-CO) between M (in its low-spin valence state) and the n CO ligands, as well as the bond energy ΔH between $M(CO)_{n-1}$ and CO.

There are two sets of experimental data with a bearing on the M-CO bond strength in $M(CO)_n$, namely, the mean bond energy E corresponding to the process

$$M(CO)_n \text{ (g)} \rightarrow M(g) + nCO(g) - nE \qquad (2a)$$

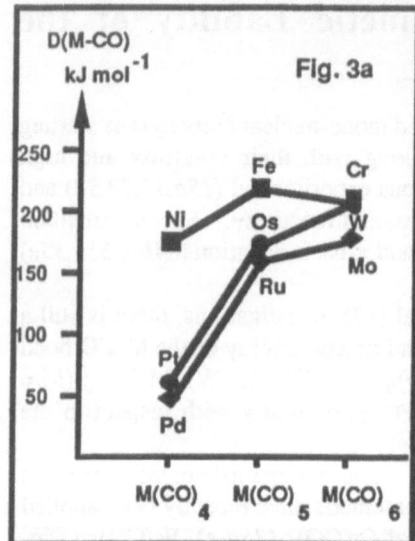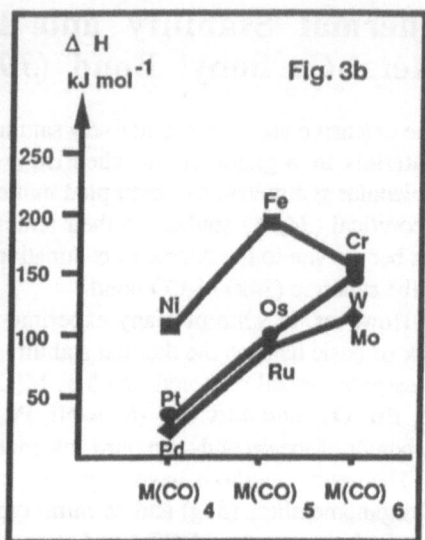

Fig. 3   Calculated bond energies for $M(CO)_6$ (M = Cr, Mo, W), $M(CO)_5$ (M = Fe, Ru, Os), and $M(CO)_4$ (M = Ni, Pd, Pt): a) intrinsic bond energies D(M-CO); b) first CO dissociation energies ΔH.

and the first bond dissociation energy ΔH corresponding to the process

$$M(CO)_n \rightarrow M(CO)_{n-1} + CO - \Delta H \qquad (2b).$$

It is important to note that E is given by

$$E = D(M\text{-}CO) - 1/n\Delta E_{prep} \qquad (3),$$

where ΔE_{prep} is the energy required to promote the metal atom from its high-spin electronic ground state to the low-spin valence configuration.   As a consequence, one can not conclude that the order of E will correspond to the order of D(M-CO) down a triad, since ΔE_{prep} might differ significantly for the three elements.

The first bond dissociation energy ΔH is on the other hand a direct measure for the strength of the M-CO bond interaction.   It is further an extremely important kinetic parameter, since the dissociation process of Eq.(2b) is assumed to be a key step in the large volume of kinetically useful substitution reactions (36)

$$M(CO)_n + L \rightarrow M(CO)_{n-1}L + CO \qquad (4)$$

where L is introduced into the coordination sphere of M by replacing one carbonyl ligand.

Our computational results for D(M-CO) and ΔH are depicted in Figure 3.   It appears from the values of D(M-CO) (Figure 3a) that strength of the M-CO bond decreases going from the 3d to the 4d metal of a triad, to increase again for the complex of the 5d metal.   The destabilization of the M-CO bond of the

4d and 5d elements, compared to their 3d counterparts increases in going from the Cr to the Ni group; in fact within the group 6 carbonyls, the M-CO bond in the W carbonyl is at a par in strength with the bond in the Cr complex. Among the systems studied, the strongest M-CO bond is assigned to the $Fe(CO)_5$ complex, whereas the weakest bond is found in the $Pd(CO)_4$ complex. It is important to note that the first dissociation bond energies ΔH (Figure 3b) follow closely the same trends observed for the intrinsic bond energies D(M-CO) (Figure 3a).

$$\pi^*_{CO} \qquad\qquad\qquad\qquad \sigma_{CO}$$

$$\underline{\textbf{5 a}} \qquad\qquad\qquad\qquad \underline{\textbf{5 b}}$$

The periodic trends discussed above can be readily rationalized in terms of the stabilizing electronic interactions and the destabilizing steric interactions which determine the strength of the M-CO bond. The electronic terms are represented by π-back-donation from occupied nd-orbitals on the metal centre to the empty π_{CO} orbital (**5a**), as well as σ-donation from the doubly-occupied σ_{CO} orbital (**5b**) to vacant nd orbitals. The steric terms are dominated by the repulsive four-electron two-orbital interactions between σ_{CO} and occupied nd orbitals on the metal centre. Our results show that electronic factors are most favourable for the pentacarbonyls where both π-back-donation and σ-donation are important, whereas the steric interactions are most favourable for the M-CO bond among the hexacarbonyls, where all nd orbitals of σ-symmetry are empty and only mild repulsive interactions between σ_{CO} and the occupied ns and np metal orbitals are present. For first-row transition metals, the repulsive interactions between occupied nd- and σ_{CO}-orbitals are still modest, since the nd-σ_{CO} overlap integrals are relatively small for the contracted 3d-orbitals compared to the more diffuse 4d and 5d orbitals. Thus, electronic factors will make the intrinsic mean bond energy larger for $Fe(CO)_5$ than for $Cr(CO)_6$ (Figure 3a). On the other hand, in carbonyls of 4d and 5d elements, where repulsive interactions between occupied nd- and σ_{CO}-orbitals are considerable, the steric factors cause the M-CO bonds in $Ru(CO)_5$ and $Os(CO)_5$ to be weaker than in $Mo(CO)_6$ and $W(CO)_6$, respectively. The tetracarbonyls, in which all interactions between the nd- and σ_{CO}-orbitals are repulsive, have weaker M-CO bonds than the corresponding hexacarbonyls and pentacarbonyls in each of the transition series (Figure 3a). Finally, relativistic effects will stabilize the 5d element carbonyl compared to the 4d metal carbonyl within a triad, see Figure 3a). Such a stabilization is sufficient to bring the strength of the M-CO bond in $W(CO)_6$ at a par with the strength of the bond in $Cr(CO)_6$. The rational given above for the variations in D(M-CO) can further be used to explain the trends in the first CO dissociation energy ΔH (Figure 3b).

Table 5   Comparison between calculated and experimental values for the mean bond energy E of several carbonyl complexes.   Calculated values for D(M-CO) and ΔE_{prep} are also given.   All values in kJ mol^{-1}.

M(CO)$_n$	D(M-CO)	$1/n \cdot \Delta E_{prep}$	E^c	E(Exp.)^c
Cr(CO)$_6$	211	100.7	110	110[a]
Mo(CO)$_6$	178	51.6	126	151[a]
W(CO)$_6$	210	54.4	156	179[a]
Fe(CO)$_5$	216.8	98.42	117	118.4
Ni(CO)$_4$	178.9	—	—	191[b]

[a] Ref. 15a.   [b] Experimental intrinsic mean dissociation energy D(M-CO), Ref. 15a.   [c] Mean bond dissociation energy.

We have calculated the promotion energy ΔE_{prep} needed to evaluate the mean bond dissociation energy E, according to Eq.(3).   In Table 5, we compare our calculated value for E with the available experimental mean bond dissociation energies.   We observe a good to fair agreement of our theoretical estimates with the experimental data; in particular, the experimental trend in the values of E down the only triad for which such data are available, the group 6 carbonyls, is correctly reproduced.   Here, see Table 5, the Cr(CO)$_6$ complex has the lowest average dissociation energy in spite of having the largest value of D(M-CO). The different trends in E and D(M-CO) are related to variations in the promotion energies ΔE_{prep} (Table 5), as the promotion energy is seen to be much larger for Cr than for Mo or W.   This result is not unexpected since ΔE_{prep} depends on exchange integrals that in general are larger for the relatively contracted 3d-orbitals of chromium than for the more diffuse 4d- and 5d-orbitals of molybdenum and tungsten.

Our calculated values for the first CO dissociation energy ΔH are also in fair to good agreement with available experimental data, see Table 6.   Finally, the indication that the M-CO bonds are fairly weak in the Pd(CO)$_4$ and Pt(CO)$_4$ complexes, inferred by our theoretical estimates of both D(M-CO) and ΔH (Figure 3), finds support in the apparent instability of these species at room temperature.

The value of the non-local contribution to the exchange in the calculated bond energies for carbonyls ranges from 109 to 138 kJ mol^{-1}.   Thus, the HFS and LSD estimates (16a,35e) of bond energies are in some case twice as large as the correct experimental values, and the non-local correction to the exchange is once again found to be essential to achieve good agreement with experiment. For calculations on transition metal complexes, the LSD/NL method must also be considered more reliable than the HF-method itself, which tends to severely under-estimate the bond energies in such systems, as attested by the prediction, given by HF-results, that the Fe(CO)$_4$ and Ni(CO)$_4$ complexes be unstable with respect to the free metal atom and free CO molecules (35a).   On the other hand, due to their slow convergency (35b), CI calculations are not at present widely applicable to transition metal complexes.   These considerations point to the

Table 6   Comparison between calculated and experimental values for the first CO dissociation energy ΔH, values in kJ mol^{-1}. Calculated values do not include geometry relaxation of the fragments $M(CO)_{n-1}$.

$M(CO)_n$	LSD/NL	Exp.	$M(CO)_n$	LSD/NL	Exp.
$Cr(CO)_6$	147	162[a]	$Fe(CO)_5$[b]	185[c]	176[d]
$Mo(CO)_6$	119	126[a]	$Ru(CO)_5$[b]	92	117[e]
$W(CO)_6$	142	166[a]	$Ni(CO)_4$	106	104[f]

[a] Ref. 37. [b] Equatorial CO dissociation energy. [c] The dissociation product is the fragment $Fe(CO)_4$ in its singlet state. [d] Ref. 32a. [e] Ref. 38. [f] Ref. 39.

LSD/NL-approximation as one of today's methods of choice in the theoretical investigation of transition metal complexes.

References

1.   Becke, A.D. *J. Chem. Phys.* **1986**, *84*, 4524.
2.   (a) Fan, L.;Versluis, L.; Ziegler, T.; Baerends, E.J.; Ravenek, W. *Int. J. Quantum Chem.* **1988**, *522*, 173.
     (b) Harrod, J.F; Ziegler, T.; Tschinke, V. *Organometallics* **1989**, *9*, 897.
     (c) Versluis, L.; Ziegler, T.; Baerends, E.J.; Ravenek, W. *J. Am. Chem. Soc.* **1989**, *111*, 2018.
     (d) Ziegler, T.; Fan, L.; Tschinke, V.; Becke, A. *J. Am. Chem. Soc.* **1989**, *111*, 9177.
3.   Baerends, E.J.; Ellis, D.E.; Ros, P. *Chem. Phys.* **1973**, *2*, 71.
4.   Snijders, J.G.; Baerends, E.J.; D.E.; Ros, P. *Molec. Phys.* **1979**, *38*, 1909.
5.   Stoll, H.; Golka, E.; Preuss, H. *Theor. Chim. Acta* **1980**, *55*, 29.
6.   Vosko, S.H.; Wilk, L.; Nusair, M. *Can. J. Phys.* **1980**, *58*, 1200.
7.   (a) Ziegler, T.; Rauk, A. *Theor. Chim. Acta* **1977**, *46*, 1.
     (b) Ziegler, T.; Snijders, J.G.; Baerends, E.J. *J. Chem. Phys.* **1981**, *74*, 1271.
8.   (a) Snijders, J.G.; Baerends, E.J.; Versnoijs, P. *At. Nucl. Data Tables* **1982**, *26*, 483.
     (b) Versnoijs, P.; Snijders, J.G.; Baerends, E.J. *Slater Type Basis Functions for the Whole Periodic System.* Internal Report; Free University: Amsterdam, 1981.
9.   Krijn, J.; Baerends, E.J. *Fit Functions in the HFS-method.* Internal Report (in Dutch); Free University: Amsterdam, 1981.
10. Ziegler, T.; Tschinke, V.; Becke, A. *Polyhedron* **1987**, *6*, 685.
11. (a) Shim, I. K. *Dan. Videmsk. Selsk. Mat.-Fys. Medd.* **1985**, *41*, 47; and references therein.
     (b) Weltner, W.; Van Zee, R.J. *Annu. Rev. Phys. Chem.* **1984**, *35*, 291.
12. Baycara, N.A.; McMaster, B.N.; Salahub, D.R. *Molec. Phys.* **1984**, *52*, 891; and references therein.
13. Moore, C.E. In *Atomic Energy Levels, Nat. Bur. Stand. (U.S.)* C**1958**, *467*; Vol. 3.

14. Ziegler, T.; Tschinke, V.; Versluis, L.; Baerends, E.J.; Ravenek, W. *Polyhedron* **1988**, *7*, 1625.
15. (a) Connors, J.A. *Top. Curr. Chem.* **1977**, *71*, 71.
     (b) Pearson, R.G. *Chem. Rev.* **1985**, *85*, 41.
     (c) Beauchamp, J.L. *Chem. Rev.* in press.
     (d) Bryndza, H.E.; Fong, L.K.; Peciello, R.A.; Tam, W.; Bercaw, J.E. *J. Am. Chem. Soc.* **1987**, *109*, 1444.
     (e) Bruno, J.W.; Marks, T.J.; Morss, L.R. *J. Am. Chem. Soc.* **1983**, *105*, 6824.
16. (a) Ziegler, T.; Tschinke, V.; Versluis, L. *NATO ASI* **1986**, *Series C 176*, 189; and references therein.
     (b) Hay, P.J.; Rohlfing, C.M. *NATO ASI* **1986**, *Series C 176*, 135.
17. For a full account on the energy decomposition of the bond energies of Cl_3M-L and L-$Co(CO)_4$, see Ref. 14.
18. (a) Roddick, D.M.; Santasiero, B.D.; Bercaw, J. *J. Am. Chem. Soc.* **1985**, *107*, 4670.
     (b) Cardin, D.J.; Lappert, C.L.; Raston, C.L.; Riley, P.L. In *Comprehensive Organometallic Chemistry*, Wilkinson, G., Ed.; Pergamon Press: New York, 1982.
19. Crabtree, R.H. *Chem. Rev.* **1985**, *85*, 245.
20. (a) Mandich, M.L.; Halle, L.F.; Beauchamp, J.L. *J. Am. Chem. Soc.* **1984**, *106*, 4403.
     (b) Armentrout, P.B.; Halle, L.F.; Beauchamp, J.L. *J. Am. Chem. Soc.* **1981**, *103*, 6501.
     (c) Connor, J.A.; Zafarani-Moattar, M.T.; Bickerton, J.; Saied, N.I.; Suradi, S.; Carson, R.; Al Tackhin, G.A.; Skinner, H.A. *Organometallics* **1982**, *1*, 1166.
21. (a) Ziegler, T.; Tschinke, V.; Baerends, E.J.; Snijders, J.G. *J. Phys. Chem.* in press.
     (b) Ziegler, T.; Tschinke, V.; Becke, A. *J. Am. Chem. Soc.* **1987**, *109*, 1351.
22. Ziegler, T.; Wendan, C.; Baerends, E.J.; Ravenek, W. *Inorg. Chem.* **1988**, *27*, 3458.
23. Halpern, J. *Acc. Chem. Res.* **1982**, *15*, 238.
24. Bruno, J.W.; Stecher, H.A.; Mors, L.R.; Sonnenberg, D.C.; Marks, T.J. *J. Am. Chem. Soc.* **1986**, *108*, 7275.
25. (a) Georgiadis, R.; Armertrout, P.B. *J. Am. Chem. Soc.* **1986**, *108*, 2119.
     (b) Aristov, N.; Armertrout, P.B. *J. Am. Chem. Soc.* **1986**, *108*, 1806.
26. Halle, L.F.; Armentrout, P.B.; Beauchamp, J.L. *Organometallics* **1982**, *1*, 963.
27. (a) Squires, R.R. *J. Am. Chem. Soc.* **1985**, *107*, 4385.
     (b) Sallans, L.; Lane, K.R.; Squires, R.R.; Freiser, B.S. *J. Am. Chem. Soc.* **1985**, *107*, 4379.
     (c) Schilling, J.B.; Goddard, W.A. III; Beauchamp, J.L. *J. Am. Chem. Soc.* in press.
     (d) Girling, R.B.; Grebenik, P.; Perutz, R.N. *Inorg. Chem.* **1986**, *24*, 31.
     (e) Ungvary, F. *Organomet. Chem.* **1972**, *36*, 363.
28. Ziegler, T.; Versluis, L.; Tschinke, V. *J. Am. Chem. Soc.* **1986**, *108*, 612.

29. Low, J.J; Goddard, W.A. III *Organometallics* **1986**, *5*, 609.
30. Ziegler, T.; Tschinke, V.; Ursenbach, C. *J. Am. Chem. Soc.* **1987**, *109*, 4825.
31. (a) Kettle, S.F.A. *Curr. Top. Chem.* **1977**, *71*, 111.
    (b) Braterman, P.S. In *Metal Carbonyl Spectra*; Academic Press: London, 1975.
    (c) Mingos, D.P.M. In *Comprehensive Organometallic Chemistry* ; Wilkinson, G.; Stone, F.G.A.; Abel, E.W., Eds.; Pergamon: New York;1982; Vol. 3, p1.
32. (a) Lewis, K.E.; Golden, D.M.; Smith, G.P. *J. Am. Chem. Soc.* **1984**, *106*, 3906.
    (b) Bernstein, M.; Simon, J.D.; Peters, J.D. *Chem. Phys. Lett.* **1983**, *100*, 241.
33. (a) Rees, B.; Mitschler, A. *J. Am. Chem. Soc.* **1976**, *98*, 7918.
    (b) Beagley, B.; Schmidling, D.G. *J. Mol. Struct.* **1974**, *22*, 466.
    (c) Hedberg, L.; Lijima, T.; Hedberg,K. *J. Chem. Phys.* **1979**, *70*, 3224.
    (d) Jones, L.J.; McDowell, R.S.; Boldblatt, M. *Inorg. Chem.* **1969**, *8*,2349.
34. (a) Guenzburger, D.; Saitovitch, E.M.B.; De Paoli, M.A.; Manela, J. *J. Chem. Phys.* **1984**, *80*, 735.
    (b) Baerends, E.J.; Ros, P. *Mol. Phys.* **1975**, *30*, 1735.
    (c) Heijser, W.; Baerends, E.J.; Ros, P. *J. Mol. Struct.* **1980**, *19*, 1805.
    (d) Bursten, B.E.; Freier, D.G.; Fenske, R.F. *Inorg. Chem.* **1980**, *19*, 1804.
    (e) Demuynk, J.; Veillard, A. *Theor. Chim. Acta* **1973**, *28*, 241.
    (f) Caulton, K.G.; Fenske, R.F. *Inorg. Chem.* **1975**, *7*, 1273.
    (g) Hubbard, J.L.; Lichtenberger, J. *J. Am. Chem. Soc.* **1982**, *104*, 2132.
    (h) Elian, M.; Hoffmann, R. *Inorg. Chem.* **1975**, *14*, 1058.
    (i) Saddei, D.; Freund, H.J.; Hohlneicher, G. *Chem. Phys.* **1981**, *55*, 1981.
    (j) Johnson, J.B.; Klemperer, W.G. *J. Am. Chem. Soc.* **1977**, *99*,7132.
    (k) Chatt, J.; Duncanson, L.A. *J. Chem. Soc.* **1953**, 2939.
    (l) Hillier, I.H.; Saunders, V.R. *Mol. Phys.* **1971**, *22*, 1025.
    (m) Vanquickenborne, L.J.; Verhulst, J. *J. Am. Chem. Soc.* **1977**, *99*, 7132.
    (n) Ford, P.C.; Hillier, I.H.; Pope, S.A.; Guest, M.F. *Chem. Phys. Lett.* **1983**, *102*, 555.
    (o) Burdett, J.K. *J. CHem. Soc., Faraday Trans.* 2 **1974**, *70*, 1599.
    (p) Penzak, D.A.; McKinney, R.J. *Inorg. Chem.* **1979**, *18*, 3407.
    (q) Osman, R.; Ewig, C.S.; Van Wazer, J.R. *Chem. Phys. Lett.* **1978**, *54*, 1341.
    (r) Serafini, A.; Barthelat, J.C.; Durand, P. *Mol. Phys.* **1978**, *36*, 1341.
    (s) Sakai, T.; Huzinaga, J. *Chem. Phys.* **1982**, *76*, 2552.
35. (a) Bauschlicher, C.W.; Bagus, P.S. *J. Chem. Phys.* **1984**, *81*, 5889.
    (b) Lüthi, H.P.; Siegbahn, P.E.M.; Almlof, J. *J. Phys. Chem.* **1985**, *89*, 2156.
    (c) Daniel, C.; Benard, M.; Dedieu, A.; Wiest, R.; Veillard, A. *J. Phys. Chem.* **1984**, *88*, 4805.
    (d) Sherwood, D.E; Hall, M.B. *Inorg. Chem.* **1983**, *22*, 93.

(e) Baerends, E.J.; Rozendaal, A. *NATO ASI* **1986**, *Series C*, *176*, 159.

(f) Rösch, N.; Jorg, H.; Dunlap, B. *NATO ASI* **1986**, *Series C*, *176*, 179.

(g) Veillard, A., Ed. *NATO ASI* **1986**, *Series C*, *176*.

(h) Rolfing, C.M.; Hay, P.J. *J. Che. Phys.* **1985**, *83*, 4641.

36. Kirtley, S.W. In *Comprehensive Organometallic Chemistry* ; Wilkinson, G.; Stone, F.G.A.; Abel, E.W., Eds.; Pergamon: New York;1982; Vol. 3, p 783.

37. (a) Angelici, R.J. *Organomet. Chem. Rev. A* **1968**, *3*, 173.

(b) Covey, W.D.; Brown, T.L. *Inorg. Chem.* **1973**, *12*, 2820.

(c) Centini, G.; Gambino, O. *Atti Acc. Sci. Torino I* **1963**, *97*, 757.

(d) Werner, H. *Angew. Chem. Int. Ed. Engl.* **1968**, *7*, 930.

(e) Graham, J.R.; Angelici, R.J. *Inorg. Chem.* **1967**, *6*, 2082.

(f) Werner, H.; Prinz, R. *Chem. Ber.* **1960**, *99*, 3582.

(g) Werner, H.; Prinz, R. *J. Organomet. Chem.* **1966**, *5*, 79.

38. Huq, R.; Poe, A.J.; Chawla, S. *Inorg. Chem. Acta* **1979**, *38*, 121.

39. Turner, J.J.; Simpson, M.B.; Poliakoff, M.; Maier, W.B. *J. Am. Chem. Soc.* **1983**, *105*, 3998.

11
DGauss: Density Functional — Gaussian Approach. Implementation and Applications

J. ANDZELM

Abstract

The DGauss program (for Density - Gaussian) is an analytical implementation of the density functional (DF) method. In this approach variational fitting to the density can be accomplished leading to exact Coulomb forces. Remaining exchange-correlation energy is a smooth function of the density and can be accurately fitted on a small, adaptive set of grid points. The DGauss program employs Gaussian basis sets. This allows one to build on the wealth of experience gained from Hartree-Fock (HF) molecular orbital calculations. HF basis sets used in DF calculations lead to accurate molecular geometries. However, energetics of reactions are reproduced much better if local spin density (LSD) optimized basis sets are used. Gradient corrected Hamiltonians have to be used in order to yield meaningful results for energetics of chemical reactions, particularly if different types of bonds and/or weak bonds are involved. The Becke-Perdew (BP) and Becke-Stoll (BSPP) DF-gradient Hamiltonians are examined in this paper for hydrogenation reactions, reactions involving single and double bonds and zinc-water complexes. The recently developed analytical energy gradient technique was used to optimize the geometries and calculate vibrational frequencies of organic and organo-metallic molecules.

Introduction

While density functional (DF) theory is practically the only first-principles, quantum mechanical approach being used in solid state physics and materials science, research in theoretical chemistry is dominated by *ab initio* (Hartree-Fock, correlated methods) and semiempirical calculations (Hehre *et al.*, 1986; Stewart, 1990). DF method was introduced to chemistry in 1970's, with the development of the Xα-SW technique (Johnson, 1973; Slater, 1974), Hartree-Fock-Slater program (Baerends *et al.*, 1973), DVM code (Averill and Ellis, 1973) and the Gaussian approach (Sambe and Felton, 1975). However, those methods could not address many important questions asked by chemists, like molecular structures and energies. Therefore, the majority of chemists were not using DF methods. With the advent of accurate gradient-corrected exchange-correlation potentials and gradient techniques for geometry optimization, together with more efficient DF software, the DF method is now becoming a valuable, popular tool for chemical research (Borman, 1990).

The DF theory as originated by Hohenberg, Kohn, and Sham (1964, 1965), and extended by Levy (1979), introduces the electron density as a basic variable which determines the ground state properties of a molecule. Moreover, the DF method constitutes a conceptual framework for an understanding the chemical bond (Parr and Yang, 1989; Parr, 1990). In practice this theory requires approximations. The fundamental approximation is Local Spin Density (LSD), which has proven to be successful in calculating properties of molecules as well as the solid state (Salahub, 1987; Wimmer et al., 1987; Jones and Gunnarsson, 1990).

There are various techniques being used to solve the Kohn-Sham equations for molecular systems. Slater-type orbitals (Baerends et al., 1973), numerical atomic basis sets (Delley and Ellis, 1982) and Gaussian-type orbitals (Sambe and Felton, 1975; Dunlap et al., 1979) have shown a great promise for the accurate and efficient application of the LSD method in chemistry. More recently, plane-wave basis sets (Teter et al., 1989) and basis set-free (Becke, 1989a) solutions of the DF equations have been introduced.

There is a growing evidence that the LSD approach provides accurate results for molecular structures and electronic properties for a variety of molecular systems. This includes systems like fluorocarbons (Dixon et al., this volume), organometallic compounds (Ziegler et al., 1989), transition metal clusters (Hoek and Baerends, 1989; Andzelm et al., 1989a), polymers (Mintmire, 1989), compounds containing heavy elements (Ziegler et al., 1989) and metal surfaces and interfaces (Wimmer et al., 1985). The LSD method is computationally less demanding than traditional ab initio methods, and therefore can be used for large molecular systems of practical importance (Dixon et al., this volume; Caldwell and Redington, this volume)

Bond energies are typically overestimated in LSD calculations and in some cases, e.g. molecules with hydrogen bonds, the LSD optimized geometry may not be accurate (Hill et al., this volume). The LSD approach is based on the homogeneous electron gas model of the density, which may be a crude approximation of the real molecular system, with its inhomogeneous electron density. One way to account for inherent LSD errors is to introduce nonlocal corrections into the Hamiltonian via density -gradient terms (Becke, 1989b). There is clearly a need for extensive research in the area of nonlocal electron-density functional approaches; the latest developments (Wilson and Levy, 1990) are most encouraging.

The DGauss Implementation

The foundation for the DGauss implementation is the LCGTO-MCP-LSD (linear combination of the Gaussian type orbitals- model core potential- local spin density) method (Sambe and Felton, 1975; Dunlap et al., 1979; Salahub, 1987; Andzelm et al., 1989; Salahub et al., this volume; Wimmer, this volume).

In the present DGauss program the "best" local spin density Hamiltonian (VWN) is being used (Vosko et al., 1980). The VWN exchange correlation potential provides an accurate approximation for the homogeneous electron gas over the wide range of densities, suitable for atoms, molecules and metals. In addition the gradient corrected exchange Hamiltonian due to Becke (Becke, 1989b) and the correlation correction due to Perdew (Perdew, 1986) can be calculated. The semiempirical exchange energy correction (Becke, 1986) and the correlation energy by Stoll (Stoll et al., 1978) are also available. In the present implementation the optimized geometries of molecules are found by minimizing the LSD energy using analytical-gradients. Based on the resulting LSD densities, the non-local corrections are then evaluated yielding improved energies. This non-self consistent approach for density-gradient corrections has been found successful in molecular calculations for diatomics (Becke, 1989b) and organometallic systems (Ziegler et al., 1989). In certain cases, eg. hydrogen bonded systems (Hill, et al., this volume) or weakly bonded complexes (Mlynarski and Salahub, 1990) a self consistent account for density-gradient corrections seems to be mandatory. This is because the gradient-corrections have an impact on both geometry and energy of the molecules.

The approach used in DGauss relies on the variational, analytical approximation to the density. In order to achieve a variational fitting to the density the second order Coulomb energy term arising from the difference between the exact and the fitted density is minimized (Dunlap, et al., 1979). This leads to analytical expressions for the fitting coefficients involving Coulomb-type three-index, two electron integrals. The remaining exchange-correlation energy is a smooth function of the density and therefore can be accurately fitted on a small, adaptive set of grid points. It is worthwhile to mention that a method for the variational fitting of the exchange correlation potential has been recently formulated (Dunlap, et al., 1990), although it is not yet implemented in the present DGauss program.

Direct LSD Method

In traditional I/O-based methods the integrals are calculated once and than reused each time the Hamiltonian matrix is created or the fitting is performed. For large systems, in order to minimize the wall clock time of calculations and to reduce the need for extensive disk storage, the direct LSD scheme is being used in DGauss (Andzelm, to be published). The inevitable increase in

CPU time does not exceed 30% of the CPU time for the traditional method. The direct scheme is also successfully being used in modern Hartree-Fock programs (Almlof et al., 1982; Haser and Ahlrichs, 1989).

In order to introduce the direct LSD approach, let us briefly summarize the underlying LCGTO (linear combination of Gaussian-type orbitals) approach. The electron density, $\rho(r)$, and the exchange correlation potential, $\mu(r)$, are approximated using Gaussian auxiliary basis sets, $g(r)$, of the form

$$\rho(r) = \Sigma_r \, \rho_r \, g_r(r)$$

$$\mu(r) = \Sigma_s \, \mu_s \, g_s(r)$$

Here ρ_r and μ_s are the fitting coefficients of the density and exchange-correlation potential, respectively. The Hamiltonian matrix can then be expressed as follow

$$H_{pq} = h_{pq} + \Sigma_r \rho_r \, [pq|r] + \Sigma_s \mu_s \, [pq|s]$$

where h_{pq} is the one-electron Hamiltonian matrix; $[pq|r]$ and $[pq|s]$ are two and one-electron integrals, respectively (Obara and Saika, 1986).

In the direct approach only the difference in the Hamiltonian matrix (δH_{pq}^{i+1}), resulting from the change in the fitting coefficients needs to be calculated. The hamiltonian matrix in the SCF iteration i+1 can be written as follows

$$H_{pq}^{i+1} = H_{pq}^i + \delta H_{pq}^{i+1}$$

$$\delta H_{pq}^{i+1} = \Sigma_r \, (\rho_r^{i+1} - \rho_r^i) \, [pq|r] + \Sigma_s \, (\mu_s^{i+1} - \mu_s^i) \, [pq|s]$$

The exchange-correlation contributions to δH_{pq}^i can be estimated quite accurately from an exponential prefactor (Obara and Saika, 1986) and the difference in fitting coefficients, $\mu_s^{i+1} - \mu_s^i$. In the case of Coulomb contributions the exact estimation based on a Schwarz inequality, (Haser and Ahlrichs, 1989) is possible.

$$(\rho_r^{i+1} - \rho_r^i) \, EST \, [pq|r] < Threshold$$

The exact estimation (EST) of the integral requires the calculation of two center integrals, if we use a convenient normalization of the density fitting functions (Dunlap et al., 1979)

$$EST \, [pq|r] \le [pq|pq] * [r|r] = [pq|pq]$$

The direct scheme can be introduced to density fits, as well. The variational method of density fitting (Dunlap et al., 1979) leads to a matrix equation for the ρ_r coefficients

$$\rho^{i+1} = S^{-1} * (t^i + \delta t^{i+1})$$

For simplicity, the normalization factor was omitted in the above equation. The change in the t vector can be written by using the density matrix P_{pq} and can be exactly computed as follows

$$\delta t_r^{i+1} = \Sigma_{pq} (P_{pq}^{i+1} - P_{pq}^i) [pq|r]$$

$$(P_{pq}^{i+1} - P_{pq}^i) \ EST \ [pq|r] < Threshold$$

We have found that at the beginning of the SCF procedure the Threshold can be rather large and we decrease it to 10^{-10} near convergence. One can expect that the mixed scheme of the traditional (disk or memory storage) and the direct algorithms will utilize supercomputer resources optimally (Haser and Ahlrichs, 1989). In the present DGauss program only the direct scheme is available.

Analytical Energy Derivatives

Analytical energy derivatives of the total LSD energy with respect to nuclear coordinates are available in the present version of DGauss. The algorithm (Andzelm et al., 1989a; Fournier et al., 1989) yields the exact Coulomb gradients and the approximate gradients of the exchange-correlation (XC) energy. A brief description of the algorithm is presented here.

Due to incompleteness of the orbital basis set the Pulay correction (Pulay, 1969) to the forces has to be used. Since the density fit basis set is used as well, the additional correction to the forces is needed. It can be calculated exactly because in this method the variational fitting of density is performed (Dunlap et al., 1979). Assuming that perfect numerical integration and fit to the XC potential can be done (Andzelm et al., 1989a), the LSD gradient formula in the LCGTO-LSD method, is as follows

$$\partial E_{LSD}/\partial x = F_{HFB} + F_D$$

F_{HFB} is the Hellman-Feynman force plus the correction for the orbital basis set dependence on the nuclear coordinate, x.

$$F_{HFB} = \Sigma_{pq} P_{pq} \{ \partial h_{pq}/\partial x + \Sigma_r \rho_r [\partial(pq)/\partial x|r] + \Sigma_s \mu_s [\partial(pq)/\partial x|s]\}$$
$$+ \partial U_n/\partial x - \Sigma_{pq} W_{pq} \partial[pq]/\partial x$$

The term F_D is the correction due to incompleteness of the density fit basis set

$$F_D = \Sigma_r \, \rho_r \, [\, \partial(r)/\partial x \, | \, (\rho - \rho^f) \,]$$

Clearly, if the "exact" density (ρ) and the fitted density (ρ^f) are the same, the term F_D vanishes.

The error due to inaccurate calculation of the XC gradient has been analyzed in details by Fournier et al., (1989). It was found, that in order to increase the accuracy of the gradients, rather large fitting sets had to be used (Fournier et al., 1989).

In the present version of DGauss this algorithm has been modified. The direct numerical integration of the XC force is performed instead. This makes calculation of forces computationally more demanding, but the results are more reliable. The accuracy of forces can be improved by eliminating spurious one-center contributions in a similar way as has been done by Versluis and Ziegler (Versluis and Ziegler, 1988). Finally, the translational invariance of the energy may be enforced by explicitly eliminating any remaining spurious translational or rotational forces (St-Amant and Salahub, 1990).

In summary, in the SCF procedure, the XC potential is calculated analytically following the method of Dunlap et al., (1989), and then, in the gradient calculations, the numerical calculation of XC gradients is done using an augmented set of grid points (Andzelm and Wimmer, 1990, Salahub et al., this volume).

The new variational method of fitting the exchange-correlation potential yields forces less dependent on the fitting procedure (Dunlap et al., 1990), and certainly will improve the accuracy of the LSD gradients.

The present accuracy of the gradients has been found satisfactory in calculations on organic and organo-metallic molecules. The typical error in locating the minimum based on minimization of the energy and the zero of the gradient is on the order of 0.001Å- 0.004Å for bond distances and less than 1 deg for bond angles. The results presented in this paper have been obtained by optimizing the geometry in Cartesian coordinates and using the BFGS method to update the inverse of the hessian matrix. Vibrational frequency calculations have been done through finite differencing of analytic energy derivatives.

Basis Sets, Model Core Potentials and Grid Selection

The exact calculation of the three-center Coulomb integrals needed for the variational fitting of the density can be accomplished only if Gaussian-type orbitals are used. This requirement brings the DGauss method into the mainstream of the *ab initio* molecular orbital calculations. In particular, one can use the same type of orbital basis sets as in modern Hartree-Fock programs and employ highly efficient techniques for integral calculations. In molecular calculations there is frequently a need to augment atomic basis sets with diffuse or several polarization functions. In those cases Hartree-Fock experience may be invaluable. On the other hand, Hartree-Fock basis sets used in

LSD calculations likely will exhibit Basis Set Superposition Error (BSSE).

In order to minimize this error LSD optimized basis sets have been developed (Andzelm, *et al.*, 1985a). The optimization procedure is based on the Tatewaki and Huzinaga algorithm (Tatewaki and Huzinaga, 1979) which was used to develop Hartree-Fock basis sets for all the elements (Huzinaga, *et al.*, 1984a). There are several advantages in using these basis sets. The high quality of the valence orbitals, as judged by comparison with the numerical orbital energies and shapes, can be achieved even with a relatively modest representation of the core. The number of primitive Gaussian type orbitals can be smaller than that used in non-contracted basis sets, but the properties of valence orbitals are comparable. The quality of molecular results improves systematically when optimized minimal basis sets are replaced by double-zeta split-valence (DZV) sets, and further, the polarization functions are added resulting in DZVP basis sets, which are the default in the DGauss program. The present version of the DGauss program has LSD-optimized basis sets for all atoms up to Xe (Godbout *et al.*, 1990). Those basis sets have a "harder core" than those previously developed (Andzelm *et al.*, 1985a). This leads to a smaller BSSE in molecular calculations. There is growing evidence (Dixon *et al.*, this volume; 1990; Hill *et al.*, this volume) that better core representation will not affect the molecular geometry results, and that the energies of chemical reactions converge within relative accuracy of 5%. Therefore, it seems that the present technique of optimizing LSD-basis sets yields high quality double-zeta-split-valence+polarization (DZVP) basis sets. Some dependence of the molecular results on the higher level orbital sets of triple-zeta quality with multiple polarization functions have been noticed (Dixon *et al.*, this volume). Performance of those higher level basis sets has yet to be assessed through systematic molecular calculations.

The advantage of using Gaussian-type functions is that the pseudopotential technique, developed for Hartree-Fock theory, can readily be used in the LSD approach. The effect of the model core potential can be exactly calculated if Gaussian-type orbitals are used. The model core potentials (MCP) allow the minimization of BSSE. They lead to accurate valence orbitals, and easily enable the inclusion of the relativistic effects. The method of generating the MCP, used in DGauss, is based on the method of Huzinaga (Huzinaga *et al.*, 1984b), which was successfully used in Hartree-Fock calculations. Applying Huzinaga's projection operators it was found that in general spin-polarized LSD calculations one can exactly separate the valence and core effects (Andzelm *et al.*,1985b). The effect of the core orbitals can then be approximated by model potentials. The shape and orbital energies of valence orbitals closely match the results of reference numerical atomic calculations. The effect of polarizable outer-core orbitals can easily be incorporated by enlarging the valence space. The valence orbitals have a nodal structure that seems to be important for the proper accounting of correlation effects (Hay and Wadt, 1985). In numerous calculations on transition metal complexes and clusters (see for example Andzelm *et al.*, 1989), semiconductors (Andzelm *et al.*, 1987; Gryko and Allen, to be published), and molecules containing main group elements (Papai *et al.*, 1990), MCP's provided consistently reliable molecular results. In the present version of DGauss, MCP are available only for some of the atoms and work is in progress to develop MCP for all atoms.

Besides the orbital basis set there is a need to introduce additional fitting basis sets to approximate the electron density and the exchange-correlation potential and energy. The even-tempered expansion of the s-, p-, and d- type uncontracted Gaussians provides a flexible enough representation of the atomic density, provided the variational method of fitting is chosen. The first term in the even-tempered expansion corresponds to the most defuse orbital function according to the procedure by Dunlap *et al.* (1979). In practice, the double-zeta d-type expansion is already accurate, at least for geometry optimization of organic molecules. Typically, we perform calculations with triple-zeta p- and d- type fitting sets, and find that quadruple-zeta sets do not change molecular results significantly (Dixon *et al.*, this volume; 1990)

The exchange-correlation potential is also expanded in Gaussian-type basis functions. However, the determination of the expansion coefficients has to be carried out numerically using a grid similar to the one used in the Numol (Becke, 1989a) and Dmol (Delley, 1990) programs.

The grid selection in DGauss is an adaptive procedure based on the approximation of the exchange-correlation energy on the angular shell around each atom. Based on the required accuracy and the maximal size of the Lebedev quadrature (Lebedev, 1976) the angular integration is performed automatically. The radial distribution of points is accomplished using the method of Becke (Becke, 1988). In order to minimize sensitivity of the total energy on the choice of the grid rotations of the sets of angular points are performed. In the detailed study on the benzene molecule at 91 configurations the largest difference in total energies was found to be 0.15 kcal/mol (Andzelm, unpublished), using the default number of grid points (~ 1000 points/atom). Once the Gaussian-expansions of the exchange-correlation potential is accomplished, further calculation can be performed analytically.

Computational Requirements

The Hamiltonian matrix elements involving three center two- and one-electron integrals are calculated using an efficient method of Obara and Saika (OS)(Obara and Saika, 1986; Andzelm *et al.*, 1989a). A highly vectorized and parallelized integral programs can be written (Andzelm and Wimmer, 1990) using the recursive method of OS in the case of LSD calculations. For gradient calculations a modified OS method due to Head-Gordon and Pople (HGP) (Head-Gordon and Pople, 1988) has been used. The LSD implementation of the HGP method yields integrals which can be calculated three times faster than the Hartree-Fock type integrals. The main gain of the LSD algorithm is that the number of molecular integrals grows with the size of basis set, N, as N^3 instead of N^4 in the case of Hartree-Fock approach.

The major computational effort of any density functional calculations is the synthesis of the electron density in the real space on all grid points. This algorithm depends on the number of basis functions, N, the number of occupied molecular orbitals, M, and the number of grid points, P, as NMP. The numerical integration of the matrix elements is a computationally even more demanding operation (it scales as N^2P), although in the present DGauss calculations it is being invoked only twice for each geometry optimization cycle. This formidable task can be accomplished using a sparse matrix algorithms, which can be fully parallelized (Andzelm, 1989b). Theoretically, the integral calculations, density synthesis, numerical integration of matrices, diagonalization and matrix inversion algorithms depend on the size of the basis set, N, at most as N^3. In practice the first three operations utilize the inherent sparsity of the Gaussian basis set and for large organic molecules (with 500 -1000 basis sets) actual scaling of $N^{2.2}$ is typically observed (Andzelm 1989b; Andzelm, and Wimmer, 1990).

All the above mentioned algorithms are fully vectorized and parallelized. It has been demonstrated that the self-consistent LSD calculations, for large molecules, can be carried out at sustained speed of 1GFLOPS on an 8-processor CRAY Y-MP supercomputer (Andzelm, 1989b; 1989c).

Applications

Molecular Geometries

Hartree-Fock optimized basis sets can be used in LSD calculations and are found to provide quite accurate geometries for various molecules including nitro compounds (Redington and Andzelm, this volume). In this paper we study the performance of the newly optimized LSD basis sets (Godbout, et al., 1990). The results are presented in Table 1.

In the calculation (A) the DZVPP basis set has been used which corresponds to the popular 6-31G** Pople's basis set (Hariharan and Pople, 1972; Hehre, et al., 1986). For atoms of the second row this orbital basis set has a pattern (621/41/1*), and the hydrogen basis set is (41/1*) (Huzinaga et al., 1984a). The corresponding fitting set of triple-zeta quality (A1) with the pattern (7/3/3) has been used for carbon-like atoms and for hydrogen a (5/1/1) set was employed. The maximum number of angular points, 110(194), and the accuracy of selecting angular shells, 10^{-8} (10^{-10}), have been used for LSD (Gradient) calculations, respectively. The number of radial shells in numerical calculations was kept at 30 (25) for second-row and hydrogen atoms, respectively (Becke, 1988). The calculation (A) corresponds to the default choice of parameters in DGauss. In calculation (B) the DZVP basis set has been used, with the lower quality set for fitting (A2) and the limited set of grid points. The DZVP orbital basis set is similar to the 6-31G* Pople basis set (Hehre, et al., 1986). The fitting set A2 has a form (7/3/2) for second row atoms and (3) for hydrogen. In the (B) -type calculations the same grid has been used for both LSD and gradient calculations. The maximum number of angular points was 110, the accuracy of selecting the angular shells was 10^{-8}, and the number of radial shells was 25 and 20 for carbon-like and hydrogen atoms, respectively.

Table 1. Comparison of LSD, HF and MP2 geometries. A(DZVPP) and B(DZVP) levels of LSD calculations are described in the text. HF, MP2(6-31G*) and experimental (Expt) data are taken from Hehre et al. (Hehre et al., 1986). Bond distances are in Å, bond angles are in degrees.

	bond,angle	LSD(A)	LSD(B)	HF	MP2	Expt
H_2CO						
	CO	1.212	1.213	1.184	1.221	1.208
	CH	1.123	1.129	1.092	1.104	1.116
	HCH	115.9	116.1	115.7	115.6	116.5
CH_2NH						
	CN	1.274	1.275	1.250	1.282	1.273
	CH_s	1.110	1.115	1.084	1.096	1.103
	CH_a	1.106	1.109	1.080	1.090	1.081
	NH	1.035	1.036	1.006	1.027	1.023
	H_sCN	125.4	125.6	124.7	125.4	123.4
	H_aCN	118.2	118.1	119.2	116.1	119.7
	HCN	111.0	110.7	111.6	109.7	110.5
CH_3NH_2						
	CN	1.453	1.456	1.453	1.465	1.471
	CH_t	1.114	1.117	1.091	1.100	1.099
	CH_g	1.105	1.106	1.084	1.092	1.099
	NH_a	1.025	1.027	1.001	1.018	1.010
	HNH	107.1	106.9	106.9	105.9	107.1
CH_3OH						
	CO	1.410	1.413	1.400	1.424	1.421
	CH_t	1.103	1.105	1.081	1.090	1.094
	CH_g	1.110	1.114	1.087	1.097	1.094
	OH	0.974	0.977	0.946	0.970	0.963
	HOC	108.6	107.8	109.4	107.4	108.0
CH_3F						
	CF	1.378	1.382	1.365	1.392	1.383
	CH	1.106	1.107	1.082	1.092	1.100
	HCH	109.5	109.8	109.8	109.8	110.6

Results from Table 1 show that the two levels of LSD calculations (A) and (B) provide very similar molecular geometries. They compare favorably with MP2 and experimental results and are better than HF results with the exception of the bond distances involving hydrogen. In those cases the LSD calculations overestimate the bond distance by typically 0.02Å. The level (A) of calculations leads to better bond distances involving hydrogen than the set (B).

Energies of Chemical Reactions

It is well known that the LSD energies of chemical bonds are overestimated and the gradient corrected DF method should be used instead of the LSD approach. The bond energies and energetics of the reactions (Dixon *et al.*, this volume) are more sensitive to the choice of proper basis set.

In Table 2 The A-H bond dissociation energies are calculated using the LSD (VWN) and Becke-Stoll (BSPP) Hamiltonians and the Pople basis set, 6-31G** (Hariharan and Pople, 1972). The results are compared with the HF and MP2 calculations using the same basis set (Hehre el at., 1986). Several calculations have been done using the Gaussian 88 program (Frisch *et al.*, 1988). For some molecules the results of the Numol program (Becke, 1989a) are also presented.

Table 2. A-H Bond Dissociation energies (in kcal/mol) at 6-31G**//6-31G* basis set level. The Numol (Becke, 1989a) calculations are in parentheses.

		DGauss (Numol):		Gaussian:	
	Expt	LDF	BSPP	HF	MP2
LiH	58	60 (61)	62 (62)	32	45
BeH(2σ)	50,56	61	58	52	52
BH	82	92 (91)	87 (86)	62	77
CH(2π)	84	91	88	55	73
CH(4σ)	67	80	62	62	68
OH(2π)	107	123	103	67	96
FH	141	154	132	93	131
FH[f]		161 (164)	138 (141)		

[f] Dunning-Huzinaga basis set

Comparison with experimental data (Huber and Herzberg, 1979) shows that LSD results are already better than HF values. However, LSD predicts the BeH bond to be stronger than the LiH bond, contrary to experimental evidence. The same is true for BH and CH2(π) molecules. The gradient corrected DF calculations with the BSPP Hamiltonian are able to correct those LSD errors. In the case of OH and FH molecules LSD considerably overestimates the dissociation energy. The BSPP calculations consistently correct the inaccuracies of the LSD model. It is worthwhile to note that the results obtained by the fully numerical, basis set free method of Becke

(Numol program), are very similar to the Gaussian basis set results (DGauss program) provided that a good basis set has been used. This is especially true for the HF molecule where the 6-31G* basis set is not adequate and the Dunning-Huzinaga set should be used instead (Dunning and Hay, 1977). The gradient corrected DF results closely approach experimental values and are superior to MP2 results.

The results of dissociation energies and hydrogenation energies of simple hydrocarbons are presented in Table 3. The LSD (VWN), Becke-Perdew (BP) and Becke-Stoll (BSPP) energies are compared with experimental data. Level A (DZVPP) of DGauss calculations has been used to study those reactions. One can notice that the gradient-corrected calculations are always superior to LSD calculations and in most cases good agreement with experimental data was found. With the exception of the dissociation energy of the ethylene molecule the BP nonlocal Hamiltonian leads to better results as compared with experimental data.

Table 3. Bond Dissociation Energies and Energies of Hydrogenation Reactions (in kcal/mol). Results of LSD and NLSD DZVPP calculations.

reaction	LSD	BP	BSPP	Expt
$CH_3\text{-}CH_3 \rightarrow 2CH_3$	115	95	82	97
$CH_2 \text{=} CH_2 \rightarrow 2CH_2$	205	177	169	164
$CH\equiv CH \rightarrow 2CH$	269	235	213	237
$C_2H_6 + H_2 \rightarrow 2CH_4$	18	19	21	19
$C_2H_4 + 2H_2 \rightarrow 2CH_4$	67	60	50	57
$C_2H_2 + 3H_2 \rightarrow 2CH_4$	132	114	94	105

The energetics of hydrogenation reactions and reactions involving different types of bonds have been calculated (see Table 4). The geometries of the reactants and products were optimized (see Table 1), then the LSD and BP energies were calculated and compared with the *ab initio* and experimental results (Hehre *et al.*, 1986). In the case of hydrogenation reactions, both local, nonlocal density functional (DF), and *ab initio* results are comparable. The DF results are actually closer to MP2 than to HF results. One can notice the similar dependence on the quality of basis set from DF and HF -type calculations.

When both single and double bonds are involved in reactions the differences between local and nonlocal DF calculations are significant. In all investigated cases the BP(DZVPP) results are close to MP2 data within 2 kcal/mol, whereas LSD results may exhibit an error of 13 kcal/mol. Particularly, the results for the reaction involving aldehyde and methyl alcohol

(Table 4) indicate the importance of gradient corrections. Clearly, in order to quantitatively evaluate the energetics of chemical bonds the gradient-corrected Hamiltonian must be used. The geometries of investigated molecules obtained on the local level are already satisfactory, although the corresponding energetics of the reactions may not be.

Table 4. Energies of Chemical Reactions (in kcal/mol). The energetics obtained using DZVPP set(A) with LSD and nonlocal BP method is compared with HF and MP2 results obtained using 6-31G** set. The results from DZVP and 6-31G* calculations are shown in parentheses.

reaction	LSD	BP	HF	MP2	Expt
$CH_3\text{-}NH_2+H_2 \rightarrow CH_4+NH_3$	-25(-23)	-26(-24)	-28(-27)	-25	-26
$CH_3\text{-}OH+H_2 \rightarrow CH_4+H_2O$	-28(-24)	-28(-25)	-30(-27)	-28	-30
$CH_3F+H_2 \rightarrow CH_4+HF$	-28(-22)	-27(-20)	-27(-23)	-26	-29
$CH_2=O+2H_2 \rightarrow CH_4+H_2O$	-68(-63)	-57(-49)	-58(-54)	-55	-59
$CH_2=NH+CH_4+NH_3 \rightarrow 2CH_3\text{-}NH_2$	-20(-21)	-8(-7)	-7(-7)	-8	-12
$CH_2=O+CH_4+H_2O \rightarrow 2CH_3\text{-}OH$	-11(-14)	0(0)	1(1)	2	1

Vibrational Frequencies

Results for vibrational frequencies of several molecules are presented in Tables 5 and 6. In order to obtain second derivatives the numerical differentiation of the gradients has been done at equilibrium geometry (see Table 1). A two-point difference formula was used and a displacement of 0.01 a.u. was applied. Due to numerical instabilities we expect the lower frequencies to be in error not greater than 2%. This is caused by the fact that the accuracy of gradients is no better than 10^{-5}. Additionally, some sensitivity of the results with respect to the SCF convergence has been observed. The rotations and vibrations have not yet been transformed out in the present calculations. The present results of vibrations show a good agreement between LSD vibrations and experimental data. Typically, high frequencies are overestimated, while low ones are smaller than corresponding experimental results. The Hartree-Fock frequencies are considerably too large. In Table 6 the measured frequencies corrected for anharmonicity are presented. The LSD frequencies underestimate measured harmonic frequencies. The average percentage deviations in calculated frequencies from measured, harmonic values are 8.5, 4.5 and 2.2% for HF/6-31G*, LSD/DZVPP and MP2/6-31G*, respectively. Similar trends in LSD results have been noticed for other molecules as well (Fan *et al.*, 1988; Papai *et al.*, 1990).

Table 5. Vibrational frequencies (in cm^{-1}) from LSD(DZVPP) calculations compared with HF/6-31G*, MP2/6-31G* and experimental data (Hehre *et al.*, 1986).

molecule/mode	LSD(A)	HF	MP2	experiment
CH$_2$NH				
NH stretch	3354	3719	3463	3297
CH$_2$ stretch	3050	3347	3254	3036
	2948	3254	3116	2924
CN strech	1690	1901	1724	1640
bend	1421	1628	1542	1453
	1296	1496	1412	1347
torsion	1130	1270	1159	1123
bend	1057	1223	1107	1063
bend	1027	1164	1100	1059
CH$_3$NH$_2$				
NH$_2$ stretch	3536	3813	3641	3427
	3446	3730	3508	3361
CH$_3$ stretch	3051	3281	3228	2985
	3006	3245	3155	2961
	2895	3156	3063	2820
NH$_2$ scis	1590	1841	1745	1623
CH$_3$ deform	1448	1665	1596	1485
	1422	1648	1539	1473
	1384	1607	1469	1430
NH$_2$ twist	1280	1479	1405	1419
CH$_3$ rock	1126	1289	1237	1130
	938	1052	915	1195
CN stretch	1084	1149	1113	1044
NH$_2$ wag	784	946	941	780
torsion	316	341	351	268
CH$_3$OH				
OH stretch	3758	4117	3785	3681
CH$_3$ stretch	3059	3305	3201	3000
	2977	3231	3140	2960
	2916	3185	3065	2844
CH$_3$ deform	1438	1664	1552	1477
	1421	1652	1562	1477
	1408	1638	1542	1455
OH bend	1311	1508	1424	1345
CH$_3$ rock	1118	1289	1160	1165
	1026	1187	1120	1060
CO stretch	1102	1164	1082	1033
torsion	305	348	250	295

Table 6. Vibrational frequencies (in cm^{-1}) from LSD(DZVPP) calculations compared with HF/6-31G*, MP2/6-31G* and experimental data: measured and corrected for anharmonicity(harmonic) (Hehre *et al.*, 1986).

molecule/mode	LSD(A)	HF	MP2	experiment harmonic	measured
H$_2$CO					
CH$_2$ stretch	2873	3231	3064	3009	2843
	2811	3159	3019	2944	2783
CO stretch	1806	2028	1786	1764	1746
CH$_2$ scis	1458	1680	1567	1563	1500
CH$_2$ rock	1203	1384	1249	1287	1249
CH$_2$ wag	1130	1336	1194	1191	1167
CH$_3$F					
CH$_3$ stretch	3042	3312	3205	3132	3006
	2947	3232	3110	3031	2930
CH$_3$ deform	1417	1653	1556	1498	1467
	1414	1652	1549	1490	1464
CF stretch	1095	1186	1102	1059	1049
CH$_3$ rock	1146	1312	1213	1206	1182

Computational Effort

In order to assess the computational effort of the two calculations (A) and (B) the CPU time of the LSD (with 10 SCF cycles, including startup time) and the gradient calculations are reported for the methylamine molecule. The size of the orbital and fitting set and the total number of grid points used is also presented.

Table 7. Computational effort of the direct DGauss (A) and (B) calculations for CH$_3$NH$_2$.

	DGauss (A)	DGauss (B)
basis[a]	DZVPP (55, 138)	DZVP (40, 71)
grid[b]	7508, 13656	6308, 6308
time(sec)[b]	28.0, 10.5	8.5, 2.6

[a] size of the orbital and fitting set is given

[b] first number corresponds to LSD calculations, second is for gradient calculations

>From Table 7 it can be seen that the level (B) calculation is almost 3.5 times more economical than the default-type calculation (A). More tests are required to evaluate fully the accuracy and performance of the level (B) computations.

The timing in Table 7 can not serve as a guide to estimate time of calculations for large molecules. The reason is that the startup time is considerable for small molecules and the small vector dimensions make the program run less efficiently than in the case of, e.g., 1000 basis functions. In recent test calculations on a 62 atom organometallic molecule with 551 orbitals (DZVP level) it took about 20 and 8 minutes to perform one SCF and gradient calculation, respectively. One processor of a CRAY YMP was used in these calculations.

Structures of Zinc-Water Complexes

Zn^{2+} complexes are important in many inorganic and biochemical reactions. Recently we have studied the properties of a zinc-insulin complex (Andzelm, 1989b; Andzelm and Wimmer, 1990). One important question to be resolved is the structure of the hydration shell around the zinc ion. In the present work we study the structure of zinc-water complexes. This problem attracted considerable attention recently (Tossell, 1990; Giessner-Prettre and Jacob, 1989; Magnera et al., 1989).

In Table 8 the geometries of zinc cation-water complexes are presented together with the average energy of the Zn-water bond. The full, analytic-gradient geometry optimization has been performed using a DZVPP basis set and the default grid and triple-zeta selection of the fitting functions in the DGauss program. For zinc the $(63321/531^*/41^+)$ orbital basis set has been used (Godbout et al., 1990). The energetics were studied using VWN and BP Hamiltonians.

It can be seen that the present approach introduces errors for bonds and valence angles involving hydrogen of about 0.02 Å and 1 deg, respectively. We can notice a trend: the O-H bond decreases and the H-O-H angle becomes smaller with the increasing number of water molecules in the complex. The bond distance Zn^{2+} - O increases from 1.85Å to 2.06Å, for the hexaaquazinc complex. The value of 2.06Å is close to recent Hartree-Fock result of 2.09Å (Tossell, 1990) and is in reasonable agreement with experimental data ranging from 2.08 to 2.14 Å (Giessner-Prettre and Jacob,1989). Typically the Zn-O distance varies from 2.05 to 2.14Å in hexaaquazinc salts (Cariati et al., 1983). The energy per bond decreases with the number of coordinated water molecules. The nonlocal BP results of the bonding energy are always smaller by about 15% in comparison to the local VWN values. In order to calibrate the method the calculation on a Zn^+- single water molecule is presented. In recent work by Magnera et al. (Magnera et al., 1989), the accurate thermochemical data for transition-metal hydrates are presented. The experimental result of 39 kcal/mol for Zn^+ is very close to the present value of 38 kcal/mol obtained using the nonlocal BP approach. Although we can not expect this excellent agreement to hold for other types of bonds, we definitely prove that the nonlocal BP approach is considerably better than local VWN method in studying the energetics of zinc-water complexes. The detailed results of the DF study on zinc complexes will be published elsewhere (Andzelm and Wimmer, to be published).

Table 8. Geometries and energetics of the zinc-water complexes. Geometry (Å, deg) and the average energy for Zn-water bond (kcal/mol) are given. Experimental data are presented in parentheses.

complex	Geometry			Energy/bond	
	Zn-O	O-H	H-O-H	VWN	BP
H_2O	-	0.977 (0.958)	105.5 (104.5)		
Zn^+-H_2O	2.004	0.989	109.2	48	38 (39.0)
Zn^{+2}-H_2O	1.845	1.006	110.6	120	106
Zn^{+2}-$(H_2O)_4$	1.95	0.989	108.4	89	76
Zn^{+2}-$(H_2O)_6$	2.06 (2.05-2.14)	0.985	108.1	72	60

Conclusions

In the present contribution the principles of the LCGTO-MCP-DF method have been reviewed and the implementation of the method in the DGauss program has been discussed. Calculations of the geometry, vibrational frequencies, and the energies of reactions were carried out for simple organic molecules and organometallic complexes. The local approach (LSD) based on the VWN potential provides accurate geometries and vibrations, at least for the molecules studied in this paper. The energetics of reactions have to be studied using the nonlocal approach, like BP, used in this paper. The energetics of reactions presented here show a remarkable improvement over LSD results and satisfactory agreement with experimental data. The LCGTO-MCP-DF method as implemented in DGauss shows a favorable scaling of close to N^2 (where N is the basis set size). Therefore we can expect that calculations on large organic and organo-metallic systems can be performed using this methodology.

Acknowledgements

The author is grateful to Erich Wimmer and all colleagues at Cray Research Inc. for fruitful collaboration on the DGauss program.

Dennis Salahub's continuous encouragements and collaboration are very appreciated. I thank Axel Becke for his calculations on A-H molecules using the Numol program, Nathalie Godbout for developing the LSD-optimized basis sets, Alain St-Amant and Brett Dunlap for help in developing the gradient program, Bernard Delley for conversations on the numerical calculations, George Fitzgerald for help in using geometry optimization techniques and Tom Ziegler for reading and commenting on the manuscript.

References

Almlof, J., Faegri, Jr.,K. and Korsell, K., 1982, *J. Comput. Chem.* 3:385.

Andzelm, J., Radzio, E. and Salahub, D.R., 1985, *J. Comput. Chem.* 6:520,533.

Andzelm, J., Radzio, E. and D.R. Salahub, 1985, *J. Chem. Phys.* 83:4573.

Andzelm, J., Russo, N. and Salahub, D.R., 1987, *J. Chem. Phys.* 87:6562.

Andzelm, J., Wimmer, E. and Salahub, D.R., 1989, *The Challenge of d and f electrons: Theory and Computation*, (Edited by Salahub, D.R. and Zerner, M.C.), ACS Symposium Series No. 394, p.228, American Chemical Society, Washington, D.C..

Andzelm, J., 1989, *Proceedings NCSA II Conference on Parallel and Vector Processing*, NCSA.

Andzelm, J., 1989, *Gigaflop report*, Cray Research Inc, p15.

Andzelm, J. and Wimmer, E., 1990, to be published.

Averill, F.W. and Ellis, D.E., 1973, *J. Chem. Phys.* 59: 6412.

Baerends, E.J., Ellis, D.E. and Ros, P., 1973, *Chem. Phys.* 2:41.

Becke, A.D., 1986, *J. Chem. Phys.* 84:4524.

Becke, A.D., 1988, *J. Chem. Phys.* 88:2547.

Becke, A.D., 1989, *Int. J. Quant. Chem. Symp.* 23:599.

Becke, A.D., 1989, *The Challenge of d and f electrons: Theory and Computation*, (Edited by Salahub, D.R. and Zerner, M.C.), ACS Symposium Series No. 394, p.166, American Chemical Society, Washington, D.C..

Borman, S., 1990, *Chemical and Engineering News*, April 9, p22.

Caldwell, D.J. and Redington, P.K., this volume.

Cariati, F., Erre, L., Micera, G., Panzanelli, A., Ciani, G. and Sironi, A., 1983, *Inorganica Chemica Acta* 80:57.

Delley, B. and Ellis, D.E., 1982, *J. Chem. Phys.* 76:1949.

Delley, B., 1990, *J. Chem. Phys.* 92:508.

Dixon, D.A., Andzelm, J., Fitzgerald, G., Wimmer, E. and Jasien, P., this volume.

Dixon, D.A., Andzelm, J., Fitzgerald, G. and Wimmer, E., 1990, to be published.

Dunlap, B.I., Connolly, J.W.D. and Sabin, J.R., 1979, *J. Chem. Phys.* **71**:3396.

Dunlap, B.I., Andzelm, J. and Mintmire, J.W., 1990, Phys. Rev. A., in press.

Dunning, T.H.Jr. and Hay, P.J.,1977, *Methods of Electronic Structure Theory*, (Edited by Schaefer, H.F.,III), Ch.1, Plenum Press, New York.

Fan, L., Versluis, L., Ziegler, T., Baerends, E.J. and Ravenek, W., 1988, *Int. J. Quant. Chem. Symp.* **22**:173.

Fournier, R., Andzelm, J. and Salahub, D.R., 1989, *J. Chem. Phys.* **90**:6371.

Frisch, M.J., Head-Gordon, M., Schlegel, H.B., Raghavachari, K., Binkley, J.S., Gonzalez, C., Defrees, D.J., Fox, D.J., Whiteside, R.A., Seeger, R., Melius, C.F., Baker, J., Martin, R., Kahn, L.R., Stewart, J.J.P., Fluder, E.M., Topiol, S. and Pople, J.A., 1988, Gaussian, Inc., Pittsburgh, PA.

Giessner-Prettre, C. and Jacob, O., 1989, *J. Comput.-Aided Mol. Des.* **3**:23.

Godbout, N., Andzelm, J., Wimmer, E. and Salahub, D.R., 1990, to be published.

Hariharan, P.C. and Pople, J.A., 1972, *Chem. Phys. Lett.* **66**:217.

Haser, M. and Ahlrichs, R., 1989, *J. Comput. Chem.* **10**:104.

Hay, P.J. and Wadt, W.R, 1985, *J. Chem. Phys.* **82**:270,284,299.

Head-Gordon, M. and Pople, J.A., 1988, *J. Chem. Phys.* **89**:5777.

Hehre, W.J., Radom, L., Schleyer, P. and Pople, J.A., 1986, *Ab initio molecular orbital theory*, John Wiley & Sons, New York.

Hill, R.A., Labanowski, J.K., Heisterberg, D.J. and Miller, D.D., this volume.

van den Hoek, P.J. and Baerends, E.J., 1989, *Surface Sci.* **221**:L791.

Hohenberg, P. and Kohn, W., 1964, *Phys. Rev.* **B. 136**:864.

Huber, K.P. and Herzberg, G., 1979, *Molecular Spectra and Molecular Structure IV: Constants of Diatomic Molecules*, Van Nostrand Reinhold, New York.

Huzinaga, S., Andzelm, J., Klobukowski, M., Radzio, E., Sakai, Y. and Tatewaki, H., 1984, *Gaussian Basis Sets for Molecular Calculations*, Elsevier, Amsterdam; and references therein.

Huzinaga, S., Klobukowski, M. and Sakai, Y., 1984, *J. Phys. Chem.* **88**:4880.

Jones, R.O. and Gunnarsson, O., 1990, *Reviews of Modern Physics* **61**:689.

Johnson, K.H., 1973, *Adv. Quantum Chem.*, **7**:143

Kohn, W. and Sham, L.J., 1965, *Phys. Rev.* **A. 140**:1133.

Lebedev, V.I., 1976, *Zh.V.Mat.Fiz.* **16**:293.

Levy, M., 1979, *Proc. Natl. Acad. Sci. (USA)* **76**:6062.

Magnera, T.F., David, D.E. and Michl, J., 1989, *J. Am. Chem. Soc.* **111**:4100.

Mintmire, J.W., 1989, *Phys. Rev.B* **39**:13350.

Mlynarski, P. and Salahub, D.R., to be published.

Obara, S. and Saika, A., 1986, *J. Chem. Phys.* **84**:3963.

Papai, I., St-Amant, A., Ushio, J. and Salahub, D.R., 1990, *Int. J. Quant. Chem.*, in press.

Parr, R,G., 1990, *Chemical and Engineering News*, July 16, p.45.

Parr, R.G. and Yang, Weitao, 1989, *Density-Functional Theory of Atoms and Molecules*, Oxford University Press, New York.

Perdew, J.P., 1986, *Phys. Rev.* **B. 33**: 8822

Pulay, P., 1969, *Mol. Phys.* **17**:197.

Redington, P.K. and Andzelm, J, this volume.

Salahub, D.R., 1987, *Ab Initio Methods in Quantum Methods in Quantum Chemistry-II*, (Edited by Lawley, K.P.), p.447, J.Wiley & Sons, New York.

Salahub, D.R., Fournier, R., Mlynarski, P., Papai, I., St-Amant, A. and Ushio, J., this volume.

Sambe, H. and Felton, R.H., 1975, *J. Chem. Phys.* **62**:1122.

Slater, J.C., 1974, *The Self-Consistent Field for Molecules and Solids,* vol. 4, McGraw-Hill, New York.

St-Amant, A. and Salahub, D.R., 1990, *Chem. Phys. Lett.* **169**:387.

Stewart, J.P., 1990, J. Comput. Mol. Des. **4**:1.

Stoll, H., Pavlidou, C.M.E. and Preuss, H., 1978, *Theor. Chim. Acta* **49**:143.

Tatewaki, H. and Huzinaga, S., 1979, *J. Chem. Phys.* **71**:4339.

Teter, M.P., Payne, M.C. and Allan, D.C., 1989, *Phys. Rev.* **B40**:12255.

Tossell, J.A., 1990, *Chem. Phys. Lett.* **169**:145.

Versluis, L. and Ziegler, T., 1988, *J. Chem. Phys.* **88**:322.

Vosko, S.H., Wilk, L. and Nusair, M., 1980, *Can. J. Phys.* **58**:1200.

Wilson, L.C. and Levy, M., 1990, *Phys. Rev.* **B. 41**:12930.

Wimmer, E., this volume.

Wimmer, E., Freeman, A.J., Fu, C.-L., Cao, P.-L., Chou, S.-H. and Delley, B., 1987, *Supercomputer Research in Chemistry and Chemical Engineering*, (Edited by Jensen, K.F. and Truhlar, D.G.), ACS Symposium Series No. 353, p.49, American Chemical Society, Washington, D.C.

Wimmer, E., Krakauer, H. and Freeman, A.J., 1985, *Advances in electronics and electron physics*, (Edited by Hawkes, P.), **65**:357, Academic Press.

Ziegler, T., Fan, L., Tschinke, V. and Becke, A., 1989, *J. Am. Chem. Soc.* **111**:9177.

Ziegler, T., Tschinke, V., Baerends, E.J., Snijders, J.G. and Ravenek, W., 1989, *J. Phys. Chem.* **93**:3050.

12
Nonlocal Correlation Energy Functionals and Coupling Constant Integration

MEL LEVY

I. Introduction

This chapter is concerned with nonlocal correlation energy functionals. To help one understand the basic concepts of density-functional theory, I shall first (Section II) discuss the "constrained-search" formulation of density-functional theory. In this formulation all variational functionals are viewed as minimizations with wavefunctions (or ensembles) which are simultaneously constrained to yield each trial density. For instance, the Hohenberg-Kohn $F[n]$ for $<\hat{T} + \hat{V}ee>$ has been identified as (Levy, 1979)

$$\min_{\Psi \to n} \langle \Psi | \hat{T} + \hat{V}ee | \Psi \rangle$$

where n is the electron density. This constrained-search expression facilitates the discovery of properties of $F[n]$ for computational purposes.

I shall then review two correlation energy functionals, $E_c^{QM}[n]$ and $E_c^{HF}[n]$, which are intended to be added to traditional Hartree-Fock calculations in quantum chemistry. With $E_c^{QM}[n]$, in Section III I point out that the exact ground-state density is not needed for one to obtain the exact ground-state energy. Indeed, we know that the exact traditional quantum mechanical correlation energy can be expressed as a functional of the Hartree-Fock density (Harris and Pratt, 1985; Levy, 1985, 1987). With $E_c^{HF}[n]$, in Section IV I point out that when the corresponding correlation potential, $\delta E_c^{HF}[n]/\delta n(\vec{r})$, is added to the Fock potential and self-consistency is achieved, then the exact ground-state energy and density are obtained (Baroni and Tuncel, 1983) and the highest-occupied orbital energy turns out to be the exact ionization energy (Levy et.al., 1987).

Most of the chapters in this book and most of my research utilize approximations to the whole F[n] where the Hartree-Fock computation is by-passed. With this in mind, in Sections (V-IX) the constrained-search formulation, coordinate scaling, and the adiabatic connection (coupling constant integration formula) of Harris and Jones (1974), Langreth and Perdew (1975, 1977), and Gunnarsson and Lundgvist (1976) are employed to construct approximate correlation energy functionals which are meant to approximate that exact correlation energy, $E_c[n]$, which is defined to be used in conjunction with the exact exchange energy, $E_x[n]$, to construct the exact exchange-correlation component, $E_{xc}[n]$, of the whole F[n].

A recently published nonlocal correlation energy functional of Wilson and Levy (1990) is revealed in Section VII. Although extremely simple, it appears to compare favorably with other more complicated correlation energy functionals.

Coordinate scaling shall play a key role in this chapter because coordinate scaling allows one to learn a significant amount about the properties of $E_c[n]$. Part of the appeal of coordinate scaling is the fact that it is often quite easy to observe immediately how a functional behaves upon coordinate scaling. (Quite often a scaling analysis will literally take just a few minutes.) For example,

$$\int n \left(|\nabla n| n^{-4/3} + n^{-1/3} \right)^{-1} d^3 r$$

simply becomes

$$\int n \left(|\nabla n| n^{-4/3} + \lambda^{-1} n^{-1/3} \right)^{-1} d^3 r$$

when n is replaced by the uniformly scaled density $n_\lambda(x,y,z) = \lambda^3 n(\lambda x, \lambda y, \lambda z)$. This result arises because

$$\int n_\lambda (|\nabla n| n_\lambda^{-4/3} + n_\lambda^{-1/3})^{-1} d^3 r =$$

$$\int n(\lambda \vec{r}) (\lambda^4 |\nabla(\lambda \vec{r}) n(\lambda \vec{r})| \lambda^{-4} n^{-4/3}(\lambda \vec{r}) + \lambda^{-1} n(\lambda \vec{r})^{-1/3})^{-1} d^3 \lambda r$$

$$= \int n(\vec{r}) (|\nabla(\vec{r}) n(\vec{r})| n^{-4/3} + \lambda^{-1} n^{1/3}(\vec{r}))^{-1} d^3 r$$

II.   New Constrained-Search Representation of Hohenberg-Kohn F[n]: Solution To Degeneracy and v-Representability Problems

Assume that we are interested in the ground-state energy and density of $\hat{H}_v$

$$\hat{H}_v = \hat{T} + \hat{V}ee + \sum_{i=1}^{N} v(\vec{r}_i) , \tag{1}$$

where $\hat{T}$ is the kinetic energy operator, $\hat{V}ee$ is the electron-electron repulsion operator, and $v(\vec{r})$ is the electron-nuclear attraction operator of interest. Then by the Hohenberg-Kohn theorem (1974) the ground-state energy, E_v^{GS}, is obtained from

$$E_v^{GS} = \min_{n} \{\int v(\vec{r}) n(\vec{r}) + F[n]\} \tag{2}$$

where F[n] is the Hohenberg-Kohn universal functional of $<\hat{T}+\hat{V}ee>$ for each trial electron density $n(\vec{r})$. Equation (2) constitutes the Hohenberg-Kohn variational theorem.

There were significant restrictions on the original Hohenberg-Kohn F[n]. Namely, the trial n had to be a <u>non-degenerate</u> <u>ground-state</u> density of some Hamiltonian H_v', where

$$\hat{H}_{v'} = \hat{T} + \hat{V}_{ee} + \sum_i v'(\vec{r}_i), \tag{3}$$

so that the formal identification of the original Hohenberg-Kohn F[n] was

$$F[n] = \langle \Psi_n^{GS} \mid \hat{T} + \hat{V}_{ee} \mid \Psi_n^{GS} \rangle, \tag{4}$$

where Ψ_n^{GS} yields n and is simultaneous the non-degenerate ground-state wavefunction of some $\hat{H}v'$. Now what if there is no Ψ_n^{GS} associated with a given trial n? Then, the original Hohenberg-Kohn F[n] of Eq.(4) is not defined for this trial n.   This constitutes the degeneracy and v-representability problems. I solved these problems (Levy, 1979) by defining a new F[n]. (In this connection, see Dunlap (1987, 1988) for other relevant treatments of degeneracies.) My work followed in the spirit of the work of Percus (1978) for noninteracting systems, even through the paper of Percus did not concern itself with the problems of degeneracy and v-representability. The Hohenberg-Kohn F is a special case of the F which I defined which is (Levy, 1979)

$$F[n] = \min_{\Psi \to n} \langle \Psi \mid \hat{T} + \hat{V}_{ee} \mid \Psi \rangle \tag{5}$$

or

$$F[n] = \langle \Psi_n^{min} \mid \hat{T} + \hat{V}_{ee} \mid \Psi_n^{min} \rangle, \tag{6}$$

where Ψ_n^{min} is that antisymetric wavefunction which minimizes $\langle \hat{T} + \hat{V}_{ee} \rangle$ and is simultaneously constrained to yield n. For this reason the F[n] in Eqs.(5) or (6), and its extensions, are often called "constrained-search" functionals or representations. It is easy to verify that

$$\Psi_n^{GS} = \Psi_n^{min} \tag{7}$$

when Ψ_n^{GS} exists, so that the F[n] in Eq.(4) equals the F[n] in Eqs.(5-6) when Ψ_n^{GS} exists. In any case, the F[n] in Eqs.(5-6) is valid even when Ψ_n^{GS} does not exist, such as when n belongs to an H_v' with degenerate ground-states. Another major advantage of the F[n] in Eqs.(5-6) is that the proof of Eq.(2) becomes trivial, as follows:

$$\min_n \{\int v(\vec{r}) n(\vec{r}) d^3r + F[n]\}$$
$$= \min_n [\min_{\Psi \to n} \langle \Psi | \hat{H}_v | \Psi \rangle] \tag{8}$$
$$= \min_\Psi \langle \Psi | \hat{H}_v | \Psi \rangle = E_v^{GS}$$

Moreover, the constrained search explicit identification of F[n] provides a natural vehicle for proving theorems concerning the properties of F[n] for computational purposes, as shall be seen in later sections of this chapter. Finally, I should emphasize that even though both the Hohenberg-Kohn and constrained-search representations of F employ wavefunctions in their definitions, both representations of F are meant to be approximated explicitly in terms of the density and not in terms of wavefunctions.

III.   Exact Ground-State Energy and Correlation Energy from Hartree-Fock Density

It has been proven independently by Harris and Pratt (1985) and by Levy (1985, 1987) that the traditional quantum mechanical correlation energy, E_c^{QM}, is a functional of the <u>Hartree-Fock</u> density. Surprisingly, the exact ground-state density is not needed. For Hamiltonian $\hat{H}_v$ simply perform a Hartree-Fock calculation to obtain the Hartree-Fock energy, E_v^{HF}, and the

Hartree-Fock density, n_v^{HF}. The exact ground-state energy is then obtained by

$$E_v^{GS} = E_v^{HF} + E_c^{QM}[n_v^{HF}] . \qquad (9)$$

The object, of course, is to approximate $E_c^{QM}[n]$.

IV.   Exact Ground-State Energy, Ionization Energy, and Density From Fock Potential and Correlation Potential

Another useful correlation energy is defined by Baroni and Tuncel (1983):

$$E_c^{HF}[n] = \langle \Psi_n^{min} | \hat{T} + \hat{V}ee | \Psi_n^{min} \rangle \\ - \langle \Phi_n^{HF} | \hat{T} + \hat{V}ee | \Phi_n^{HF} \rangle , \qquad (10)$$

where Φ_n^{HF} is that single determinant (Levy, 1979; Payne, 1979) which yields n and minimizes $\langle \hat{T} + \hat{V}ee \rangle$. When the corresponding correlation potential, $\delta E_c^{HF}[n]/\delta n(\vec{r})$, is added to the ordinary Fock potential for $\hat{H}_v$ and self-consistency is achieved (Baroni and Tuncel, 1983), it turns out that $\Phi_{n_v^{GS}}^{HF}$ is generated, where n_v^{GS} is the exact (not Hartree-Fock) density of H_v. Further, the highest-occupied orbital energy turns out to be the negative of the exact ionization energy (Levy et al., 1987) of $\hat{H}_v$. (Note that the final Fock potential is expressed in terms of the orbitals of the single determinant $\Phi_{n_v^{GS}}^{HF}$.) Finally, E_v^{GS} is obtained from (Baroni and Tuncel, (1983):

$$E_v^{GS} = \langle \Phi_{n_v^{GS}}^{HF} | \hat{H}_v | \Phi_{n_v^{GS}}^{HF} \rangle + E_c^{HF}[n_v^{GS}] . \qquad (11)$$

V.   Partitioning of F[n]

For the purpose of achieving an accurate approximation to F[n] for computational purposes, it is often convenient to partition F[n] according to

$$F[n] = T_s[n] + U[n] + E_{xc}[n] , \qquad (12)$$

where

$$T_s[n] = \langle \Phi_n^{min} | \hat{T} | \Phi_n^{min} \rangle , \qquad (13)$$

where U[n] is the classical electron-electron repulsion energy, and where E_{xc} is the universal exchange-correlation energy which may be decomposed as

$$E_{xc}[n] = E_x[n] + E_c[n] , \qquad (14)$$

with

$$E_x[n] = \langle \Phi_n^{min} | \hat{V}ee | \Phi_n^{min} \rangle - U[n] \qquad (15)$$

so that

$$E_c[n] = \langle \Psi_n^{min} | \hat{T} + \hat{V} | \Psi_n^{min} \rangle \\ - \langle \Phi_n^{min} | \hat{T} + \hat{V}ee | \Phi_n^{min} \rangle . \qquad (16)$$

Here Φ_n^{min} is that single determinant which yields n and minimizes just $\langle \hat{T} \rangle$. To simplify the presentation, this chapter shall restrict itself to Φ_n^{min} which is a non-degenerate ground-state of some noninteracting Hamiltonian with a local (multiplication) one-body potential (Kohn-Sham (1965) potential.)

Although there are obviously other ways to partition E_{xc}, the partitioning as given by equations (14) and (15) is quite convenient and popular amond density functional formalists (Harris and Jones, 1974; Sahni et. al., 1982), and it is the "correlation energy", $E_c[n]$, as defined in Eq.(16) which shall be featured in the remaining part of this chapter. With this in mind, it has been observed and reasoned that

$$E_c[n] \sim E_c^{QM}[n] \sim E_c^{HF}[n] \qquad (17)$$

for many atoms and molecules, especially when the number of electrons is small.

The "exchange energy" defined by Eq.(15) exhibits very simple homogeneous coordinate scaling (Levy and Perdew, 1985). Namely,

$$E_x[n_\lambda] = \lambda E_x[n] \qquad (18)$$

where

$$n_\lambda(x, y, z) = \lambda^3(\lambda x, \lambda y, \lambda z), \qquad (19)$$

and where λ is a coordinate scale factor. Note that the prefactor λ^3 is present to conserve the norm to N electrons. The gradient expansion for $E_x[n]$ takes the form

$$E_x[n] = a\int n^{4/3}(\vec{r}) \, d^3r + gradient \ terms. \qquad (20)$$

The familiar exponent of 4/3 is consistent with Eq.(18) and with the exchange energy as defined by Eq.(15). Indeed, the simple homogeneous scaling in Eq.(18) does not apply for the exchange as defined with the Hartree-Fock determinant.

In essence, the $E_c[n]$ in Eq.(16) is that piece of E_{xc} with the complicated coordinate scaling. For the purpose of approximating $E_c[n]$, I shall now review a few of the recently dervied limiting coordinate scaling properties of $E_c[n]$. I shall then present the non-local correlation energy functionals of Wilson and Levy (1990). These functionals were constructed to satisfy most of these constraints.

VI.   Limiting Coordinate Scaling Properties of $E_c[n]$

The following limiting uniform coordinate scaling
constraints have recently been derived (Levy, 1989; Levy, 1990 ab) for $E_c[n]$:

$$\lim_{\lambda \to \infty} E_c[n_\lambda] = a[n] \tag{21}$$

$$\lim_{\lambda \to 0} E_c[n_\lambda] \sim \lambda b[n] \tag{22}$$

where a[n] is <u>bounded</u> from below.   In contrast, the local density approximation for the left-hand-side of Eq.(21) is unbounded because it goes as $-\log(\lambda)$. See, for example; the VWN functional (Vosko et al., 1980). The local density approximation for E_c, however, does correctly scale linearly with λ as $\lambda \to 0$.

Recently derived limiting nonuniform coordinate scaling conditions include (Levy, 1990 b)

$$\lim_{\lambda \to \infty} E_c[n_\lambda^x] \gtrsim -\int n(\vec{r}_1) n(\vec{r}_2) |x_1 - x_2|^{-1} d^3 r_1 d^3 r_2 \tag{23}$$

$$\lim_{\lambda \to 0} \lambda^{-1} E_c[n_\lambda^x] \geq -\int n(\vec{r}_1) n(\vec{r}_2) |x_1 - x_2|^{-1} d^3 r_1 d^3 r_2 \tag{24}$$

where

$$n_{\lambda}^X(x, y, z) = \lambda n(\lambda x, y, z).$$ (25)

In contrast, the local density approximation for the left-hand-side of Eq.(23) is incorrectly <u>unbounded</u> from below because it goes as $-\log(\lambda)$. Moreover, the local density approximation for the left-hand-side of Eq.(24) is also unbounded from below because it goes as $-\lambda^{-2/3}$, for $\lambda \to 0$. In fact no local form, no matter how complicated, can simultaneously satisfy conditions (22) and (24).

VII.   New Nonlocal Correlation Energy Functionals

Quite recently Wilson and Levy (1990) have published the following non-local correlation energy functional as an approximation to the exact E_c:

$$E_c^{WL}[n_{\lambda}] = \int \frac{(an + b|\vec{\nabla}n|/n^{1/3})}{(c + d|\vec{\nabla}n|/(n/2)^{4/3} + \lambda^{-1}r_s)} d^3r$$ (26)

where $r_s = (3/4\pi n)^{1/3}$, and where a= -0.74860, b= 0.06001, c= 3.60073, and d= 0.90000. (Note that $E_c^{WL}[n]$ is obtained by setting $\lambda = 1$ on the right-hand-side of Eq.(26).)

Observe that $E_c^{WL}[n_{\lambda}]$ satisfies conditions (21) and (22) in that $E_c^{WL}[n_{\lambda}]$ is bounded from below as $\lambda \to \infty$ and that $E_c^{WL}[n\lambda]$ goes linearly in λ as $\lambda \to 0$. Further, the gradient terms are needed to insure that $E_c^{WL}[n_{\lambda}]$ does not, unreasonably, approach the same value [(constant)N] regardless of the density. Also, the gradient terms are present, in accordance with the work of Ou-Yang and Levy (1990), to insure that, generally,

$$E_c^{WL}[n_\lambda^X] \neq E_c^{WL}[n_\lambda^Y] \neq E_c^{WL}[n_\lambda^Z] \qquad (27)$$

for non-spherical densitites. An equality in Eq.(27) for all non-spherical densities, as given by any completely local form, would be incorrect (Levy and Ou-Yang, (1990)). Moreover, the gradient terms are present in both the numerator and denominator to guarentee that $E_c[n_\lambda^\eta]$ is bounded from below as $\lambda \to \infty$ [condition (23)]. However, it is easy to verify that $E_c^{WL}[n_\lambda^\eta]$ does not satisfy condition (24). It turns out that $E_c^{WL}[n_\lambda^\eta]$ is formed by replacing λ by $\lambda^{1/3}$ on the right-hand-side of Eq.(26). Hence, $\lambda^{-1}E_c^{WL}[n_\lambda^\eta]$ shares the LDA misfortune of blowing up by going as $-\lambda^{-2/3}$ as $\lambda \to o$. In fact, I am not aware of any present functional which satisfies condition (24). [Condition (24) was not involved in the construction of E_c^{WL} because this condition was unknown to us at the time of the inception of E_c^{WL}].

Following is our (Wilson and Levy, 1990) spin-density counterpart to Eq.(26):

$$E_c^{WL}[n_\alpha, n_\beta] = \int \frac{(an+b|\nabla n|/n^{1/3})\,(1-\zeta^2)^{1/2}}{[c+d(\dfrac{\nabla n_\alpha}{n_\alpha^{1/3}} + \dfrac{\nabla n_\beta}{n_\beta^{1/3}}) + r_s]}\, d^3r \qquad (28)$$

where $\zeta = (n_\alpha - n_\beta/n)$. Here, α signifies up spin and β signifies down spin. The $(1-\zeta^2)^{1/2}$ factor makes $E_c^{WL}=o$ for all one-electron system, which is the correct result. However, E_c^{WL} is always zero in the feromagnetic limit, regardless of the number of electrons. This is probably a bit drastic. In any case, for improvements we should consider making further use of the squares of the spin densities (Dunlap, 1988).

The numerical results of Eqs.(26) and (28) as exhibited , in part, in Tables I and II appear to compare favorably with those of other recent correlation energy functionals.

VIII.   Coordinate Scaling in the Adiabatic Connection Formula for $E_c[n]$

The following adiabatic connection formula (coupling constant integration formula) provides a powerful tool for understanding and approximating E_c:

$$E_c[n] = \int_o^1 V_{ee}^{C,\alpha}[n]\, d\alpha \qquad (29)$$

where

$$V_{ee}^{C,\alpha}[n] = \langle \Psi_n^{min,\alpha} | \hat{V}_{ee} | \Psi_n^{min,\alpha} \rangle - \langle \Phi_n^{min} | \hat{V}_{ee} | \Phi_n^{min} \rangle , \qquad (30)$$

and where α is the coupling constant. Here $\Psi_n^{min,\alpha}$ yields n and minimizes $\langle \hat{T} + \alpha \hat{V}_{ee} \rangle$. It should be clear that $\Psi_n^{min,1} = \Psi_n^{min}$ and $\Psi_n^{min,o} = \Phi_n^{min}$. The density is held fixed on the right-hand-side of Eq.(29). Hence, the particular formula of Langreth and Perdew (1975, 1977) and Gunnarson and Lundgvist (1976) is employed. The density changes in the integrand of Harris and Jones (1974). See Harris (1984) for a recent analysis of the adiabatic connection.

From Eq. (29), it follows that knowledge of the behavior of $V_{ee}^{\alpha}[n]$ dictates knowledge of $E_c[n]$. With this in mind, I now list several known properties of $V_{ee}^{\alpha}[n]$:

$$V_{ee}^{C,o}[n] = o, \qquad (31)$$

$$\frac{\partial V_{ee}^{C,\alpha}[n]}{\partial \alpha} \leq 0, \tag{32}$$

$$V_{ee}^{C,\alpha}[n] \geq - \langle \Phi_n^{min} | V_{ee} | \Phi_n^{min} \rangle \tag{33}$$

or

$$V_{ee}^{C,\alpha}[n] \geq - E_x[n] - U[n] \geq - U[n] , \tag{34}$$

and the following important coordinate scaling equality (Levy, 1990 ab)

$$V_{ee}^{C,\alpha}[n] = \alpha V_{ee}^{C,1}[n_\lambda] ; \quad \lambda = \alpha^{-1} . \tag{35}$$

Equation (31) has been noted by Becke (1988), Eq.(32) has been proved by Levy and Perdew (1985), and Eq.(33) follows from the requirement of a non-negative repulsion energy (Levy, 1990 b).

Becke (1988) has recently made nice use of Eq. (29) in the development of his nonlocal correlation energy functional. Here I simply illustrate the use of Eq.(29) which I feel will play a large role in the future. For illustrative purposes, I now write down what is perhaps the simplest local form to satisfy equations (31), (32), and (35). This form is

$$\tilde{V}ee^{c,\alpha}[n] = \int [\frac{\alpha n(\vec{r})}{a + \alpha b n^{-1/3}(\vec{r})}] d^3 r,$$

where a and b are negative constants. However, the above $\tilde{V}ee^{c,\alpha}$ violates Eq.(34) when $n = n_\beta^x$ with $\beta \to o$. This violation of condition (34) is shared (Wilson and Levy, unpublished) by Becke's functional and probably by most, if not all, of the present correlation energy functionals. In any case, integration of the above equation gives

$$\tilde{E}_c[n] = \int d^3r \int_o^1 [\frac{\alpha n(\vec{r})}{a+\alpha bn^{-1/3}(\vec{r})}] d\alpha$$

or

$$\tilde{E}_c[n] = \int d^3r [b^{-1}n^{4/3} + ab^{-2}n^{5/3}\log(\frac{a}{a+bn^{-1/3}})] .$$

IX.   Constrained-Search Derivation of Adiabatic Connection Formula (Coupling Constant Integration Formula)

From the definitions of $\Psi_n^{min,\alpha}$, Ψ_n^{min}, and Φ_n^{min}, it should be clear that

$$E_{xc}[n] + U[n] = \int_o^1 \frac{\partial}{\partial \alpha} <\Psi_n^{min,\alpha}|\hat{T}+\alpha\hat{V}ee|\Psi_n^{min,\alpha}> d\alpha$$

$$(36)$$

so that

$$E_{xc}[n] + U[n] = \int_o^1 [\frac{\partial}{\partial \alpha} <\Psi_n^{min,\alpha'}|\hat{T}+\alpha\hat{V}ee|\Psi_n^{min,\alpha'}>]_{\alpha'=\alpha} d\alpha$$

$$(37)$$

$$+ \int_o^1 [\frac{\partial}{\partial \alpha} <\Psi_n^{min,\alpha}|\hat{T}+\alpha'\hat{V}ee|\Psi_n^{min,\alpha}>]_{\alpha'=\alpha} d\alpha$$

Now, the second term in Eq. (37) is zero because $\Psi_n^{min,\alpha}$ minimizes $<\hat{T} + \alpha\hat{V}ee>$. Hence

$$E_{xc}[n] + U[n] = \int_{o}^{1} \langle \Psi_{n}^{min,\alpha} | \hat{V}ee | \Psi_{n}^{min,\alpha} \rangle d\alpha \quad (38)$$

which is the coupling constant integration formula for exchange-correlation. For correlation alone, combine equations (14), (15), (38) and (30) to form Eq. (29).

TABLE 1:   Correlation energies E_c of closed-shell species. Energies are in atomic units.

Species	P^a	B^b	LYP^c	WL^d	$Expt.^e$
He	-0.044	-0.042	-0.0437	-0.0420	-0.0420
Be^{2+}	-0.049	-0.055	-0.0490	-0.0452	-0.0443
Be	-0.095	-0.092	-0.095	-0.095	-0.094
B^+		-0.107	-0.107	-0.100	-0.111
Ne	-0.395	-0.391	-0.383	-0.383	-0.387
Mg	-0.471	-0.466	-0.459	-0.444	-0.444
Ar	-0.810	-0.785	-0.751	-0.787	-0.787

[a]Perdew's functional (1986).

[b]Becke's functional (1988).

[c]Lee-Yang-Parr functional (1988).

[d]Wilson-Levy functional (1990).

[e]These values refer to the exact $E_c^{QM}[n_v^{HF}]$.

TABLE II.   Correlation energies E_c of open-shell species. Energies are in atomic units.

Species	P[a]	B[b]	LYP[c]	WL[d]	Expt.[e]
H	-0.002	-0.000	-0.000	-0.000	-0.000
Li	-0.054	-0.055	-0.053	-0.046	-0.046
B	-0.130	-0.129	-0.128	-0.129	-0.125
C	-0.168	-0.166	-0.161	-0.160	-0.157
N	-0.206	-0.202	-0.193	-0.188	-0.189
Na	-0.421	-0.419	-0.408	-0.399	-0.398
P	-0.608	-0.590	-0.566	-0.554	-0.553

[a]Perdew's functional (1986)

[b]Becke's functional (1988).

[c]Lee-Yang-Parr functional (1988).

[d]Wilson-Levy functional (1990).

[e]These values refer to the exact $E_c^{QM}[n_v^{HF}]$.

References

Baroni, S., and Tuncel, E., 1983, *J. Chem. Phys.* 79: 6140.

Becke, A. D., 1988, *J. Chem. Phys.* 88: 1053.

Dunlap, B. I., 1987, *Advan. Chem. Phys.* 69:287.

Dunlap, B. I., 1988, *Chemical Physics* 125:89.

Gunnarsson, O., and Lundgvist, B. I., 1976, *Phys. Rev.* B13:4274.

Harris, J., and Jones, R. O., 1974, *J. Phys.* F4: 1170.

Harris, J., 1984, *Phys. Rev. A* 29: 1648.

Harris, R. A., and Pratt, L. R., 1985, *J. Chem. Phys.* 83: 4024.

Hohenberg, P., and Kohn, W. , 1964, *Phys. Rev.* 136: B864.

Kohn, W., and Sham, L. J., 1965, *Phys. Rev.* 140: A 1133.

Langreth, D. C., and Perdew, J. P., 1975, *Solid State Commun.* 17: 1425.

Levy, M., 1979, *Proc. Natl. Acad. Sci. (USA)* 76: 6062; *Bull-Amer. Phys. Soc.* 24:626.

Levy, M., and Perdew, J. P., 1985, *Phys. Rev. A* 32: 2010.

Levy, M., 1985, Lecture given at the "A. J. Coleman Symposium on Density Matrices and Density Functionals", Kingston, Canada, August 1985.

Levy, M., Yang, W., and Parr, R. G., 1985, *J. Chem. Phys.* 83: 2334.

Levy, M., 1987, *Density Matrices and Density Functionals,* (Edited by Erdahl, R. and Smith, Jr., V. H.), Reidel, Boston. (Proceedings of "Coleman Symposium", 1985).

Levy, M., Pathak, R. K. Perdew, J. P., and Wei, S., 1987, *Phys. Rev. A* 36:2491.

Levy, M., 1989, *Int. J. Quantum Chem.* S23: 617.

Levy, M., and Ou-Yang, H., 1990, *Phys. Rev. A* 42:651.

Levy, M., 1990a, *Bull. Amer. Phys. Soc.* 35:822.

Levy, M., 1990b, *Phys. Rev. A*, submitted for publication.

Ou-Yang, H., and Levy, M., 1990, *Phys. Rev. A* 42:651.

Payne, P. W., 1979, *J. Chem. Phys.* 71:490.

Percus, J. K., 1978, *Int. J. Quantum Chem.* 13:89.

Perdew, J. P., 1986, *Phys. Rev. B* 33:8822.

Sahni, V., Gruenbraum, J., and Perdew, J. P., 1982, *Phys. Rev. B* 26:4371.

Vosko, S. J., Wilk, L. and Nusair, M., 1980, *Can. J. Phys.* 58:1200.

Wilson, L. C., and Levy, M., 1990, *Phys. Rev. B* 41:12930.

WIlson, L. C., and Levy, M., unpublished.

13

A Simplified Self-Interaction Correction Method for Covalently Bonded Solids: Application to trans-Polyacetylene

JOSEPH G. HARRISON

I. Introduction

Density functional theory (DFT) presents a practical alternative to wave-function based methods in molecular and solid state systems primarily because of the utility of the local approximation. While the success of this approximation was viewed with considerable skepticism in early applications to atoms and molecules, much has been done in the past two decades to construct a sound theoretical basis for that success (Gunnarsson et al., 1976, 1979; Langreth, et al., 1977; Sahni et al., 1988). Nevertheless, there remains a quite justified concern about how far the local approximation can be pushed. The recently developed gradient corrections (Perdew, 1986; Becke, 1988) to the local density approximation (LDA) and a nonlocal, scale-consistent correlation functional (Wilson et al., 1990) are timely advances which should allow approximate DFT to maintain a marketable advantage in the face of concurrent advances in ab-initio methodology.

From the vantage point of Hartree-Fock (HF) theory, we could characterize the errors incurred in the local approximation as either being self-interaction errors or correlation errors, but such a distinction is untenable because of the lack of transformation invariance of the so-called self-interaction term (Heaton et al., 1985). Alternatively, we may characterize the errors as orbital-approximation errors; the self-interaction error being associated with that departure from a point-particle model which is inherent in the orbital representation and the correlation error being related to the lack of a proper representation of the pair distribution by a single-determinant of orbitals. Of course, the HF method eliminates the spurious self-interaction of an "orbitally-smeared" electron by explicit cancellation of the self-Coulomb and self-exchange terms. From a viewpoint more consistent with DFT, we may follow Perdew and Zunger (Perdew et al., 1981) by formulating the self-interaction (SI) in terms of the orbital densities and the DFT energy functional. The SI error may be expected to account for a substantial part of the local-approximation error. To what extent gradient-corrected self-interaction corrections (GC-SIC) are needed or are even useful is an issue we can not address in this article, although an SIC exchange-correlation functional involving gradient corrections could be formulated in straightforward fashion as outlined by Perdew and Zunger (Perdew et al., 1981) and reviewed below. In this article we simply focus on the implementation of SIC within a standard LDA functional. We will begin with a review of previously developed SIC formalism for atoms and molecules. Then we review the bandstructure-adapted formalism and point out the difficulty in a direct implementation of that formalism in systems with hybrid bands. Then we discuss our simplified method for approximate incorporation based on a Mulliken-weighting scheme. As we will try to show, this method provides a way to incorporate SIC with very minor modifications in any standard energy-band code employing local orbitals such as the the method of linear combinations of atomic orbitals (LCAO) or Gaussian orbitals (LCGTO). Finally, we present our application of this method to trans-polyacetylene.

II. The Self-interaction Correction (SIC)

A. The self-interaction error

We begin with the total energy functional of DFT which we write as (Perdew, et al., 1981)

$$E[\rho\uparrow,\rho\downarrow] = T_0 + V_{ext} + U_c[\rho] + E_{xc}[\rho\uparrow,\rho\downarrow] \qquad (1)$$

where,

T_0 is the noninteracting kinetic energy,

V_{ext} is the external interaction,

U_c is the mutual Coulomb interaction associated with the density, ρ,

E_{xc} is the exchange-correlation (XC) energy.

In the limit of a single electron (described by an orbital density, $\rho_{i\sigma}$), we should have

$$E[\rho_{i\sigma}, 0] = T_0 + V_{ex} \tag{2}$$

since point-electrons do not self-interact.   This then requires that the Coulomb and XC functionals have the property that

$$U_c[\rho_{i\sigma}] + E_{xc}[\rho_{i\sigma}, 0] = 0. \tag{3}$$

In the local spin density (LSD) approximation, we replace the exact XC functional with

$$E_{xc}^{LSD} = \int \rho(r) \; \epsilon_{xc}[\rho\uparrow, \rho\downarrow] d^3r \tag{4}$$

which does not satisfy Eq. (3) in general,

$$U_c[\rho_{i\sigma}] + E_{xc}^{LSD}[\rho_{i\sigma}, 0] \neq 0 \tag{5}$$

B. Manifestations

The spurious self-interaction implied in Eq. (5) manifests itself in a variety of ways in the results of LDA calculations. The orbital eigenvalues in the Kohn-Sham (KS) equations (Kohn et al., 1964) have no approximate interpretation as physical removal energies analogous to the Koopmans' interpretation of the HF orbital eigenvalues. The exception to this is the KS eigenvalue of the highest occupied level, which corresponds to the negative of the first ionization potential for the system.   A related problem is that calculated energy-band gaps are too small by typically 30 - 40%. Negative ions admit of no self-consistent solutions in LDA without the imposition of Watson-sphere potentials or

Latter (Latter, 1955) corrections (Shore, et al., 1977; Schwarz, 1978). Another less obvious manifestation is in the difference in the HF and LDA solutions for the free-electron gas. The vanishing of the density of states at the Fermi level is a well known failure of the HF treatment of this model for a metal. No such pathological behavior is found in the LDA solution. Thus, even in the free electron gas, we find a manifestation of the residual, non-physical self-interaction error of LDA (Harrison, 1987a).

C. Orbital SIC

The orbital SIC of Perdew and Zunger (PZ) is simply and elegantly formulated as an orbital-by-orbital subtraction of the residual SI embodied in Eq. (5). We write the SIC-LSD energy functional as

$$E^{SIC\text{-}LSD} = E^{LSD} + U^{SIC} \qquad , \qquad (6)$$

where E^{LSD} is the LSD energy functional (Eq. (1) and (4)) and

$$U^{SIC} = -\sum_{\sigma} \sum_{i}^{N\sigma} \{ E_{xc}^{LSD}[\rho_{i\sigma}, 0] + U_c[\rho_{i\sigma}] \} \quad . \qquad (7)$$

Quite a number of orbital SIC forms have been proposed in the past (e.g. Cowan, 1967; Lindgren, 1971; also see references in Perdew, et al., 1981), but the PZ form gives a universal prescription for formulating a correction for any approximate energy functional, including those with gradient-correction terms.

A difficulty with Eq. (7) that has been repeatedly noted for all such orbital SIC forms, is the lack of transformation invariance to other orbital sets even for closed-shell systems. In fact, the very observation that the correction expressed in Eq. (7) vanishes for Bloch orbitals was taken as grounds for dismissing such a correction for LDA energy-band calculations (Slater, et al., 1971). Related questions about "size-consistency" were also raised but have since been resolved (Perdew, 1989). PZ suggested the existence of some "optimally localized" orbital set to be used in the evaluation of Eq. (7), but the procedure for obtaining such a set was not spelled out.

D.   The localization equations

Pederson, Heaton and Lin (Pederson et al., 1985) derived a variational procedure for identifying an optimal local-orbital set. The set was optimized in the sense that it minimized the the SIC-LSD energy within norm-conserving variations of the density in terms of unitarily equivalent orbital sets used to construct the density. A useful result from their work was that an explicit set of coupled Schroedinger-like equations for the optimum orbital set was derived along with some auxiliary conditions (localization equations) on the orbitals which serve to uniquely identify the appropriate orbital set from all the unitarily equivalent possibilities. Those equations can be written as

$$(h_\sigma^{LSD} + V_{i\sigma}^{SIC}) \; \Phi_{i\sigma} = \Sigma_j \; \lambda_{ji}^{\sigma} \; \Phi_{j\sigma} \qquad , \qquad (8)$$

$$< \Phi_{i\sigma} \mid (V_{i\sigma}^{SIC} - V_{j\sigma}^{SIC}) \mid \Phi_{j\sigma} > = 0 \qquad , \qquad (9)$$

where h_σ^{LSD} is the LSD Hamiltonian and the SIC potentials are

$$V_{i\sigma}^{SIC} (r) = -\int \rho_{i\sigma}(r') \; |r - r'|^{-1} \; d^3r' - V_{xc}[\rho_{i\sigma},0;r] \qquad (10)$$

Pederson went further to identify a canonical set of orbitals for the SIC-LSD functional which diagonalized the SIC-LSD local-orbital Lagrange multiplier matrix, Λ^σ . These satisfy the equations

$$(h_\sigma^{LSD} + \Delta V_{i\sigma}^{SIC}) \; \Psi_{i\sigma} = \; \epsilon_i^{\sigma} \; \Psi_{i\sigma} \qquad , \qquad (11)$$

where

$$\Delta V_{i\sigma}^{SIC} = \Sigma_j \; M_{ji}^{\sigma} \; V_{j\sigma}^{SIC} \; \Phi_{j\sigma} < \Phi_{j\sigma} \mid \Psi_{i\sigma} >/ \; \Psi_{i\sigma} \} \qquad (12)$$

is the appropriate (non-local) SIC potential for the canonical orbitals derived from the unitary transformation (M_{ji}^{σ}) between the the solutions to Eq. (8-9) and the canonical set. In applications to atoms, the orbital SIC has been shown to greatly improve the accuracy of approximate DFT results for total energies, exchange energies, correlation energies, ionization energies, and electron affinities (Perdew et al., 1981; Harrison, 1983, 1987b; Harrison, et al., 1983).

II. SIC for Energy-Band Structures

A. Crystal orbitals.

We may briefly review the elements of energy-band theory needed to clarify the SIC formalism for this application and to establish our notation for the following sections. A full discussion is not appropriate for this contribution but may be found in any standard solid state physics textbook (e.g. Ashcroft et al., 1976; Ziman, 1972). We take as our crystal orbitals the solutions to the KS equations expanded in Bloch sums, $|b_{ik}^{\mu}>$, as

$$h^{LSD} \ | \ \Psi_{nk}> = \epsilon_n(k) \ | \ \Psi_{nk}> \tag{13}$$

$$| \ \Psi_{nk}> = \Sigma_{\mu} \ \Sigma_i \ c_{in}^{\mu}(k) \, |b_{ik}^{\mu}> \tag{14}$$

where μ covers the atoms in the unit cell (crystal basis), i goes from 1 to M_{μ}, the number of basis functions for atom μ, and the Bloch sum is given by

$$|b_{ik}^{\mu}> \ = \Sigma_{\alpha} \ \exp(i \ k \cdot R_{\alpha}) \ \Phi_{i\mu}(r - R_{\alpha} - t_{\mu}) \tag{15}$$

where $\Phi_{i\mu}$ is an atomic or gaussian-type orbital in the usual implementation (Heaton et al., 1980).

Since the reader may very well have his formal training in quantum chemistry rather than solid state physics, we will digress slightly to give a brief, qualitative discussion of the symbol "k" used in the above expressions. k can be simply viewed as a translational-symmetry label, and its significance can be understood by analogy with other symmetry labels such as those found in atomic systems. For a spherical atom, we have as a consequence of $[L^2 \ , \ H] = 0$ and $[L_z \ , \ H] = 0$, that ℓ and m_{ℓ} are good symmetry labels, and we may use them to label our states in terms of energies ($\epsilon_{n\ell}$) and wave functions ($\Psi_{n\ell m}$). In perfect crystals we have symmetry operators, T_{α}, corresponding to the translational symmetry of these systems which commute with H ($[T_{\alpha}, \ H] = 0$). We ignore any additional point-group symmetry in the full space group for this discussion. Our symbol k thus is our good symmetry label in analogy with ℓ and m_{ℓ}, and we may likewise use it to label our states in terms of energies (or energy bands, $\epsilon_n(k)$) and wave functions ($|\Psi_{nk}>$). In the discussion to follow, we will be examining energy-band structures, so a simple analogy to a similar display of information in atomic systems may be helpful. In displaying the range of orbital energies corresponding to bound states in atomic systems, we could simply draw short horizontal lines spaced according to some chosen energy scale as sketched in Fig. 1a. A more useful way to display this information would be to make separate columnar arrays of these lines, each column corresponding to a value of the quantum number, ℓ, as sketched in Fig. 1b. In close analogy to this we have the alternative ways to represent the energy bands in solids sketched in Fig. 2. The band structure gives more information about the relative energies of states with the same translational symmetry label, k.

(a)   (b)

Fig. 1   Atomic Energy-Level Diagram
         with (b) and without (a) symmetry information

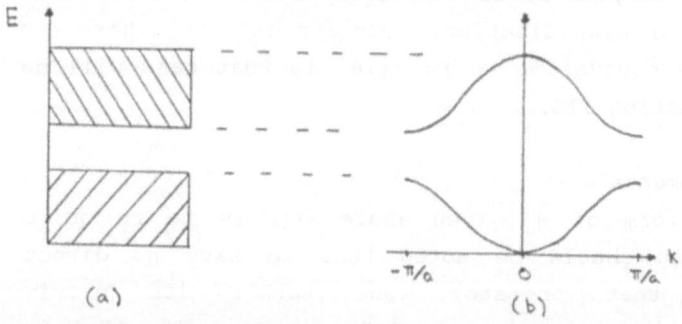

Fig. 2   Energy Bands
         with (b) and without (a) symmetry information

B. Unified Hamiltonian, H_u

The formalism for incorporating the SIC for energy-band calculations has been given elsewhere (Heaton et al., 1983) and we refer the reader to that article for further details. We summarize the relevant equations below in terms of the projection operators P_{nk}, O_k, and the unified Hamiltonian, H_u, which satisfies KS-like equations of the same formal structure as Eq. 13:

$$P_{nk} \equiv |\Psi_{nk}\rangle\langle\Psi_{nk}| \quad , \quad O_k \equiv 1 - \Sigma_{nocc} \, P_{nk} \tag{16}$$

$$H_u = h_0 - \Sigma_{n,m}' \, P_{nk} \, h_0 \, P_{mk} + \Sigma_n \, P_{nk} \, \Delta V_{nk} \, P_{nk}$$

$$+ \Sigma_n \, [\, O_k \, \Delta V_{nk} \, P_{nk} + P_{nk} \, \Delta V_{nk} \, O_k \,] \quad , \tag{17}$$

where h_0 is the LSD Hamiltonian, the prime indicates omission of m=n, all sums run up to the highest occupied band, and

$$\Delta V_{nk} = N^{-\frac{1}{2}} \, \Sigma_\mu \, \Sigma_i^{M\mu} \, C_{in}^{\ \mu}(k) \, \Sigma_\alpha \, \exp(i \, k \cdot R_\alpha) \, V_{n\mu}^{SIC}(r-R_\alpha-t_\mu)$$

$$\times \, \Phi_{i\mu}(r-R_\alpha-t_\mu)/ \, \Psi_{nk} \quad , \tag{18}$$

In Eq. (18) the summation over the basis set for each atom ($i=1,\ldots,M_\mu$) was introduced (Heaton et al., 1983) to render a local orbital whose density (sum over k) exhibited the same degree of localization as a true Wannier density which would be much more difficult to generate. As noted by Heaton et al., this procedure becomes more complicated and less acceptable for complex bands especially for bands exhibiting a high degree of hybridization. Our approximate scheme for generating the equivalent to Eq. (18) in that case will be outlined in section IID.

C. Matrix elements

Although the form of H_u given above appears to be quite complicated, it should be noted that we have no direct interest in that operator, but rather its matrix representation in our basis set of Bloch sums since that is the key to finding solutions in the finite-basis method. Those matrix elements are in fact quite simple and we summarize the relevant expressions needed to set up the secular equation for any standard eigensystem processing code.

The basic unified-Hamiltonian matrix element may be written as

$$< b_{\alpha k}^{\tau} \mid H_u(k) \mid b_{\beta k}^{\mu} > = H_{\alpha\beta}^{\tau\mu} \qquad . \qquad (19)$$

We first note that the effect of the projection operators on one of the Bloch sums can be written as

$$P_{nk} \mid b_{\beta k}^{\mu} > = S_{\beta n}^{\mu} \mid \Psi_{nk} > \qquad , \qquad (20)$$

$$O_k \mid b_{\beta k}^{\mu} > = \mid b_{\beta k}^{\mu} > - \Sigma_n S_{\beta n}^{\mu} \mid \Psi_{nk} > \qquad , \qquad (21)$$

where

$$S_{\beta n}^{\mu}(k) \equiv < \Psi_{nk} \mid b_{\beta k}^{\mu} > \qquad . \qquad (22)$$

We can thus reduce $H_{\alpha\beta}^{\tau\mu}$ to a combination of three basic types of integrals:

$$< b_{\alpha k}^{\tau} \mid h_o(k) \mid b_{\beta k}^{\mu} > \qquad \{ \text{ usual LSD-LCAO integrals} \}$$

$$< b_{\alpha k}^{\tau} \mid \Delta V_{ik} \mid \Psi_{ik} > \qquad \{ \text{ reduce to 3-center integrals} $$

$$< \Psi_{ik} \mid \Delta V_{ik} \mid b_{\beta k}^{\mu} > \qquad \text{(Heaton et al., 1983)} \} \qquad .$$

We may write the unified Hamiltonian in matrix form by first introducing some auxiliary matrices that are normally calculated in standard LSD-LCAO energy-band codes. For a basis set of M Bloch sums they are the MxM matrices

$$S_{\alpha\beta}^{\tau\mu} = < b_{\alpha k}^{\tau} \mid b_{\beta k}^{\mu} > \qquad \text{overlap matrix}$$

$$H_{\alpha\beta}^{0} = < b_{\alpha k}^{\tau} \mid h_o(k) \mid b_{\beta k}^{\mu} > \qquad \text{LSD Hamiltonian matrix}$$

$$c^f = [C_1 \, C_2 \, \cdots \, C_M] \qquad \text{matrix of eigenvectors.}$$

From these we derive the following matrices

$$c = [C_1 \, C_2 \, \cdots \, C_{nocc}] \qquad \text{first nocc columns of } c^f$$

$$D = C^+ S \qquad \text{(cf. Eq. (20))}$$

$$F^+ = [\ F_1^+\ F_2^+\ \ldots\ F_M^+\ ] \qquad F_{an} = <\ b_{ak}^{\tau}|\ \Delta V_{nk}\ |\ \Psi_{nk}>$$

$$A = C^+ F \qquad A_{ij} = <\ \Psi_{ik}|\ \Delta V_{jk}\ |\ \Psi_{jk}\ >.$$

Now, defining two matrices

$$K^0 \equiv C^+ H^0 C \quad \text{and} \quad E_{ij} \equiv (A_{ii} + K^0_{ii})\ \delta_{ij}$$

we may finally write the matrix representation of the unified Hamiltonian in the Bloch-sum basis as

$$H_u = H^0 - D^+[K^0 + A + A^+ - E]D + F D + D^+ F^+ \qquad (23)$$

D. Mulliken-weighted SIC

We recall first the correct expression for the SIC potential based on Wannier functions (WF), $w_{n\mu}(r-R_\alpha)$ (Heaton et al., 1983)

$$\Delta V_{nk} = \Sigma_\mu\ \Sigma_\alpha\ V_{n\mu}^{SIC}(r - R_\alpha - t_\mu)\ \exp(i\ k \cdot R_\alpha)\ w_{n\mu\alpha}\ /\ \Psi_{nk}. \qquad (24)$$

Although the rigorous application of this form would appear to be intractable because of the difficulty of generating WF, it should be noted that a variational method for generating WF has been introduced by Pederson and Lin (Pederson et al., 1987) which may in fact may such a procedure practical in the near future. We base our simplified method on Eq. (24) and replace the WF with k-dependent local orbitals given by

$$u_{nk}^\mu(r-R_\alpha-t_\mu) = \Sigma_i\ C_{in}^\mu(k)\ \Phi_{i\mu}(r-R_\alpha-t_\mu) \qquad (25)$$

(Heaton et al., 1983) and replace the orbital SIC potential with the approximate form

$$V_{n\mu}^{SIC} = \Sigma i\ N_{in}^\mu\ V_i^{SIC}\ (atom) \qquad (26)$$

where N_{in}^μ is the Mulliken population of atomic orbital i of atom-species μ for state nk. We obtain the populations by diagonalization of the Hamiltonian in a minimal basis and use the SIC potential derived from the atomic orbitals for each atom in our basis. The use of Eq. (25 - 26) simplifies the SIC calculation tremendously in that no explicit generation of Wannier densities is carried out. While we use fixed atomic SIC potentials in Eq. (26), the Mulliken populations fold in some hybridization effects. This could in fact be relaxed with some attendant complication by carrying out a local decomposition of the crystal potential and using those local potentials to generate a new set of "atomic" orbitals for use in generating the atomic SIC potentials. We did not explore that possibility in this work. It should be noted that this simplified method for incorporating SIC requires only minor modifications of an existing LCAO band structure code. Finally, we note one satisfying property of our Mulliken-weighted scheme; namely, that it correctly represents the dissociation limit as that of SIC-LDA atoms. In applying this method to trans-polyacetylene, we were careful to maintain the angular features of the atomic SIC potentials, i.e. we maintained different SIC potentials for the densities derived from the carbon $2p_x$, $2p_y$, and $2p_z$ orbitals. The Coulomb part of the SIC potential for these orbitals can be shown to reduce to the common form

$$V_c^\ell(r) = P_\ell \, h(r) + q(r) \tag{27}$$

where P_ℓ is just x^2, y^2, or z^2, and $h(r)$ and $q(r)$ are spherical functions derived from radial integrations of the orbital densities. These functions are displayed in Fig. 6 along with the SIC potentials for the 1s and 2s carbon orbitals.

III. Application to trans-Polyacetylene (CH_x)

A. Equilibrium LDA structure
Trans-CH_x has been the subject of quite a number of HF and LDA studies (e.g. Mintmire et al, 1983, 1987; Springborn, 1986 and references therein). In the LDA results there remains some questions regarding the equilibrium structure which we will attempt to summarize. It should be noted, however, that our application to this system did not have as a primary aim the

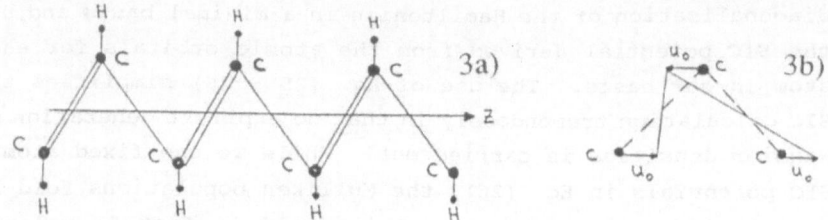

Fig. 3   Structure of Trans-CH$_x$
         (a) Bond-alternate structure   (b) dimerization coordinate, u$_0$

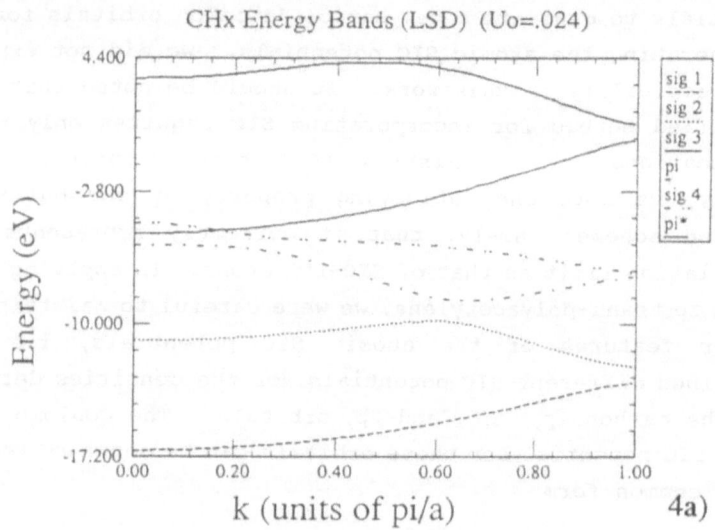

4a)

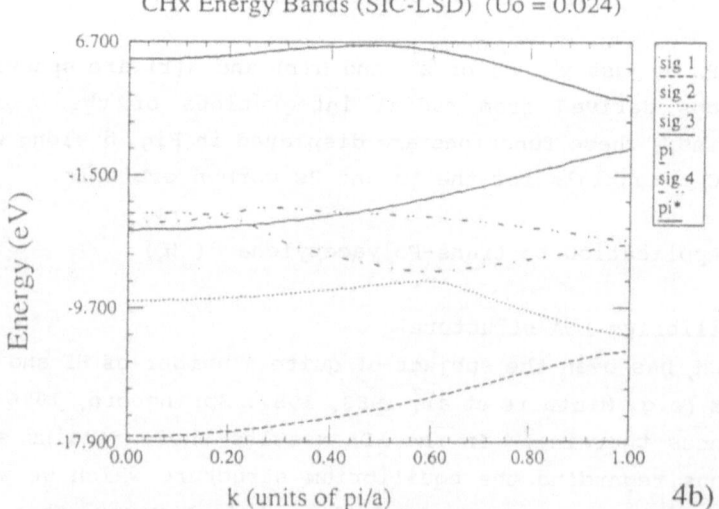

4b)

Fig. 4   Energy Bands of Trans-CH$_x$ (u$_0$ = 0.024 Å)
         (a) LSD   (b) SIC-LSD

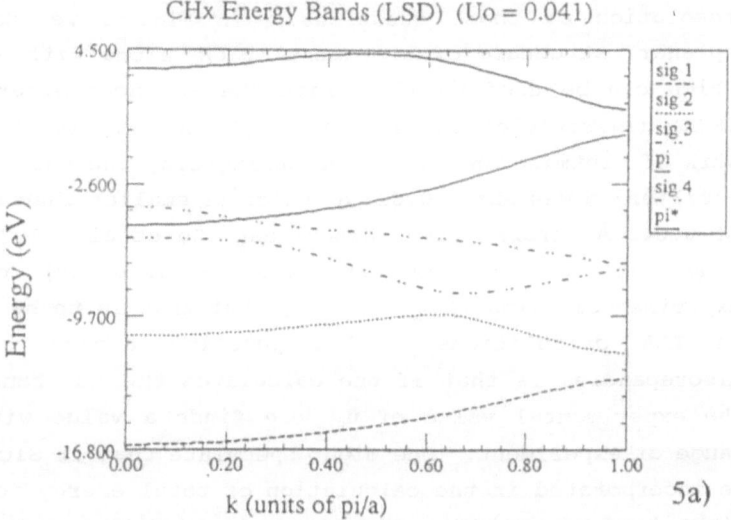

5a)

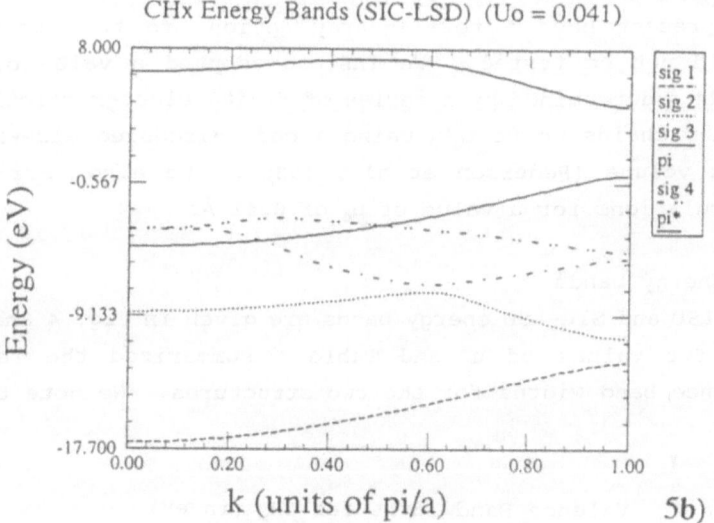

5b)

Fig. 5 Energy Bands of Trans-CH$_x$ (u$_0$ = 0.041 Å)
         (a) LSD   (b) SIC-LSD

resolution of those questions.   In Fig. 3 we sketch the (planar) structure of a segment of CH_x along with a diagram which can be used to illustrate the Su, Schrieffer, Heeger (SSH) dimerization coordinate, u_0 (Su et al., 1979).   In the work of Mintmire and White and Springborn, the value of u_0 at equilibrium was about 0.025 Å, which is smaller than the value of 0.042 Å inferred from experiment (Su et al., 1979).   The band gap is also smaller (0.96 - 0.26 eV) than the experimental value (1.4 - 1.8 eV), but that is to be expected in LDA calculations.   The puzzling aspect of these discrepancies is that if one calculates the LDA band gap at the experimental value of u_0, one finds a value within the range of experiment.   One might speculate that if SIC were to be incorporated in the calculation of total energy for CH_x it might lead to a different value of u_0 as well as resolve the band gap problem.   The present codes used in this work do not at present have a total-energy option, so this conjecture could not be tested.   We instead adopted a value of u_0 of 0.024 Å determined by a series of finite-cluster calculations on CH_x chains up to C_8H_8 using a code discussed elsewhere in this volume (Pederson et al., 1988).   We also carried out calculations for a value of u_0 of 0.41 Å.

B. Energy bands

The LSD and SIC-LSD energy bands are given in Fig. 4 and 5 for the two values of u_0 and Table 1 summarized the relevant valence band widths for the two structures.   We note that σ

Table 1.   Valence Bandwidths for CH_x (in eV)

u_0	0.024 Å		0.041 Å	
band	LSD	SIC-LSD	LSD	SIC-LSD
σ_1	3.67	5.25	3.51	5.17
σ_2	2.45	3.29	2.10	3.02
π	5.03	4.90	4.82	4.65
Total	16.8	17.6	16.5	17.3

band widths increase with the SIC while the π band width decreases. Also the relative position of the upper σ and π bands near k=0 changes. It would be of interest to compare the SIC-LSD band with HF results for these values of u_0 but we are not aware of any in the literature. They do seem to follow the trend toward larger valence band widths found in HF calculations in general.

The question of energy band gap poses an interesting problem of conduction band models which we have discussed in some earlier work (Heaton et al., 1983). If we define the band gap in terms of a model in which we account for the energy required to ionize from the top of the valence band (TVB) and the energy gained in adding an electron to the bottom of the conduction band (BCB), assuming a completely delocalized hole, then a reasonable way to calculate the gap as an eigenvalue difference would be to take the negative of the SIC-LSD TVB eigenvalue and add to it the eigenvalue for the BCB derived from an LDA calculation, $\epsilon_{BCB} - \epsilon_{TVB}$, i.e., apply no SIC to the conduction band states. We refer to this as the electron affinity (EA) model. Alternatively, we could adopt an excitonic (Reynolds et al., 1981) model or improved-virtual-orbital (IVO) method (Hunt et al., 1969) for dealing with the conduction band. In either approach, we solve for the conduction band states for an N-1 electron Hamiltonian. We may approximate this by the SIC-LSD Hamiltonian for the valence-band states, using a procedure for modifying the unified Hamiltonian discussed in earlier work (Harrison et al., 1983; Heaton et al., 1983).

We find that in the EA model, we get band gaps that are too large for either value of u_0 (3.6 and 4.4 eV for u_0 values of 0.024 Å and 0.041 Å, respectively). The corresponding LSD gaps are 0.8 and 1.3 eV. If we adopt the IVO model we find gaps very close to the LDA values (0.8 and 1.4 eV, respectively). Here the implication may be that the IVO model assumes too strong a localization for the hole in using the SIC-LSD Hamiltonian for the valence band states. It would be interesting to devise a model excited-state Hamiltonian based on some other measure of localization, e.g., the localization of solitons in this lattice. On the other hand, we have not investigated the possibility that the larger value of u_0 may in fact be the equilibrium value in SIC-LSD. If so, the IVO band gap is in good agreement with experiment, as is the LSD value, but the equilibrium u_0 in LSD is much smaller. Thus further resolution of the band gap question must await a total-energy study using SIC-LSD.

SIC Atomic Potentials (C-A Correlation)

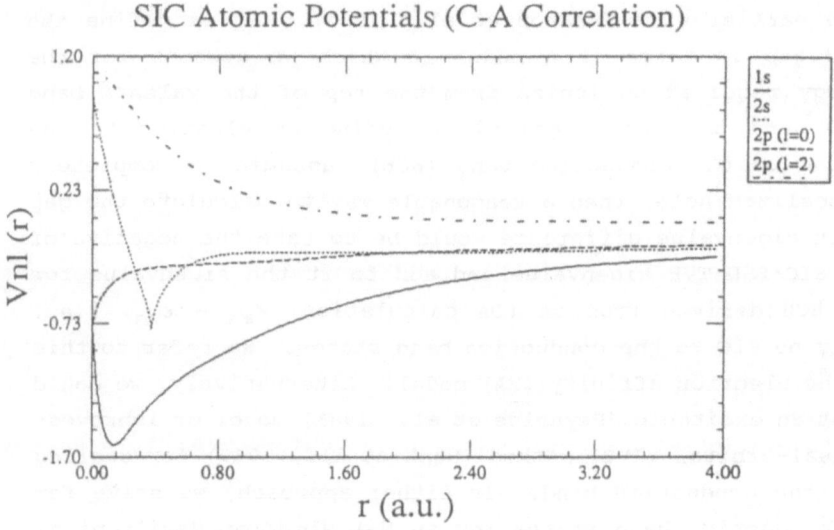

Fig. 6   SIC Atomic Potentials (C-A Correlation).

References

Ashcroft, N. W., and Mermin, N. D., 1976, Solid State Physics Holt, Rinehart and Winston, New York.
Becke, A. D., 1988, J. Chem. Phys. 88:1053.
Cowan, R. D., 1967, Phys. Rev. 163:54.
Gunnarsson, O., and Lundqvist, B. I., 1976, Phys. Rev. B 13:4274.
Gunnarsson, O., Jonson, M., and Lundqvist, B. I., 1979, Phys. Rev. B 20:3136.
Harrison, J. G., 1983, J. Chem. Phys. 78:4562.
Harrison, J. G., Heaton, R. A., and Lin, C. C., 1983, J. Phys. B 16:2079.

Harrison, J. G., 1987a, Phys. Rev. B 35:2273.
Harrison, J. G., 1987b, J. Chem. Phys. 86:2849.
Heaton, R. A., and Lin, C. C., 1980, Phys. Rev. B 22:3629.
Heaton, R. A., Harrison, J. G., and Lin, C. C., 1983, Phys. Rev. B 28:5992.
Heaton, R. A., Harrison, J. G., and Lin, C. C., 1985, Phys. Rev. B 31:1077.
Kohn, W., and Sham, L. J., 1964, Phys. Rev. 140:A1133.
Langreth, D. C., and Perdew, J. P., 1977, Phys. Rev. B 15: 2884.
Latter, R., 1955, Phys. Rev. 99:510.
Lindgren, I., 1971, Int. J. Quantum Chem. 5:411.
Mintmire, J. W., and White, C. T., 1983, Phys. Rev. B 28: 3283.
Mintmire, J. W., and White, C. T., 1987, Phys. Rev. B 36: 4180.
Pederson, M. R., Heaton, R. A., and Lin, C. C., 1985, J. Chem. Phys. 82:2688.
Pederson, M. R., Klein, B. M., and Broughton, J. Q., 1988, Phys. Rev. B 38:3825; also, see article by Pederson in this volume.
Perdew, J. P., and Zunger, A., 1981, Phys. Rev. B 23:5048.
Perdew, J. P., 1986, Phys. Rev. B 33:8822.
Perdew, J. P., 1989, in Density Functional Theory of Many-Fermion Systems, (Edited by Trickey, S. B.). Advances in Quant. Chem., Academic Press, New York.
Sahni, V., Bohnen, K.-P., and Harbola, M. K., 1988, Phys. Rev. A 37:1895.
Schwarz, K., 1978, Chem. Phys. Lett. 57:605.
Shore, H. B., Rose, J. H. and Zaremba, E., 1977, Phys. Rev. B 15:2858.
Slater, J. C., and Wood, J. H., 1971, Int. J. Quantum Chem. 4 :3.
Springborn, M., 1986, Phys. Rev. B 33:8475.
Su, W. P., Schrieffer, J. R., and Heeger, A. J., 1979, Phys. Rev. Lett. 42:1698.
Wilson, L. C., and Levy, M., 1990, Phys. Rev. B (in press).
Ziman, J. M., 1972, Principles of the Theory of Solids, Cambridge University Press, New York.

14
Correlation Contributions from Density Functionals

Andreas Savin

Dedicated to Prof. Dr. H. Preuss on his 65th birthday.

Introduction

How is it possible to improve a density functional calculation? One way is to refine the density functional. This route has already been pointed out by Hohenberg and Kohn (1964) and much progress has been made along this line (see, e.g., the contributions of Becke, Levy, or Parr to this volume). On the other hand, it is well-known that wavefunctions can be improved systematically by increasing the number of Slater determinants used. Several attempts have been made to couple the latter approach with density functional theory. The purpose of this paper is to describe such couplings.

Correlation energy density functionals

We will consider first the case of adding a density functional (DF) to the energy obtained with a one-determinant wavefunction (Hartree-Fock, HF).

The foundation for such an approach was put forward by Kohn and Sham (1965). The simplest explanation was given by Levy (1979) in the following way. The ground state energy DF is given by:

$$E_0[\rho] = F_0[\rho] + \int v\rho \tag{1}$$

where ρ is the density, v the external potential and F_0 the universal DF defined by

$$F_0[\rho] = \min_{\Psi \to \rho} \langle \Psi | T + V_{ee} | \Psi \rangle \tag{2}$$

(Ψ is a wavefunction giving ρ, T and V_{ee} are the operators for the kinetic energy and the interelectronic interaction, respectively). If Ψ is restricted to one-determinant wavefunctions, then the equation obtained in analogy to (1) is:

$$E_{HF}[\rho] = F_{HF}[\rho] + \int v\rho \tag{3}$$

The difference between $E_0[\rho]$ and $E_{HF}[\rho]$ is a universal functional of the density and can be used to define a correlation energy DF:

$$E_c[\rho] = F_0[\rho] - F_{HF}[\rho] \tag{4}$$

Other definitions are possible, too (see, e.g., Levy, 1987). The usual definition of the correlation energy (Löwdin, 1959; Wigner, 1934) , $E_0[\rho_0] - E_{HF}[\rho_{HF}]$, gives values which lie between $E_c[\rho_{HF}]$ and $E_c[\rho_0]$ (ρ_0 is the exact ground-state density, ρ_{HF} the HF ground-state density; Savin et al., 1986). In general, only $E_c[\rho_{HF}]$ is computed: the errors made in the approximation of $E_c[\rho]$ are larger than the difference between $E_c[\rho_0]$ and $E_c[\rho_{HF}]$.

In practice, an expression for $E_c[\rho]$ is needed. The simplest approach is to make the local approximation:

$$E_c[\rho] \approx \int \rho \, \varepsilon\big(\rho(r)\big) \, d^3r \tag{5}$$

Here ε is an as yet undetermined function. The universality of $E_c[\rho]$ suggests that ε could be determined once and for all in a different type of calculation in some reference system, and then be used for all other systems. The most natural choice for determining ε is the homogeneous electron gas. The most widely used analytical expression describing $\varepsilon(\rho)$ is due to Vosko, et al. (1980). It gives differences smaller than 1 per cent with the Monte Carlo calculations of Ceperley and Alder (1980) and has the correct asymptotic behaviour for $\rho \to 0$ and $\rho \to \infty$.

As mentioned in the introduction, it is possible to refine the density functional. An important improvement is due to the introduction of a supplementary dependence into ε, namely on the spin polarization ($\zeta=(\rho_\uparrow-\rho_\downarrow)/\rho$; $\rho_\uparrow$ and $\rho_\downarrow$ are the spin-up and spin-down densities, respectively; Stoddard and March, 1971; von Barth and Hedin, 1972; Pant and Rajagopal, 1972). A further correction is the introduction of the dependence on the gradient of the density (Langreth and Mehl, 1981,1983; Hu and Langreth, 1985; Perdew, 1986). Some other extensions of the local approximation are quoted in Stoll and Savin, 1985. More recent functionals are those of Becke, 1988; Lee, et al., 1988; and Wilson and Levy,1990.

The effect of these corrections is illustrated with the values calculated for the dissociation energies of two molecules (in eV):

Method	Li_2	F_2
Hartree-Fock	0.2	-1.1
+ local DF	0.5	-0.8
+ local DF + spin-polarization	1.0	-0.5
+ local DF + spin-polarization + gradient correction (Perdew)	0.9	-0.1
exact (Huber and Herzberg, 1979)	1.1	1.7

Two trends appear in these data:
- the HF results are improved with correlation energy DF;
- the improvement may not be sufficient.

These features are documented by a large amount of data - mostly atomic ionization potentials and electron affinities (see, e.g., Lagowski and Vosko, 1988; Savin et al.,1983) or molecular dissociation energies (see, e.g., Clementi et al., 1989; Miehlich et al., 1989; Moscardo, et al., 1989). Many applications stress the importance of the first trend. They range from the prediction of the stable negative ions of the alkaline-earth-metal atoms (see, e.g., Froese Fischer et al., 1987) to the discussion of the f-occupancy in cerocene (Dolg et al., 1990). A long-time domain of application has been that of clusters (see, e.g., Flad et al.,1984; Savin et al., 1988; Fantucci et al., 1990). Correlation energy density functionals have also been applied for correcting solid-state HF calculations (see, e.g., Causa et al., 1987) or for the construction of water-water interaction potentials used for the Monte-Carlo simulation of liquid water (Caravetta and Clementi, 1984).

Theoretical justification for a coupling of CI with DF

It has sometimes been suggested that the correlation energy be redefined (see, e.g., Clementi,1965). This comes from the observation that often a few ('near-degenerate') configurations have important energy contributions. Well-known examples are the Be-series or the H_2 molecule at large inter-atomic distances. It is not a problem to deal with a few more configurations in a wavefunction treatment. One should have, however, a definition which permits the use of a density functional which does not include the correlation energy already introduced by the multi-determinant wavefunction in the configuration interaction (CI) calculation.

The most transparent way to define a DF for a part of the correlation energy closely follows equations (1)-(4). There $F_{HF}[\rho]$ was defined by restricting Ψ to one-determinant wavefunctions. If Ψ is restricted to the set of wavefunctions which can be generated within a well-defined set of orbitals, then another universal DF can be generated: $F_r[\rho]$. (F_{HF} is a special case of F_r). The correlation energy which can*not* be generated within the chosen set of orbitals is, in analogy to (4):

$$E_{c,r}[\rho] = E_0[\rho] - E_r[\rho] = F_0[\rho] - F_r[\rho] \qquad (6)$$

Here $E_r[\rho] = F_r[\rho] + \int v\rho$. $E_{c,r}$ depends not only on ρ, but also on the chosen set of orbitals.

In order to make use of such a definition the set of orbitals has to be defined. Many choices are possible. Although energy-optimized

(Multi-Configuration Self-Consistent-Field, MCSCF) orbitals were already used with success (Miehlich, 1990; Miehlich, et al.,1990) this paper will emphasize the use of natural orbitals (NO: they diagonalize the first-order density matrix, γ; the eigenvalues are their occupation numbers, v_i; Löwdin, 1955). A density matrix can be generated - in the present theoretical formulation - by Ψ minimizing $F_0[\rho]$. (Another possibility would be to use functionals of γ; Levy, 1979).

One reason for using natural orbitals is related to the experience showing that NO with large v_i describe important, molecule-specific effects. Another reason is seen by analysing the homogeneous electron gas, where the NOs are plane waves (Davidson, 1972): with small v_i they have large momentum and are used to describe short-range effects. It thus seems natural to include orbitals with large v_i in the CI calculation for E_r, while a density functional (local approximation) might work for $E_{c,r}$.

The local characterization of the separation into the two sets of orbitals (one for E_r the other for $E_{c,r}$) is still not specified. Here there are several possibilities too, and only more theoretical work and more numerical experience can show the best choice. For example, one could select the largest contribution to ρ of a NO not included in E_r, $v_i|\varphi_i|^2$, and take the corresponding $v_i \equiv v$ as an indicator of the separation. With this definition of the orbital sets, the symbol $E_{c,v}$ will be used instead of $E_{c,r}$.

In order to use equation (6) an approximation for $E_{c,r}$ has to be found. It is hoped that a local approximation would work better if near-degeneracy effects are removed by including them in E_r.

Approximations in the coupling of DF with CI

The purpose of the present section is to show how a local approximation is generated for $E_{c,v}$ and to how molecular calculations are performed.

In analogy to equation (5),

$$E_{c,\nu}[\rho] \approx \int \rho \ \varepsilon(\rho) \ \varphi(\rho,\nu) \qquad (7)$$

where ε has retained the meaning of equation (5) and the function φ has to be determined. As usual, a homogeneous electron gas calculation is used to this end. The total correlation energy per particle is known (ε); the contribution of the plane waves with occupation number larger than ν (momentum smaller than k_ν) to the correlation energy per particle (ε_r) must be computed; the difference $\varepsilon - \varepsilon_r$ is $\varepsilon\varphi$.

In order to obtain k_ν, the momentum distribution of Pajanne and Arponnen (1982) has been used. A comparison made by these authors with unpublished Monte-Carlo data of D. Ceperley underlines its reliability. For ε_r the coupled cluster calculation of Freeman (1977) was followed. It gives errors of a few mhartree in ε. Finally φ was obtained from $1 - \varepsilon_r/\varepsilon'$, where ε' is the value for ε obtained by Freeman. A simple analytic formula which fits the comuted values for φ acceptably is:

$$\varphi(\rho,\nu) \approx (\nu/\nu_1)^{0.329} \qquad (\nu < \nu_1) \qquad (8)$$

where $\nu_1 = (1. + 8.45/r_s)^{-1}$, $r_s = [3/(4\pi\rho)]^{1/3}$ and atomic units are used. With this formula the errors in $\varepsilon'\varphi$ are less than 2 mhartree for $0.2 \leq r_s \leq 10$ and $\nu > 0.0001$. ν_1 approximates ν for k_ν approaching the Fermi momentum. As $\varepsilon\varphi$ should not be larger than ε, one can use $\varphi = 1$ for $\nu > \nu_1$.

For molecular calculations natural orbitals have to be produced. Presently, CI calculations are used to this end, but there is reason to hope that simpler methods could be used. Afterwards, a decision has to be made about the space to be treated in a wavefunction calculation (to give E_r). Often chemical intuition helps to detect near-degeneracy; experience shows that $\nu \approx 0.01$ can be recommended for neutral systems. A good test for the choice seems to be the function φ itself. The integral over the density in the region of space where ν is larger than ν_1 (i.e., larger than permitted by the density in the corresponding homogeneous electron gas) should be zero if equation (7) is applied.

This integral gives a 'number' of electrons which are not properly described within the local approximation. For $E_{c,\nu}$ equations (7) and (8)
are used.

The end of this section will be devoted to a few remarks on the
approximate procedure presented here.

1. The method does not contain any empirical parameter.
2. $E_{c,\nu}$ does not depend on the spin polarization.
3. A system consisting of isolated electrons has no correlation energy.
4. For $\nu \to 0 \Rightarrow E_{c,\nu} \to 0$, and thus the *exact* correlation energy is obtained
   in principle (through E_r).
5. Approximations are made in practice, for E_r, too: limitations in the
   basis sets and classes of excitations considered. Thus, results may
   become worse if ν is reduced to much.
6. With a local DF the energy is still bounded from below. The bound
   may differ from the exact energy.

Different couplings of DF with CI

The first contribution to a combined DF and CI method is due to Lie
and Clementi (1974). They indicated that chemical intuition is not sufficient for choosing the relevant determinants for CI and faced the
problem of the double counting of the correlation energy. (With their
procedure a DF contribution is added to the CI energy, even if the latter is practically exact.)

Colle and Salvetti (1979, 1983) have described a procedure which
ensures that no density functional contribution is added to the correlation energy when the wavefunction is exact. This was achieved by
using the behaviour of the two-particle density matrix for short interelectronic distances. This kind of approach is appealing and several
publications exist which document its good quality (Colle and Salvetti,
1979; Amaral and McWeeny, 1984; Montagnani et al., 1984; Amaral, 1985;
Moscardo et al., 1989; San-Fabian, et al., 1990).

The derivation of the approach of Colle and Salvetti has, however, been criticised. They assume that the density functional is simulating the behaviour of a wavefunction with correlating factor. This ansatz has been shown to give very poor results in a variational Monte-Carlo calculation (Moskowitz et al., 1982): 3 per cent of the correlation energy of the Be atom (the DF gives nearly 100 per cent). A different derivation is given by Cohen et al., 1980, but it assumes that the Hartree-Fock first-order density *matrix*, γ, is a good approximation to the exact one. Furthermore, with the DF of Colle and Salvetti no correlation energy is obtained for a fully polarized system. As these difficulties will certainly be overcome in the future it may be concluded the coupling scheme of Colle and Salvetti deserves much more attention than it has received up to now.

Several other attempts to couple CI with density functionals were less successful. Savin et al. (1984) tried to define pair energies with DF but the differences between different possible definitions often exceeded 0.01 hartree.

Fritsche (1986) has suggested to use Kohn-Sham orbitals for CI calculations, but the short-range effects present in the DF are thus lost.

Stoll and Savin (1985) have suggested the use of a modified two-particle interaction in the wavefunction calculation, and then correcting it by a density functional. The modified electron-electron interaction should have a local character, and this is inconvenient in CI calculations.

Roos et al. (1987) have used the idea of a modified Hamiltonian which depends on the averaged density and the second-order density matrix, following Colle and Salvetti. The authors conclude that the Hamiltonian should depend locally on the density, but that this is not feasible in practice.

The scheme of Ziegler et al. (1977) to calculate multiplet splittings in the Hartree-Fock-Slater method, has been applied to correlation energy density functionals (Stoll and Savin, 1985). This corresponds to a shift of the diagonal elements of the Hamiltonian matrix with density functional contributions. Like the new method of Colle and Salvetti (1990), which also adds DF contributions to the non-diagonal elements of the Hamiltonian matrix, it does not allow for a proper balance of the CI and DF contributions.

Good results have been obtained using variants of the procedure presented in the preceding section. It is possible to define a global ν (Savin, 1988), for example by assuming that all $|\varphi_i|^2$ have similar values. This is reasonable as long as the NO are used to describe correlation in a given region of space. (Usually ν is much larger for the valence region than for the core region.) To compensate for this approximation a gradient correction (following Perdew, 1986) was necessary in equation (7). Using a global ν reduces the sensitivity to the quality of the NOs. Results with this approach are published (Savin, 1988,1989) and some will be presented later.

Another possibility for $E_{c,r}$ is to use only the information needed for obtaining E_r. This is convenient when MCSCF orbitals are used. Such an approach has been tested (Miehlich, 1990; Miehlich et al., 1990) and found to give accurate results when a gradient correction is included. The results could be further improved by using the two-particle density matrix.

Results obtained by coupling DF with CI

In many cases near-degeneracy effects are not important. In these cases the correlation energy is expected to be represented well by a DF alone. $E_{c,\nu}$ will in general be smaller than the correlation energy in the usual local approximation, because in these cases ν is expected to be smaller than v_1. A typical example is given by the *He series*, where the local approximation gives correlation energies which increase logarithmically (Perdew, et al., 1981). Calculated $E_{c,v}$ values (up to Ne^{8+}) are around 1 eV, as they should be, while the usual local approximation gives 3 eV for He and 5.5 eV for Ne^{8+}.

A typical case where near-degeneracy is present is that of the *Be series* (Linderberg and Shull, 1960). The nearly linear increase of correlation energy with Z , due to the near-degeneracy of the 2s- and 2p-orbitals, is not reproduced by density functionals. The effect can be seen also on the NO occupation numbers. For O^{4+} $v_{2p} \approx 0.04$ while the next v_i is less than 0.001. If the 2p-orbital is used in the calculation of E_r, the energy is lowered by 98 mhartree, with respect to Hartree-Fock. Further 55 mhartree are given by $E_{c,v}$, giving good agreement with the 'exact' value of 154 mhartree. If the 2p-orbital is not included in the calculation of E_r (but in that of $E_{c,r}$) $v>v_1$ for one third of the electrons and the correlation energy obtained is too large by 67%.

It is surprising that the functional of Colle and Salvetti does not seem to behave correctly in the Be series. San-Fabian et al. (1990) have shown that after including the near-degeneracy effect into the wavefunction calculation, the DF adds only 36 mhartree to the correlation energy of O^{4+} (it gave 42 mhartree for O^{6+}, and 53 mhartree for Be). A plot of the values obtained for the correlation energy by San-Fabian et al.(1990) shows that the nearly linear increase of the correlation energy with Z, which is present in E_r, is lost after the addition of the Colle-Salvetti density functional.

Other well-known near-degeneracy effects are present in *diatomic molecules*. For example, the $3\sigma_u$-orbital not occupied in the HF-calculation of the O_2 molecule has an occupation number of ≈ 0.04 ([5s,4p,2d] basis set of Dunning, 1989). If it is included in the calculation of $E_{c,r}$ $\approx$ 20% of the electrons are not properly described ($v>v_1$). This error is eliminated after the inclusion of the $3\sigma_u$-orbital into the calculation of E_r which gives now 75% of the dissociation energy. By adding the DF contribution the dissociation energy is over-estimated by $\approx 6\%$. Another example is given in figure 1 for the F_2 potential curves. An error of ≈ 0.01 hartree in DF+CI (like in this examples) was found to be typical for the first-row dimers when the global approach was used and the CI contribution came from NOs with occupation numbers ≥ 0.01 (Savin, 1988).

Very good results could be obtained for diatomic molecules with the method of Colle and Salvetti (H_2, Li_2: Colle and Salvetti, 1979; Na_2, K_2: Montagnani, et al., 1984; LiH: Amaral and McWeeny, 1984; FH: Amaral, 1985).

There are cases where it is difficult to decide whether orbitals should be included in the wavefunction space or not. Let us consider the *electron affinity* (EA) of O. At the Hartree–Fock level its energy is higher than that of the O atom, while the experimental EA is 54 mhartree. A complete active space calculation including a supplementary set of p-orbitals for both O and O^- stabilizes the negative ion with respect to the neutral atom by 22 mhartree. Thus the supplementary set of p-orbitals seems essential. CI calculations on top of the MCSCF calculations above give 41 mhartree for the EA. (The 13s,8p basis set of vanDuijneveldt (1968) was used, after extension with diffuse s- and p-functions and the 2d,1f-functions of Dunning, 1989). This CI calculation was used to generate the NOs. If the space used for MCSCF defines that for E_r, the CI+DF calculation gives an EA of 65 mH and less than 1 per cent of the electrons are not properly described by the local approximation (in O^-, using the criterion $v > v_1$). What happens if the supplementary p-set is eliminated from E_r? The test value related to $v > v_1$ stays at 0 for O, but climbs up to 0.6 for O^-. Thus, it can be expected that the HF set of orbitals may be sufficient only for O. If this calculation is performed, E_r increases by 39 mH, while $E_{c,v}$ decreases by 43 mhartree and only a minor change occurs in the EA.

Sometimes the appearance of near-degeneracy configurations

Figure 1. Potential energy curves for the F_2 molecule. The curves were shifted to a common zero for $R \to \infty$.
　1: CI contribution to DF+CI (not to distinguish from a two configuration MCSCF calculation)
　2: NO-generating CI (no energy contribution in DF+CI)
　3: 'exact' (Lie and Clementi, 1974)
　4: DF+CI

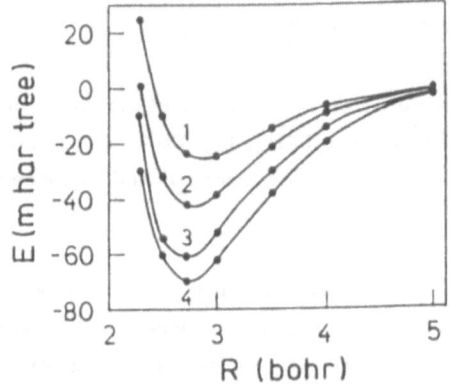

Figure 2. $d^n s^2 \to d^{n+1} s^1$ (Sc,Ni) and $d^{n+1} s^1 \to d^{n+2}$ (Ni, Cu^+) transition energies. Comparison between experimental, CI+DF and NO-generating (CEPA-1) values.

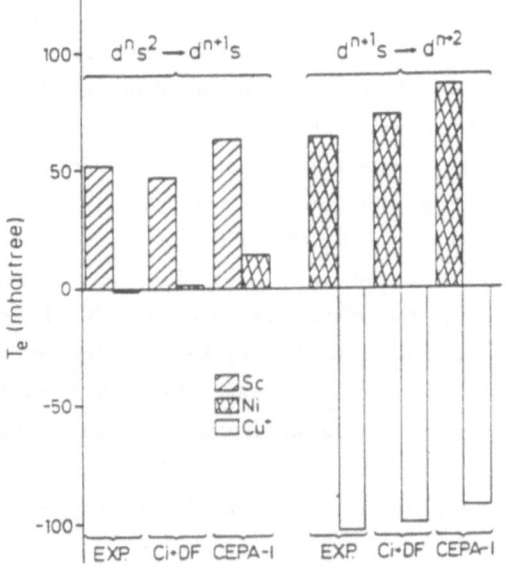

seems surprising, as it is for the s^2p^5d-configuration of Ne^+ which lowers the sp^6-cofiguration by ~ 2 eV. It was shown that the coupling of DF with CI also works reasonably in this case, giving a good value for the $s^2p^5 \rightarrow sp^6$ *transition energy* (Savin, 1989). This is a case where exchange-correlation density functionals are known to give an error of a few eV (Gunnarsson and Jones, 1985).

As a final example let us consider some $s \rightarrow d$ transitions, shown in figure 2. They were obtained by using the quasi-relativistic pseu-dopotentials and basis sets given by Dolg et al., 1987, and the global cutoff for DF+CI. The NOs were generated in a Coupled Electron Pair Approximation (CEPA-1, Meyer, 1975), with results similar to those of Werner, 1984. Ni calculations were done also using MCSCF calculations, obtaining similar results. Only a supplementary set of p-orbitals was considered for inclusion into E_r for the $d^n s^2$-configuration (mainly correlating the s-pair). The calculated $d^{n+1}s \rightarrow d^{n+2}$ transition energy has a pure DF correlation energy. For the $d^n s^2 \rightarrow d^{n+1}$ transition, the DF corrected the E_r result by 0.02 and 0.08 hartree (Sc and Ni, respecti-vely).The result is much improved over usual DF (cf. Lagowsky, Vosko, 1988; Cortona, this volume).

Although it is evident that the DF brings a significant correction to the CI contribuition to E_r, it is less clear that an improvement is present when the NO-generating CI is considered. In all the examples presented here acceptable CI calculations were attempted in order to reduce the sources of errors. It turned out, however, that in most cases the use of DF+CI reduced the errors present in the NO-genera-ting CI. In the examples of the dissociation energies of the diatomic molecules this effect gave 0.01 hartree (or an improvement by a factor of two).

How far is it possible to reduce basis sets and the CI expansion and still get a reasonable result? This question can be answered only after much more testing is done. Several calculations (global cutoff) seem to indicate that it is possible to go quite far. A three-configura-tion calculation for generating NOs in H_2 ($1\sigma_g$, $1\sigma_u$ and $2\sigma_g$) gives

DF+CI results close to that of a large expansion. It seems that the number of basis functions needed for correlation can also be reduced. It has been observed, however, that differences between calculations of different quality can give rise to errors larger than 0.01 hartree.

Conclusion

The use of density functionals permits a good description of correlation energy differences, as long as near-degeneracy effects are treated by wavefunction methods.

The typical error of the combined DF+CI procedures lies within 0.01 hartree. No attempt was made to reduce this limit further because it seems - at present - to be related to an unjustifiable increase in computational effort.

The orbital coupling leaves room for many new ideas. Maybe it is worthwile in this context to reconsider the use of functionals of the first-order density matrix. Further information might be introduced in the density functional via the second-order density matrix, as was done by Colle and Salvetti, 1979, and by Miehlich et al., 1990.

Acknowledgements

I would like to thank Prof. H. Preuss (Universität Stuttgart) who has guided my work for many years. I have to acknowledge the contribution of Prof. H. Stoll (Universität Stuttgart) who has encouraged me to find a coupling between DF and CI. I would also like to thank Prof. O. K. Andersen and Prof. H.G. von Schnering (both at Max-Planck-Institut für Festkörperforschung, Stuttgart) for their support.

I am grateful to Prof. R. Ahlrichs (Universität Karlsruhe), Dr. J. Andzelm (Cray Research, Mendota Heights), Dr. M. Casida (University of British Columbia, Vancouver), Prof. E. Clementi (IBM, Kingston), Dr. J. Flad, F.X. Fraschio (both at Universität Stuttgart), Prof. D. L. Freeman (University of Rhode Island), Dr. O. Gunnarsson (Max-Planck-Institut für Festkörperforschung, Stuttgart), Prof. M. Levy (Tulane University, New Orleans), Dr. C. Marian (Universität Bonn), B. Miehlich (Universität Stuttgart), Prof. E. Pajanne (University of Helsinki), Prof. R. G. Parr (University of North Carolina, Chapel Hill), Prof. D. Salahub (Universite de Montreal) and Prof. S. H. Vosko (University of Toronto) for stimulating discussions or correspondence. I would like to thank Prof. R. M. Pitzer (Ohio State University, Columbus) for comments on the manuscript.

This work would not have been possible without the SCF and CI programs (MELD, MOLPRO) which were made available to the Institut für Theoretische Chemie der Universität Stuttgart by Prof. E. R. Davidson (Indiana University) and Prof. H.-J. Werner (Universität Bielefeld), and were installed on the Cray by Prof. H. Stoll, Dr. U. Wedig and B. Miehlich. I would like to stress the contribution of B. Miehlich in the development of density functional programs at the Institut für Theoretische Chemie.

The financial support of the Deutsche Forschungsgemeinschaft is acknowledged.

References

The following programs were used in this work:

MELD, HF and CI: McMurchie, L.E., Elbert, S.T., Langhoff, S.R., and Davidson, E.R.;

MOLPRO, HF: Meyer, W., and Werner, H.-J.;

MCSCF: Werner, H.-J., and Knowles, P.J., 1985, *J. Chem. Phys.* 82:5053;

Knowles, P.J., and Werner, H.-J., 1988, *Chem. Phys. Letters* 115:259;

CI: Werner, H.-J., and Knowles, P.J., 1988, *J. Chem. Phys.* 89:5803;

Knowles, P.J., and Werner, H.-J., 1988, *Chem. Phys. Letters* 145:514.

MERO, DF: B. Miehlich and A. Savin.

228 A. Savin

Amaral, O.A.V., 1985, *Theor. Chim. Acta* 67:193.

Amaral, O.A.V., and McWeeny, R., 1984, *Theor. Chim. Acta* 64:171.

Becke, A.D., 1988, *J. Chem. Phys.* 88:1053.

Caravetta, V., and Clementi, E., 1984, *J. Chem. Phys.* 81:2646.

Causa, M., Dovesi, R., Pisani, C., Colle, R., Fortunelli, A., 1987, *Phys. Rev. B* 36: 891 (1987).

Ceperley, D.M., Alder, B.J., 1980, *Phys. Rev. Letters* 45:566.

Clementi, E., 1965, *J. Chem. Phys.*, 42:2783.

Clementi, E., Chakravorty, S.J., Corongiu, G., and Caravetta, V., 1989, *Modern Techniques in Computational Chemistry:MOTECC-89* (edited by Clementi, E.), p. 589. ESCOM, Leiden.

Cohen L., Santhanam, P., and Frishberg C., 1980, *Int. J. Quantum Chem.* 14: 143.

Colle R., and Salvetti, O., 1979, *Theor. Chim. Acta* 53:55.

Colle R., and Salvetti, O., 1983, *J. Chem. Phys.* 79:1404.

Colle R., and Salvetti, O., 1990, *J. Chem. Phys.* 93:534.

Davidson, E.R., 1972, *Phys. Rev.* 44:451.

Dolg, M., Fulde, P., Küchle, W., Neumann, C.-S., and Stoll, H., 1990, *J. Chem. Phys.*, submitted.

Dolg, M., Wedig, U., Stoll, H., and Preuss, H., 1987, *J. Chem. Phys.* 86:866.

Fantucci, P., Polezzo, S., Bonacic-Koutecky, V. and Koutecky J., 1990, *J. Chem. Phys.* 92: 6645.

Flad, J., Igel-Mann, G., Preuss, H., and Stoll, H., 1985, *Surface Science* 56: 379.

Freeman, D. L. ,1977, *Phys. Rev. B* 15:5512.

Fritsche, L. 1986, *Phys. Rev. B* 33:3976.

Froese Fischer, C., Lagowski, J., and Vosko, S.H., 1987, *Phys. Rev. Letters* 59:2267.

Gunnarsson, O. and Jones, R.O., 1985, *Phys. Rev. B* 31:7588.

Hohenberg, P., and Kohn, W., 1964, *Phys. Rev.* B136:864.

Huber, K.P., and Herzberg, G., 1979, *Molecular Spectra and Molecular Structure*, vol.4. Van Nostrand, New York.

Kohn, W., and Sham L., 1965, *Phys. Rev.* A140:1133.

Lagowski, J.B., and Vosko, S.H., 1988, *J. Phys. B.* 21:203.

Langreth, D.C., and Mehl, D.J., 1981, *Phys. Rev. Letters* 47:446.

Langreth, D.C., and Mehl, D.J., 1983, *Phys. Rev. B* 28:1809; erratum

1984, $\underline{29}$:2310.

Lee, C., and Parr, R.G., 1987, *Phys. Rev.* $\underline{35}$:2377.

Lee, C., Yang, W. and Parr, R.G., 1988, *Phys. Rev. B* $\underline{37}$:785.

Levy, M., 1979, *Proc. Natl. Acad. Sci. (USA)* $\underline{76}$:6062.

Levy, M. 1987, *Density Matrices and Density Functionals*, (edited by Erdahl, R., and Smith, V. H.) p. 479. Reidel, Dordrecht.

Lie, G.C., and Clementi, E., 1974, *J. Chem. Phys.* $\underline{60}$:1274, 1288.

Linderberg J., and Shull H., 1960, *J. Mol. Spectroscopy* $\underline{5}$:1.

Löwdin, P.-O., 1955, *Phys. Rev.* $\underline{97}$:1474.

Löwdin, P.-O., 1959, *Adv. Chem. Phys.* $\underline{2}$:207.

Meyer, W., 1975, *J. Chem. Phys.* $\underline{58}$:1017.

Miehlich, B., 1990, Diplomarbeit, Universität Stuttgart.

Miehlich, B., Savin, A., Stoll, H., and Preuss, H., 1989, *Chem. Phys. Letters* $\underline{157}$:329.

Miehlich, B., Savin, A., Stoll, H., and Preuss, H., 1990, to be published.

Montagnani, R., Pisani, P., and Salvetti O., 1984, *Theor. Chim. Acta* $\underline{64}$:171.

Moscardo, F., Perez-Jorda, J., and San-Fabian, E., 1989, *J. chim. phys.* $\underline{86}$:853.

Moskowitz, J.W., Schmidt, K.E., Lee, M.A., and Kalos, M.H., 1982, *J. Chem. Phys.* $\underline{76}$:1064.

Pajanne, E., and Arponnen J., 1982, *J. Phys. C* $\underline{15}$:2683; I am grateful to Prof. E. Pajanne for making available tabulated data.

Pant, M.M., and Rajagopal, A.K., 1972, *Solid State Comm* $\underline{10}$:1157.

Perdew, J.P., 1986, *Phys. Rev. B* $\underline{33}$:8822.

Perdew, J.P., McMullen, E.R., and Zunger, A., 1982, *Phys. Rev. A* $\underline{23}$:2785.

Roos, B.O., Szulkin, M., Jaszunski, M., 1987,*Theor. Chim. Acta* $\underline{71}$: 375.

San-Fabian, E., Perez-Jorda, J., and Moscardo, F., 1990; *Theor. Chim. Acta* $\underline{77}$:207.

Savin, A., Schwerdtfeger, P., Preuss, H., Silberbach, H., Stoll, H., 1983, *Chem. Phys. Letters* $\underline{92}$:226.

Savin, A., Stoll., H. and Preuss, H. 1984, *Local Density Approximations in Quantum Chemistry and Solid State Physics* (edited by Dahl, J. P., and Avery, J.) p. 263. Plenum, New York.

Savin, A., Stoll., H. and Preuss, H. 1986, *Theor. Chim. Acta* 70:207.

Savin, A., Vogel, K., Preuss, H., Nesper, R., and von Schneering, H.G., 1988, *J. Am. Chem. Soc.* 110: 373

Stoddard, J.C., and March N.H., 1971, *Ann. Phys.* (NY) 64:174.

Stoll, H. and Savin, A., 1985, *Density Functional Methods in Physics*, (edited by Dreizler, R.M. and da Providencia, J) p. 177. Plenum, New York.

van Duijneveldt,F.B., 1971, *IBM Res. Rep.* 945.

Vokso, S.H., Wilk, L., and Nusair M., 1980, *Can. J. Phys.* 58:1200.

von Barth, U., and Hedin,L., 1972, *J. Phys. C* 5:1629.

Werner, H.J., 1984, *Faraday Symp. Chem. Soc.* 19:202.

Wigner, E., 1934, *Phys. Rev.* 46:1002.

Wilson, L.C., and Levy, M., 1990, *Phys. Rev. B* 41:12930.

Ziegler, T., Rauk, A., and Baerends, E.J., 1977, *Theor. Chim. Acta* 43:261.

15

Accurate Intramolecular Forces Within Gaussian Orbital Local-Density Framework: Progress Towards Real Dynamics

MARK R. PEDERSON AND KOBLAR A. JACKSON

INTRODUCTION

With the results of a multitude of numerical tests of the Hohenberg-Kohn-Sham local spin density (LSD) approximation (Hohenberg et al., 1964, Kohn et al., 1965) in hand, there is a growing consensus amongst researchers concerning the long term viability of the LSD approximation over other <u>ab initio</u> techniques for applications to large molecules. Coupled with the ever increasing numerical data base of results is an enhanced formal understanding of the relative attributes and shortcomings of the existing approximations to the universal functional. The consensus that has emerged is that the existing LSD approximation is expected to give very accurate information about ground state molecular geometries (eg bond lengths and angles), molecular vibrational phenomena, and qualitatively correct energetics. While improvements of LSD based molecular energetics is currently a field of active research, there is confidence within the chemical and physical communities that slight improvements to the approximations will lead to quantitatively correct energetics as well. For this reason, many researchers have turned toward merging classical molecular dynamics and quantum mechanical electronic structure techniques by incorporating atomic forces, calculated within the LSD framework, into dynamical or quasidynamical algorithms. The primary focus of this paper will be on some of our recent work toward this goal. To motivate this discussion, in Sec. II we briefly outline the formalism for calculating the electronic structure and forces. In Sec. III, we turn to our work on a new, highly accurate and automatic method for the numerical evaluation of integrals requisite to quantum mechanical calculations. In Sec. IV, we discuss several different methods for the automated derivation of molecular geometries. A summary and discussion of outstanding problems follows in Sec. V.

231

FORMALISM

Within the LSD approximation and the Born-Oppenheimer approximation, the total energy for the ground state of a system of electrons and nuclei is determined by variationally adjusting the electronic charge density, $\rho(\mathbf{r})$, and the positions of the nuclei, $\{\mathbf{R}_\nu\}$ to minimize the energy functional,

$$E_t = 2\sum_i q_i \langle\psi_i| -\tfrac{1}{2}\nabla^2 + V_{nuc}|\psi_i\rangle + \tfrac{1}{2}\int d\mathbf{r}\,d\mathbf{r}'\frac{\rho(\mathbf{r})\rho(\mathbf{r}')}{|\mathbf{r}-\mathbf{r}'|} + \int d\mathbf{r}\rho(\mathbf{r})\epsilon_{xc}[\rho] + \tfrac{1}{2}\sum_{\nu\mu}'\frac{Z_\nu Z_\mu}{|\mathbf{R}_\nu-\mathbf{R}_\mu|}. \quad (1)$$

subject to the constraints that the density is constructed from a sum of the Kohn-Sham orbital densities $[\rho(\mathbf{r})=2\sum q_i|\psi_i(\mathbf{r})|^2]$ and that the Kohn-Sham orbitals are orthonormal.   The total energy consists of the electronic kinetic energy, the nuclear-electronic Coulomb energy, the electron-electron Coulomb energy, the exchange correlation energy and the nucleus-nucleus Coulomb energy respectively. For notational simplicity, Eq. (1) assumes a spin unpolarized system. The term $\epsilon_{xc}[\rho]$, which is in general a function of the electronic spin densities, has been parametrized from the results of Monte Carlo calculations on the uniform electron gas (Perdew and Zunger, 1981). For many ground state systems, the occupation numbers $\{q_i\}$ are unambiguosly either 0 or 1. However, for systems with degeneracies at the Fermi level, the lowest energy corresponding to Eq. (1) consists of fractionally occupied states with degenerate eigenvalues.   In order to find the ground state of Eq. (1), the Kohn-Sham orbitals are expanded in terms of a fixed basis according to:

$$\psi_i(\mathbf{r}) = \Sigma_j\, c_{ij}\, \phi_j(\mathbf{r}) \quad (2)$$

The wavefunction expansion coefficients, the c_{ij}'s, must be chosen such that the eigenvalue problem

$$\langle\phi_j|-\tfrac{1}{2}\nabla^2 + V_{eff}(r)-\lambda_i|\psi_i\rangle = \langle\phi_j|H_o-\lambda_i|\psi_i\rangle = 0 \quad (3)$$

is satisfied self-consistently. To ensure that the system is at a local geometrical minimum the Hellmann-Feynman forces on the nuclei must also vanish:

$$\int d\mathbf{r}\,\rho(\mathbf{r})\,\nabla_\nu V_{nuc}(\mathbf{r}) + \nabla_\nu\sum_{\mu\neq\nu}\frac{Z_\nu Z_\mu}{|\mathbf{R}_\nu-\mathbf{R}_\mu|}$$
$$+ 2\Sigma_{ij}\, q_i\, [c_{ij}\langle\psi_i|H_o-\lambda_i|(\nabla_\nu\phi_j)\rangle + \text{complex conjugate}] = 0. \quad (4)$$

The first two terms in the above equation correspond to the classical electrostatic force on a given nucleus due to the electronic charge density and the remaining nuclei. The third term, referred to as the Pulay correction (Pulay, 1977) arises if the basis (such as the fixed Gaussians used here) contain explicit dependence on the nuclear positions. The algorithms for solutions of Eqs. (2-4) are now discussed.

The codes described here allow for an *exact* (to any desired numerical precision) solution of Eqs. (2-4). The above equations are valied for any basis set and a judicious choice of a basis set is still a matter of considerable discussion. Due to the attractive analytical properties of the Gaussian-type functions, many researchers have employed these functions to perform LSD-based calcultions. One of the earlier works, due to Lin and Lafon, was performed twenty-five years ago (Lafon *et al*, 1966). While many researchers employ such functions, the numerical details used differ substantially from one code to the next and continue to evolve. We start by expanding the wavefunctions in terms of a set of Gaussian-type orbitals centered on each atomic site and self-consistently solve Schroedinger's equation. The first step consists of constructing the effective potential on a mesh of points. To obtain the Coulomb contributions, the charge density is broken up into short-range components which are zero in the region outside non-overlapping spheres centered about each atom and a long-range component. The decomposition is carried out analytically by capitalizing on the fact that the wavefunctions are constructed in terms of short and long ranged Gaussian type functions. The short range parts of the Coulomb potential can then be obtained by a spherical harmonic decomposition of the atom centered short range densities followed by one dimensional radial integrations using well known techniques (Erwin *et al*, 1990). The long-range parts of the Coulomb potential are calculated analytically (Shavitt, 1963) by exploiting the fact that the long-range part of the density consists of a linear sum of products of Gaussian functions multiplied by cartesian polynomials. By calculating the Coulomb potential exactly, we avoid difficulties and potential inaccuracies associated with least-square fitting techniques. However, while the method is exact, calculation of the long-range part of the potential is time consuming and presently accounts for approximately fifty percent of the computer time expended for a given calculation. To obtain the full potential, we add the nuclear Coulomb and electronic exchange-correlation potential to the electronic Coulomb potential. Given the potential on a mesh of points, we then construct the Hamiltonian matrix by numerical integrations according to:

$$\langle\phi_j|\text{Veff}|\phi_i\rangle = \Sigma_k \, \omega_k \, \phi_j(\underline{r}_k)\text{Veff}(\underline{r}_k)\phi_i(\underline{r}_k) \tag{5}$$

For Gaussian-type functions, the kinetic contributions may be calculated either numerically or analytically. The Hamiltonian matrix is then diagonalized, leading to new wavefunctions, and the entire process is repeated until self-consistency is achieved. The ability to perform accurate and efficient numerical evaluations (Pederson et al, 1990) of the above integrals allows us to bypass a least-square fit of the potential and is of tantamount importance, particularly in regard to calculating forces. Our methods for performing such integrations are discussed in a later section.

Before turning to the calculation of forces, we discuss several other tangential but important points regarding the self-consistent procedure. In practice, a variety of techniques may be used to obtain (at least) an overall order of magnitude savings in computer time for a given calculation. First, it is standard practice (often necessary) to iteration average the Hamiltonians to speed up the self-consistency convergence. We use an implementation of Broyden's method (Johnson, 1988) which we find to be approximately two or three times more efficient than simple mixing. Second, since most molecular systems are inherently symmetric, substantial savings may be made by using group theory (Cotton, 1971). In addition, to substantial computational savings achieved by block diagonalization of the Hamiltonian matrix, the use of symmetry allows construction of the effective potential and Hamiltonian matrix to proceed N(G) times faster, with N(G) the total number of point-group symmetry operations. Finally, we have achieved considerable time savings by ensuring that the rate determining steps are vectorized.

In order to obtain highly accurate forces, consistent with the Born-Oppenheimer approximation, careful attention must be paid to any and all explicit dependencies of the total energy on the atomic locations. For the Gaussian-type basis sets used in this work, the Pulay correction arises due to the fact that the basis functions are constrained to follow their parent nuclei as the atoms are moved through space (Jackson et al., 1990). Another possible source of inconsistencies and errors in the Hellmann-Feynman forces may result if the representation of the effective potential is dependent on the atomic locations and if this dependence is not systematically controllable. For example, in our earlier work, rather than employing our numerical integration techniques discussed below, we fit the effective potential to a linear combination of atom-centered Gaussian-type functions. Although this method led to the accurate calculation of densities of states, cohesive energies, and simple vibrational modes, even with a great deal of effort we were unable to automate it in a way that allowed for the accurate calculation of forces. We generally found that forces obtained with our least-square representation of the potential were in error by at least ten percent. In order to obtain greater accuracy, some researchers (Fournier et al, 1989) have employed variational fitting techniques along the lines originally suggested by Dunlap et al (Dunlap et al, 1979). Others (Versluis et al, 1988) have also taken careful steps to circumvent problems due to least-square fits. Others, including ourselves, have employed a variety of numerical integration schemes. In Sec. III, we discuss our algorithm for deriving an accurate and efficient numerical integration mesh. For the remainder of this section we assume that any desired numerical accuracy is indeed possible and discuss issues concerning convergence of the forces.

In calculating forces (Jackson et al, 1990), it is important to distinguish between the rate of convergence and the level of convergence for the Hellmann-Feynman forces. We note that it is possible to evaluate the Pulay correction term in Eq. (4) in several ways which are equivalent when self-consistency is achieved. In standard LSD calculations, the iteration cycle proceeds schematically as $\psi^N \rightarrow H^N \rightarrow \underline{H}^N \rightarrow \psi^{N+1}$, which indicates that the density due to a set of old wavefunctions $\{\psi^N\}$ is used to construct a Hamiltonian H^N which is subsequently iteration averaged to form $\underline{H}^N$. Upon diagonalization of $\underline{H}^N$, a new set of wavefunctions $\{\psi^{N+1}\}$ is obtained. Implicit in the derivation of the Hellmann-Feynman force and Pulay correction is the assumption of self-consistency [$H=H^N=\underline{H}^N$ and $\psi=\psi^N=\psi^{N+1}$ etc.]. Thus it is not immediately obvious which iteration-dependent quantities are appropriate for use in Eq. (4), particularly regarding the iteration dependent eigenvalue. We find that the forces converge substantially faster if the Pulay correction is calculated using iteration-consistent Hamiltonians and wavefunctions (H^N and ψ^N).   On these grounds we expect that the optimal iteration dependent "eigenvalue" would best be chosen as $\lambda = <\psi^N|H^N|\psi^N>$ but there is no compelling numerical evidence that this choice is better or worse than the eigenvalues obtained from the diagonalization of the previous iteration's averaged Hamiltonian.

The preceding arguments concern the rate of convergence of the forces. The absolute level of convergence depends, on the other hand, upon the convergence of the total energy.   While the error in the total energy is second-order with respect to errors in the charge density, the errors in the forces are linear with respect to such errors. Thus, forces converge to roughly half the precision of total energies.   We find that if the energy is converged to 10^{-8} Hartrees that the forces are converged to approximately 10^{-4} Hartrees/bohr.

To illustrate the accuracy of our forces, we discuss results from a challenging test case, the Mo_2 molecule. For these calculations, we have used a Gaussian basis set consisting of 18 single gaussians, contracted to 8 s-type, 7 p-type and 5 d-type functions (each d-type function adds an r^2 spherical function as well). In Fig. 1, the total energy and force for the Mo_2 dimer is presented as a function of atomic separation.   The equilibrium separation occurs at 3.67 bohr in good agreement with the experimental result of 3.65 and an earlier theoretical LDA value of 3.69±0.09 bohr (Delley et al, 1983).   The forces calculated from a polynomial fit of the total energy are in excellent agreement with those

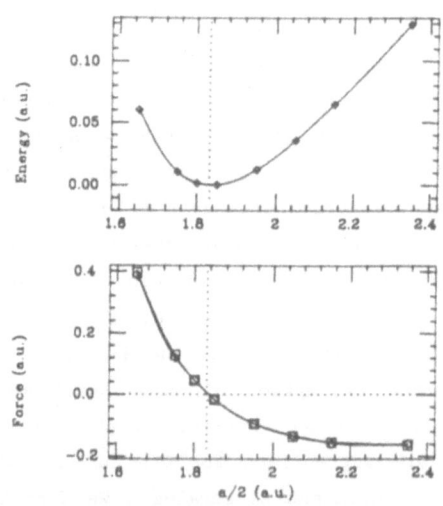

Figure 1. Forces and Total Energies for Mo_2

obtained from the Hellmann-Feynman-Pulay force (HFP).   The discrepancies between the two force curves are in fact due to the numerical schemes used for obtaining forces from the polynomial fit of our total energies- not to errors in the HFP force. To illustrate further the accuracy at the automatically determined ground state geometry (determined using the conjugate gradient technique), we obtain a force at the equilibrium separation of 0.002eV/bohr! While this accuracy is easy to obtain, it is approximately an order of magnitude better than required.

Before continuing, we note that it is indeed possible to alleviate the need for Pulay corrections by relaxing the constraint that the basis functions must follow their parent atoms and allowing each basis function to variationally float to minimize the total energy. A molecular-dynamical method for performing such calculations, with applications to Li_2, has been presented (Pederson et al, 1988). While Pulay corrections do not appear in the definition of the force, similar terms must be calculated to variationally float the functions.

VARIATIONAL MESH FOR QUANTUM-MECHANICAL SIMULATIONS

Our method for numerical integrations has been discussed in detail in a recent publication (Pederson et al, 1990). Here we simply present the salient features. As shown schematically in Fig. 2, we start by enclosing all the atoms in a "universal box" which is sufficiently large that all contributions to the wavefunctions, and thus any integrands, outside the universal box vanish. The universal box is then partitioned

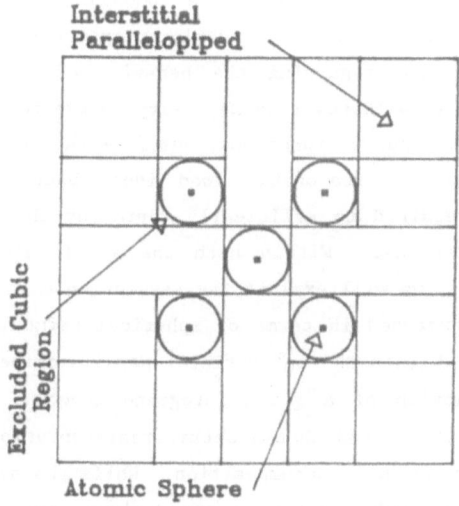

Figure 2. Schematic drawing of spatial subdivisions
into a set of parallelopipeds which contain either one or no atoms. The parallelopipeds which contain atoms are, without loss of generality, constrained to be perfect cubes. The atomic cubes are further divided by enclosing each atom by the largest atomic sphere which fits in the atomic cube. This partitioning scheme leads to three generic regions of space requiring integration meshes: the atomic spheres, the excluded cubic

regions and the interstitial parallelopipeds. The spatial tesselation described here is qualitatively similar to that used in a variety of electronic structure algorithms which include the linearized augmented plane wave algorithm (Anderson, 1975), the linear combination of atomic-orbitals method (Erwin et al, 1990), and other integration schemes which have recently been reported in the literature (Boerrigter et al, 1988; Averill and Painter, 1989). However, several aspects of the variational procedure used to choose integration meshes within each of these regions are new (Pederson et al, 1990). We now discuss this method.

Providing that the singularities in an integrand can be isolated, there are in fact an infinite number of one dimensional integration schemes which can be combined to perform any three dimensional integral to any desired accuracy. Many of these numerical integration meshes are derived by finding N points and N volume elements which are guarateed to exactly reproduce the integrals of 2N orthogonal functions. Two examples of such meshes include the real space mesh points and weights associated with Fourier transforms and those obtained from Gaussian quadrature (GQ) procedures. In addition to knowing that all of the singularities occur at the center of the atomic spheres, a prerequisite to performing efficient numerical integrations of the Hamiltonian matrix elements is to have some intuitive idea regarding the behavior of their respective integrands. Although it will never be necessary to expand the integrands in terms of an alternate set of functions, our formulation capitalizes on the fact that we have a reasonably good idea about what kinds of functions would be required to efficiently represent the integrands in each of the three regions. Within both the atomic spheres and the excluded cubic regions, we will exploit the assumption that the integrand is most efficiently expanded in terms of spherical harmonics multiplied by radial functions. Within the intersticial parallelopipeds, we assume that a harmonic expansion of a given integrand about a single atomic center would be inefficient and that a better representation would be in terms of a multicenter harmonic decomposition. While the assumptions are expected to be good, in order to ensure any desired accuracy we wish to combine them with numerical integration schemes that lead to errors which decrease exponentially with the number of mesh points. Our approach is illustrated by way of example.

Suppose we are interested in numerically evaluating the contributions of a given matrix element, $[F(\underline{r})=\phi_j(\underline{r})Veff(\underline{r})\phi_i(\underline{r})]$, in one of the intersticial parallelopipeds by combining three one dimensional integrations. That is,

$$\iiint_{p.p.} dxdydz \; F(x,y,z) \; = \; \Sigma_{ijk} w_i(a_x,b_x) w_j(a_y,b_y) w_k(a_z,b_z) F(x_i,y_j,z_k)$$

$$= \; \Sigma_i \; \Omega_i(p.p.) \; F(\underline{r}_i) \tag{6}$$

To obtain the mesh points and weights for each one dimensional integration, we start by refining our assumptions about the function $F(r)$ by assuming that it may be decomposed in terms of a multicenter sum of polynomials multiplied by Gaussians centered on each atom. This provokes us to look for a set of mesh points and weights which lead to a value of the objective function

$$D \; = \; \underset{np\alpha_k}{Max} \; W_{np\alpha_k} | \; I_{np\alpha_k} - \; \Sigma_i w_i(a,b) \; (x_i - X_n)^p \exp[-\alpha_k(x_i - X_n)^2] \; |. \tag{7}$$

less than some prescribed tolerence. In the above equation, X_n is the x-component of the position of the nth atom, and w_i and x_i are the volume elements and mesh points associated with a given "trial mesh". The test integrands are taken to be Gaussian functions multiplied by polynomials of degree $0,1,\ldots,6$. $I_{np\alpha_k}$ is the "exact" value of the integral which in practice is calculated using a very large GQ mesh. Since there are many one-dimensional integration meshes with different regimes of validity, the optimization proceeds by systematically seeking a one-dimensional mesh that provides the prescribed accuracy using the fewest number of mesh points. Once a given N-point mesh is found to be accurate enough, that mesh is chosen. The exponential convergence with respect to the number of one-dimensional mesh points is guaranteed by allowing for the choice of one-dimensional GQ integration meshes. While the convergence properties of GQ meshes are good, we have found that the total number of mesh points may be decreased by at least factor of two if a new class of GQ schemes which depend continuously on a variational parameter are employed.

The one dimensional GQ mesh has been discussed in many places. Here, we simply note that an N-point GQ mesh may be used to exactly integrate any integer power (p) of x (p<2N) according to the formula:

$$\int_a^b dx \; x^p \; = \; \Sigma_{i=1}^N \; w_i(a,b) \; x_i^p \tag{8}$$

In the above equations, the N mesh points $\{x_i\}$ coincide with the zeros of the Legendre polynomial of degree N and the weights $\{w_i(a,b)\}$ may be obtained by requiring that the Legendre polynomials of degree $0,1,\ldots,N-1$ are correctly integrated on this mesh. While it is obvious that the resulting mesh will exactly reproduce the integrals for all polynomials of degree N-1 or lower, the judicious selection of the mesh points actually guarantees that the mesh will exactly integrate all polynomials of degree 2N-1. Now, by making the coordinate substitution, $x = \exp(-\gamma y)$, Eq. (8) may be rewritten as:

$$\int_{-\ln(b)/\gamma}^{-\ln(a)/\gamma} dy \ [\ e^{-\gamma y}]^{p+1} = \Sigma_{i=1}^{N} \ (\ w_i(a,b)/\gamma x_i) \ [e^{-\gamma y_i}]^{p+1} \tag{9}$$

$$= \Sigma_{i=1}^{N} \ w_i'(a,b;\gamma) \ [e^{-\gamma y_i}]^{p+1} \tag{10}$$

which states that any polynomial of $\exp(-\gamma y)$ between degree 1 and degree 2N may be exactly integrated over the interval $[-\ln(b)/\gamma \ , \ -\ln(a)/\gamma \ ]$ by a suitable transformation of the GQ points and weights. For applications to Eq. (6), we recognize that y is a dummy variable and choose b such that $a_x = -\ln(b)/\gamma$ and a such that $b_x = -\ln(a)/\gamma$. In addition to the fact that the transformed GQ mesh is clearly intuitively better for the integrands relevant to an electronic-structure calculations, it contains an additional variational parameter (γ) which allows one to fine-tune the transformed GQ mesh to match the atomic environment. The transformed GQ mesh discussed here is but one of a growing number of adaptive grids for numerical integrations. Other adaptive grids include normal GQ meshes with arbitrary envelope functions (Gauss-Laguerre and Gauss-Hermite), a pseudo-spherical integration scheme (Averill et al, 1988) and the recent energy-minimizing mesh (Levine et al, 1989). Finally we should mention in passing that a somewhat different approach to numerical integrations of matrix elements has recently been introduced (Becke, 1988).

To illustrate the exponential convergence properties of the transformed GQ mesh, in Fig. 3 we show the average fractional error obtained when calculating the incomplete gamma functions of order 0, 1 and 2,

$$F_m(t) \ = \ \int_0^1 \ dr \ r^{2m} \ e^{-tr^2}, \quad (0.001 < t < 50000.0) \tag{11}$$

with this mesh and for Slater type integrands $[\exp(-tr)]$ with similar length scales. The exponential decay in errors is immediately clear.

We have used similar ideas to determine integration meshes in the atomic spheres and excluded cubic regions. Here we simply discuss our approach to these regions qualitatively. The integrations within the atomic spheres are performed by first carrying out angular integrations for each radial point and then performing one-dimensional radial

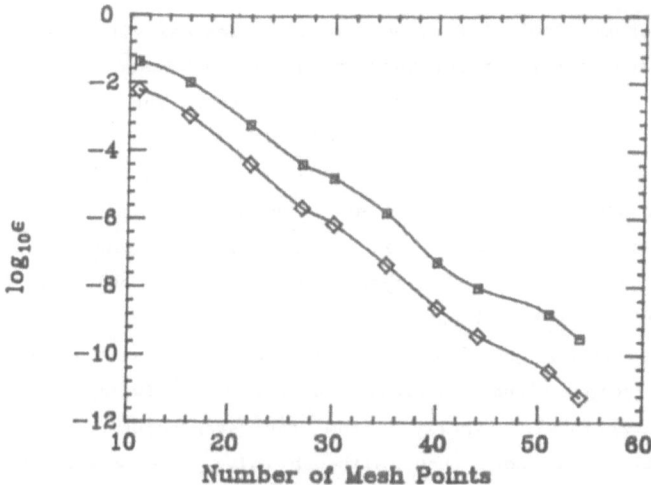

Figure 3. Average integrations errors versus number of mesh points.

integrations.   This amounts to separately choosing a two-dimensional angular mesh and a one dimensional radial mesh.   We use the two-dimensional angular meshes (Stroud, 1971) which are guaranteed to exactly integrate a polynomial of degree L or less over the surface of a sphere. The radial mesh is derived using the transformed GQ procedure of the previous paragraph.   For integrations over the excluded cubical volume $(V=8L^3-4\pi L^3/3)$, we start by writing the integral over $1/48^{th}$ of the excluded cubic region $(r>L>x>y>z>0)$ as an iterated integral in spherical coordinates. One dimensional angular meshes and the transformed GQ mesh are then combined to obtain mesh points and volume elements for this region.   Each mesh point is subsequently rotated to the 48 equivalent points in the excluded cubic region.

In applications to real systems, we find that extremely high accuracy calculations are capable with less than 2000 mesh points per inequivalent atom.   For example, to accurately calculate the charge to 0.0002 au or better, we need a total of 941 points for Li_2, 1102 points for C_2, 1015 points for As_2.   Comparison of the number of mesh points required for the highly covalent C_2 molecule to that of As_2 demonstrates that the number of points required is dependent more on the non-spherical valence region rather than the number of core states.   This is expected since the number of points near the nucleus only increases as $\ln(Z)$.

The numerical integration method discussed here starts with the assumption that it is possible to expand the effective potential in terms multicenter multipole expansion, which is basically the same assumption that we used in our earlier work which employed least square fits of the potential to Gaussian-type functions. However, the resulting numerical integrals, and ultimately forces and total energies, are more accurate than those obtained from least-square fits since the additional formalism guarantees accurate integrations even when the assumptions are not valid. An additional advantage is that we are able to include an effectively continuous range of Gaussian-type functions in our objective functions. Aside from problems with numerical instabilities, it is not feasible to use a continuous range of Gaussian-type functions for a least square fit since the computational complexity scales as N^3 (with N the number of fitting functions). Finally, from the standpoint of computer-time considerations, we note that, although naive scaling arguments suggest that construction of the Hamiltonian matrix should be an N^3 problem, for localized basis sets such as the ones employed here, the scaling is linear with the number of atoms. To see this it is only necessary to note that, in the many-atom limit, the longest range function on a given atom is zero outside a sphere with a radius of approximately ten or twenty bohr. Therefore the Hamiltonian matrix is sparse and one only needs to perform function evaluations at the mesh points contained within a relatively small sphere around each atom.

AUTOMATED GEOMETRICAL OPTIMIZATION OF MOLECULAR STRUCTURES

Given an all-electron local-orbital based method for the accurate calculation of forces, we now discuss several techniques which we have employed for the automated optimization of molecular geometries. Pioneering work in this field is due to Car and Parrinello, who recognized that within a pseudo-potential plane-wave electronic structure formalism, geometrical and electronic structure optimizations could be performed simultaneously by the introduction of a ficticious Lagrangian which depends on both electronic and nuclear degrees of freedom (Car et al, 1985). Shortly thereafter, many researchers recognized that there are a large class of iterative minimization schemes similar to the ficticious Lagrangian approach of Car and Parrinello. The performance of each of these schemes is somewhat dependent on the characteristics of the individual problem and we continue to seek alternatives to optimize our iterative minimization scheme to meet our needs. To date, we have found that the conjugate-gradient method with full quenches to the Born-Oppenheimer surface after each parameter move is most efficient for our codes (Pederson, 1988). The conjugate gradient method requires precisely the same information per "time step" or parameter move as the Car-Parrinello technique but it systematically builds up second derivative information as is necessary, ideally minimizing a quadratic function of N nuclear degrees of freedom with about 3N SCF calculations. We find that with a ficticious Lagrangian scheme, a substantially larger number of SCF calculations are required to perform the minimization.

Although it is reasonable to expect that the number of SCF calculations required to optimize a geometry by a ficticious Lagrangian scheme will always be greater than that required by the conjugate gradient method, there are reasons to believe that a robust time-dependent approach may still prove to be more efficient. The crux of the matter is that the total number of SCF iterations and not the number of SDF calculations determines the total amount of computer time which is used. One aspect of a time-dependent approach is that the Hamiltonians evolve smoothly in time. We find that after the first three or four time steps we are able to extrapolate the SCF Hamiltonian matrices forward in time accurately enough that only two or three SCF iterations, compared to the ten or fifteen iterations required for less accurate starting points, are required to converge the total energy and forces (10^{-7} Hartrees and 10^{-3} Hartees/bohr respectively). Hence, for a ficticious Lagrangian technique requiring M time steps to reach the minimum energy geometry, roughly 3M SCF iterations are performed. On the other hand, a conjugate gradient line minimization requires roughly 30 SCF iterations, so that a geometry optimization for N degrees of freedom requires 30N SCF iterations using a conjugate gradient algorithm. The relative costs are 30N for conjugate gradient and 3M for a ficticious Lagrangian method with M greater than N. So, a time dependent method which guarantees that M<10N may be faster than the conjugate gradient approach.

DISCUSSION

Before concluding with a discussion on future directions and outstanding problems, it may be appropriate to review briefly the problems which have recently been studied using the codes discussed here. Vacancy formation energies and vacancy induced lattice relaxation have recently been calculated for alkali metal clusters containing 9, 15, and 27 atoms (Pederson et al, 1989). Simulations have been done on the important interactions between hydrocarbon radicals and the dangling bonds on the <111> surface of a hydrogenated diamond crystallite [$C_{10}H_{21}$]

and bond lengths and angles for methylene, methane, acetylene, ethylene, ethane, methylacetylene, benzene (Pederson et al, 1989). The properties of nitrogen and phosphorus substitutional defects in diamond-like complexes [C_5H_{12} and $C_{17}H_{36}$] have been studied (Jackson et al, 1990). F-Center properties in 26 atom MgO clusters and substitutional Cu defects in LiCl clusters have been looked at also (Jackson et al, 1990). Automated electronic and nuclear optimizations of systems similar to these now proceed with very little user intervention.

While much has been accomplished, there are still outstanding problems which need to be addressed to extend the present capabilities of the local density approximation. From the standpoint of energetics, additional improvements along the lines of self-interaction corrections (Perdew et al, 1981; Pederson et al, 1985), gradient corrections (Langreth et al, 1983) and self-energy corrections (Pickett et al, 1984) need to be fully implemented and tested to ascertain which methods are best for obtaining accurate cohesive energies and (possibly) excitation energies. For applications to large systems with little symmetry, the occurence of accidental degeneracies at the Fermi level are expected to manifest themselves more frequently and better numerical algorithms are required to efficiently and automatically converge the total energies and forces for these systems. In connection with the Fermi-level problem, a reexamination of the meaning of LSD solutions with fractional occupation and correlated many-electron wavefunctions may be in order. Finally, it is important to note that the new numerical techniques discussed in this work come very close to eliminating the need for a Gaussian basis set. Future calculations which are either entirely numerical or include numerical atomic-like functions, Slater-type orbitals and plane-waves with Gaussian-type orbitals should indeed be possible.

ACKNOWLEDGEMENTS

Over the last decade, the authors have had many illuminating discussions with Drs. L. Boyer, J. Q. Broughton, P. Edwardson, S. C. Erwin, J. G. Harrison, R. A. Heaton, D. D. Johnson, B. M. Klein, C. C. Lin and W. E. Pickett in connection with this work. This work was supported in part by the Office of Naval Research. Computational resources were provided by the National Science Foundation at the Pittsburgh Supercomputing Center. One of us (KAJ) acknowledges partial support from the National Research Council.

REFERENCES

Andersen O. K., (1975) Phys. Rev. B 12, 3060.

F. W. Averill and G. S. Painter, (1989) Phys. Rev. B 39 8121.

Becke A. D., J. Chem. Phys. 88, 2547 (1988).

Boerrigter P. M., te Velde G., and Baerends E. J., (1988)
      Int. J. Quantum Chem. 33 87.

Car R. and Parrinello M., Phys. Rev. Lett. 55, 2471 (1985).

Cotton F. A., (Wiley Interscience, New York 1971) Chemical Applications
      of Group Theory.

Delley B.,Freeman A. J. and Ellis D. E., (1983)
      Phys. Rev. Lett. 50, 488.

Dunlap B. I., Connolly J. W. D., and Sabin J. R., (1979)
      J. Chem. Phys. 71, 3396.

Erwin S. C., Pederson M. R. and Pickett W. E., (1990)
      Phys. Rev. B. 41,   10437.

Fournier R., Andzelm J. and Salahub D. R., (1990) J. Chem. Phys. 90 6371.

Hohenberg P. and Kohn W, (1964) Phys. Rev. B 136, 864.

Jackson K. and Pederson M. R., (1990) Phys. Rev. B 42.

Jackson K. A. and Pederson M. R., (1990) Phys. Rev. B (To appear).

Jackson K. A. and Pederson M. R., (1990) Phys. Rev. B (To appear).

Johnson D. D., (1988) Phys. Rev. B 38, 12807.

Kohn W and Sham L. J., (1965) Phys. Rev. 140 A1133.

Lafon E and Lin C. C., (1966) Phys. Rev. 152, 579.

Langreth D. D. and Mehl M. J. (1983) Phys. Rev. B 28, 1809.

Levine Z. H. and Wilkins J. W., (1989) J. Comp. Phys. 83, 361.

Pederson M. R., Heaton R. A., and Lin C. C., (1985)
      J. Chem. Phys. 82, 2688.

Pederson M. R. et al, (Plenum Publishing Corporation, 1989), Atomistic
      Simulation of Materials, 79.

Pederson M. R., Jackson K. A. and Pickett W. E., (Mat. Res. Soc.,
      Pittsburgh 1989) Technology Update of Diamond Films.

Pederson M. R. and Jackson K. A., (1990) Phys. Rev. B 41 7453.

Perdew J. P. and Zunger A., (1981) Phys. Rev. B 23, 5048.

Pickett W. E. and Wang C. S., (1984) Phys. Rev. B 30, 4719.

Shavitt I., (Academic Press 1963), Methods of Computational Physics.

A. H. Stroud, (Prentice-Hall, Englewood Cliffs, NJ 1971) Approximate
      Calculations of Multiple Integrals.

Versluis L. and Ziegler T., (1988) J. Chem. Phys. 88, 322.

Wei S.-H and Krakauer H., (1985) Phys. Rev. Lett. 55, 1200.

16
Relativistic DV-Xα Studies of Three-Coordinate Actinide Complexes

WILLIAM F. SCHNEIDER, RICHARD J. STRITTMATTER, BRUCE E. BURSTEN, AND DONALD E. ELLIS

The chemistry of the actinide elements continues to present challenges to both experimental and theoretical chemists. Actinide compounds have been found to possess a diversity of structures and reactivities that are not only extensions of but significant additions to those recognized for the transition metal elements.[1] From a theoretical perspective, the number of electrons and the importance of relativistic corrections in actinide systems pose several problems: theoretical rigor is more difficult to maintain, calculations are more computationally demanding, and results are more difficult to interpret than those obtained from non-relativistic calculations. Nonetheless, significant advances have been made in the application of molecular electronic structure methods to actinide compounds. Local density functional (LDF) methods in particular have several features that make them attractive for the study of heavy element structure and bonding. We present here a brief description of a computational scheme that employs the LDF formalism to address questions in actinide electronic structure and then demonstrate its application to one system in particular: three-coordinate actinide(III) compounds.

Theoretical Model and Computational Methods

The need to incorporate relativistic corrections into the molecular Hamiltonian for heavy element systems is well documented.[2] Various schemes have been employed for this purpose, including *ab initio* relativistic effective core potentials and perturbational approaches. We describe here a fully-relativistic variational scheme based on the non-relativistic discrete variational Xα (DV-Xα) method.

The fundamental one-electron operator of relativistic quantum mechanics is the Dirac operator (1), in which $\underline{\alpha}$ and $\underline{\beta}$ represent the four 4 x 4 Dirac matrices,

$$h_D(r) = c\,\underline{\alpha}\cdot p(r) + \underline{\beta}c^2 + V(r) \qquad (1)$$

p is the momentum operator, and V(r) is any external scalar potential, such as the Coulombic electron-nucleus attraction.[3] Because the appropriate procedure for forming a corresponding many-body Hamiltonian is not known, an approximate construction must be employed.[4] In the Dirac-Fock approach the many-electron Hamiltonian is formed by combining a series of independent-particle Hamiltonians (1) with a non-relativistic electron-electron repulsion term (2). Following the traditional Hartree-Fock treatment,[5] the molecular wavefunction is represented as a Slater determinant over four-component one-

$$V_{ee} = \sum_{i>j} \frac{1}{r_{ij}} \tag{2}$$

electron wavefunctions.   Application of the variation principle to this antisymmetrized wavefunction yields the Dirac-Fock one-electron equations, in analogy to the non-relativistic case.[4] Calculations based on the Dirac-Fock treatment have been performed for many atomic[6] and an ever-increasing number of molecular systems.[2a,4c]

The Dirac-Fock equations contain both relativistic one-electron terms of the form shown in (1) and non-relativistic two-electron terms corresponding to electrostatic electron-electron repulsions. As in traditional LDF theory, the two-electron terms can be replaced by a Coulomb repulsion operator and an approximate local-density-functional exchange-correlation operator (3). Various

$$h_{DFS}(r) = c\ \underline{\alpha} \cdot p(r) + \beta c^2 + V_{ne}(r) + V_c(r) + V_{LDF}(r)$$

$$V_{ne}(r) = \sum_a \frac{Z_a}{|r - R_a|} \qquad V_c(r) = \int \frac{\rho(r')}{|r - r'|}\, dr'$$

$$V_{LDF}(r) = f[\rho(r)] \tag{3}$$

choices can be made for the last term—the simplest, and the one that was utilized in these investigations, is Slater's $X\alpha$ potential (4).[7] This choice for

$$V_{X\alpha}(r) = -3\alpha \left(\frac{3\rho(r)}{8\pi} \right)^{1/3} \tag{4}$$

the exchange potential defines the Dirac-Fock-Slater (DFS) Hamiltonian, which is a computationally convenient starting point for the investigation of molecular systems in which both relativistic effects and electron-electron repulsion interactions are significant.[8] In these calculations $\alpha = 0.70$.

To solve the DFS equations, the molecular orbitals are expressed as an LCAO expansion in a basis of symmetry-adapted four-component atomic functions. The atomic functions take the form (5). The large (P) and small (Q)

$$\psi_{nkm}(r) = \frac{1}{r} \left\{ \begin{array}{l} P_{nk}(r)\ \chi_{km}\ (\theta,\phi) \\ iQ_{nk}(r)\ \chi_{-km}(\theta,\phi) \end{array} \right\}$$

$$\chi_{km}(\theta,\phi) = |j\ m\rangle = \sum_{m'=-1/2}^{1/2} \langle l\ m-m';\ 1/2\ m'\ |jm\rangle\ Y_l^{m-m'}(\theta,\phi)\ s(m')$$

$$s(1/2) = \begin{bmatrix} 1 \\ 0 \end{bmatrix} \qquad s(-1/2) = \begin{bmatrix} 0 \\ 1 \end{bmatrix} \tag{5}$$

radial components are obtained from numerical atomic DFS calculations and are

Table I. Dimensions of atomic basis sets and well dimensions used in forming the molecular basis. The well depths were uniformly set to 3.0 a.u.

	Basis size	Inner well dimension (Å)	Outer well dimension (Å)
Th, U, Pu	8s, 7p, 6d, 5f	2.850	3.563
N	3s, 3p	2.325	2.907
H	2s, 2p[a]	1.800	2.250

[a] 2p only employed in the hydride calculations.

tabulated on a logarithmic grid; the angular parts χ_{km} are obtained as linear combinations of coupling coefficients and spherical harmonics. The angular index k takes on all non-zero integer values and is related to the j quantum number by relationship (6). Because χ_{km} and χ_{-km} have the same

$$k = \begin{cases} -(j+1/2) & \text{when } j = l+1/2 \\ (j+1/2) & \text{when } j = l-1/2 \end{cases} \qquad (6)$$

transformation properties under the molecular double group (and in fact have the same angular shapes),[9] it is possible to employ double group projection operators to generate symmetry adapted linear combinations of the $\psi_{nkm}(\mathbf{r})$ functions.[8]

The radial functions obtained from single atomic calculations seldom provide a sufficiently flexible molecular basis. In order to augment this set of functions, atomic virtual levels are included in the molecular basis. The levels are localized by placing the atom in a spherical well.[10] Both the occupied and virtual orbitals are very sensitive to the choice of well dimensions, so care must be taken to insure an even-handed treatment of atoms when forming basis sets. Because atomic DFS calculations are used to form the radial basis, the large and small radial components are coupled and the problem of variational collapse encountered in some Dirac-Fock investigations is avoided.[11] The basis set dimensions and well parameters employed in these calculations are summarized in Table I.

Once the molecular basis set is defined, a secular matrix can be formed from integrals of the basis over the DFS Hamiltonian.[8] The non-linear nature of the exchange potential makes analytical evaluation of these matrix elements difficult, and instead a numerical integration method is used. We employ a Diophantine integration scheme with 30·Z grid points per atomic center, identical to the non-relativistic case.[12] Further, in order to reduce the computational expense of evaluating Coulomb integrals, a least-squares fit of the charge density to the atomic basis functions is used.[13]

Three-Coordinate Actinide(III) Compounds

The electronic structure of three-coordinate actinide(III) compounds is of interest for several reasons. First, the +3 oxidation state is common to almost all the actinide elements. Second, an entire series of homoleptic uranium(III) complexes UX_3, X = alkyl,[14] amide,[15] aryloxide,[16] and cyclopentadienide,[17] is known, thus facilitating comparison of the influences of different ligand sets on an actinide metal in pseudo-isostructural environments. Third, the $(\eta^5\text{-Cp})_3An$ moiety is ubiquitous in the organometallic chemistry of the actinides. A thorough understanding of the orbital interactions in this fragment is essential to an understanding of its chemical behavior.[18] Fourth, the competition between spatially- and energetically-accessible metal 7s, 7p, 6d, and 5f orbitals in the frontier region is of fundamental interest.

In this study we focus on a set of small, three-coordinate actinide models. Our goal is to gain a qualitative understanding of the influences of this ligand configuration on actinide electronic structure. The degree of covalent interaction between an actinide atom and its associated ligands, and the relative participation of the various metal orbitals in forming bonds is evaluated. The similarities and differences between metal-ligand σ and π interactions are analyzed. The importance of relativistic corrections, including the spin-orbit effect, in dictating the orbital interactions is also explored.

Two classes of model complexes, containing either hydride (H^-) or amide (NH_2^-) ligands, are discussed here. The hydride substituent is ideal for modelling the σ-only interaction between a metal and a ligand. The simplicity of the hydride leads to both computational expediency and interpretational ease. We have performed calculations on a set of three trigonal planar AnH_3 molecules, An = Th, U, Pu, in order to study the σ-only actinide-ligand interactions. Similarly, to gain a fundamental understanding of π interactions between actinide elements and ligands it is important to employ a ligand set that has a well-defined and readily interpreted π character. Amide ligands are ideal for this purpose, and calculations have been performed on the trigonal-planar models $An(NH_2)_3$, An = Th, U, Pu, with the amide ligands oriented both normal to and coincident with the molecular plane. This set of calculations provides us with an overall picture of fundamental actinide-ligand interactions in the three-coordinate geometry, from which an understanding of more complex structures can be derived.

Actinide-Ligand σ Effects

The metal-ligand bonding in AnH_3 compounds is restricted to a σ-only interaction because of the orbital limitations of the hydride ligand. It is useful to consider first the consequences of molecular symmetry for these An–H interactions.[19] The hydride ligand donor orbitals span the D_{3h} representations a_1' and e'. Inspection of the D_{3h} character table reveals that the s, p, and d metal manifolds provide ample symmetry matches for the ligands—s and $d\sigma$ transform as a_1', and $p\pi$ and $d\delta$ as e'.[20] There is no symmetry-driven demand for the

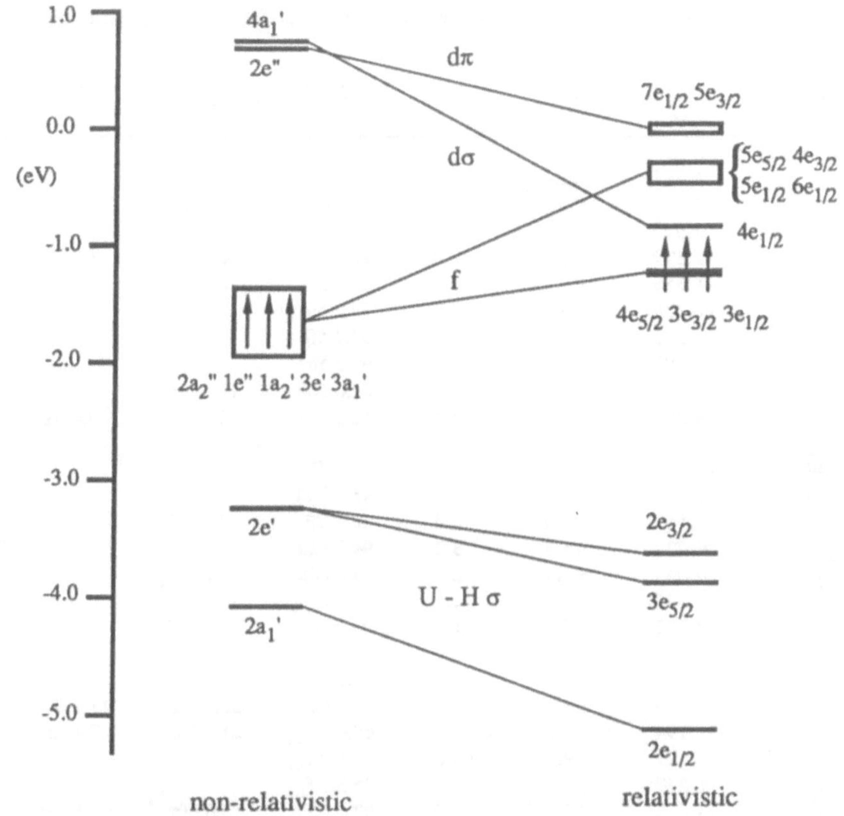

Figure 1. Comparison of non-relativistic and relativistic energy levels for UH_3.

incorporation of f orbitals into σ bonds. Metal f orbitals of both a_1' and e' symmetry exist, and the former has a particularly favorable angular orientation for the formation of an An 5f–H 1s bond, as shown in **1**. It has generally been held that f orbitals participate to a lesser extent in ligand bond formation than do the d orbitals, however.[18,21]

1

Figure 1 and Table II contain the results of non-relativistic and relativistic calculations on UH_3.[22] Comparison of these two calculations is quite instructive and confirms earlier conclusions about actinide-ligand bonding.[18] We consider first the non-relativistic results. The metal-ligand bonding orbitals ($2a_1'$ and 2e') are dominated by uranium 6d and 7s contributions but also contain considerable 5f character. Correspondingly, ligand-induced destabilization of the metal-based 6d and 7s orbitals is substantial [note in Table II, for instance, the

Table II. Molecular orbital energies and compositions for non-relativistic and relativistic UH₃.

Orbital	Energy (eV)	%U	%H	Primary Metal Contributions		
Non-relativistic UH₃:						
4e'	2.443	82.6	17.4	43.4 7pπ	38.0 6dδ	
4a₁'	0.713	100.2	-0.2	56.8 6dσ	42.6 7s	
2e"	0.676	96.5	3.5	95.3 6dπ		
3a₁'	-1.379	92.4	7.6	80.1 5fφ	6.5 7s	6.0 6dσ
3e'	-1.655	93.0	7.0	83.4 5fπ	8.6 6dδ	
1a₂'	-1.728	99.8	0.2	99.2 5fφ		
1e" [b,c]	-1.803	99.9	0.1	99.4 5fδ		
2a₂" [a]	-1.887	99.9	0.1	99.3 5fσ		
2e'	-3.246	43.7	56.3	24.0 6dδ	15.1 5fπ	
2a₁'	-4.083	36.5	63.5	16.5 5fφ	10.5 6dσ	9.4 7s
Relativistic UH₃:						
6e₃/₂	1.745	82.3	17.7	43.4 7pπ	36.6 6dδ	
6e₅/₂	1.165	91.6	8.4	48.4 7p̄	40.8 6dδ	
5e₃/₂	0.024	96.9	3.1	94.7 6dπ		
7e₁/₂	-0.063	97.7	2.3	78.0 6dπ	15.1 5f₇/₂	
6e₁/₂	-0.310	98.9	1.1	69.3 5f₇/₂	23.8 6d₁/₂	
5e₁/₂	-0.382	99.0	1.0	90.0 5f₅/₂		
4e₃/₂	-0.439	99.4	0.6	92.4 5f₃/₂		
5e₅/₂	-0.472	99.3	0.7	96.2 5f₁/₂		
4e₁/₂	-0.849	98.7	1.3	46.6 6dσ	32.0 5fφ	17.7 7s
3e₁/₂ [a,c]	-1.194	99.8	0.2	69.6 5f̄₅/₂	21.1 7s	7.4 6dσ
3e₃/₂ [a]	-1.197	99.4	0.6	97.1 5f̄₃/₂		
4e₅/₂ [a]	-1.247	98.9	1.1	95.3 5f̄₁/₂		
2e₃/₂	-3.643	40.8	59.2	32.6 6dδ	5.0 5fπ	
3e₅/₂	-3.870	34.7	65.3	23.6 6dδ	5.8 7pπ	5.1 5fπ
2e₁/₂	-5.126	36.9	63.1	27.1 7s	6.7 6dσ	5.0 5fφ

[a] Singly occupied orbital. [b] Doubly occupied orbital. [c] HOMO.

1.8 eV destabilization of the 6dδ (4e') orbital above the non-interacting 6dπ (2e")], while splitting of the 5f manifold is less. The highest-energy 5f-based molecular orbital is the fφ orbital diagrammed in 1. The relatively small splitting of the 5f manifold is a consequence of the low energy and small size of the f orbitals in non-relativistic atomic uranium.

The effects of spin-orbit coupling make the relativistic results somewhat more difficult to interpret. The number of distinct symmetry representations is reduced in the double group, and the jj-coupled basis functions (5) contain admixtures of σ and π, π and δ, and δ and φ character.[2b,9a] As pointed out by Boerrigter et al., this mixing is not difficult to decipher and can in fact yield additional information about molecular bonding.[23] For instance, the 2e₁/₂ orbital, the relativistic counterpart of the 2a₁' σ-bonding level, contains 6d̄₁/₂ and 6d₁/₂ contributions in a ratio clearly diagnostic of a dσ orbital. Not only does this correlation clarify interpretation of the orbital, but it also indicates that interactions between the metal d manifold and the ligands are significant. In the relativistic calculation, the molecular orbitals that contain substantial uranium

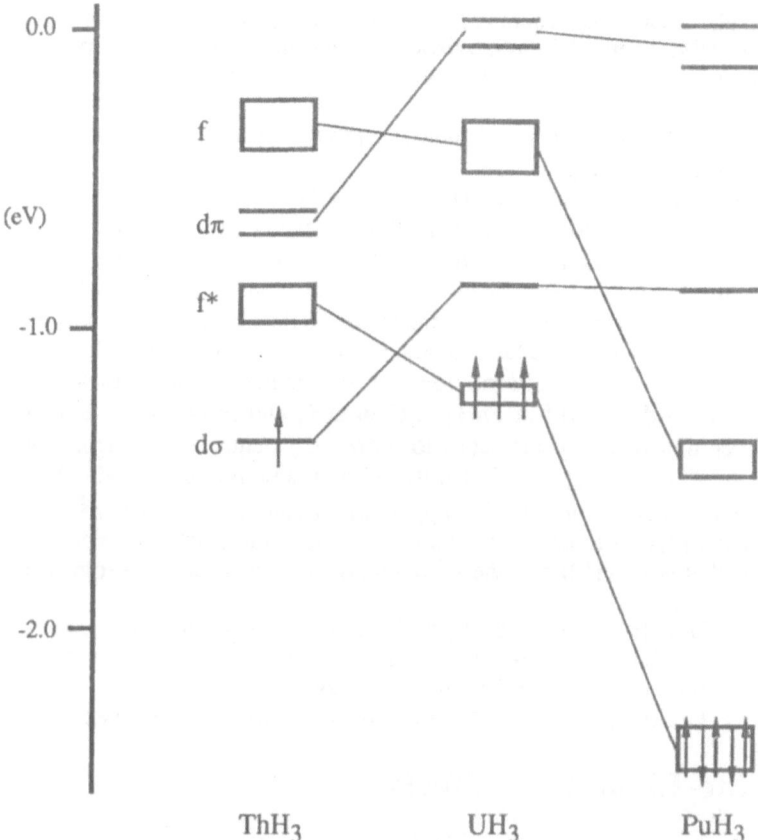

Figure 2. Comparison of DFS molecular orbital energies for three actinide trihydrides.

6d and ligand character have a 6d composition characteristic of non-spin-orbit-coupled atomic orbitals; the ligand field quenches spin-orbit coupling in the 6d manifold.[23] The data presented in Table II reflect this interpretation of the atomic compositions of molecular orbitals. In contrast, the 5f manifold exhibits only spin-orbit splitting and no ligand-field-induced splitting. The 5f orbital contributions do not appear in ratios indicative of non-spin-orbit coupled orbitals but rather as pure jj-coupled functions. The 5f orbitals do not interact substantially with the hydride ligands, and the 5f contribution to the U–H bonding levels is virtually nil.

The differences in the non-relativistic and relativistic results point out the dichotomy in the nature of the s, p, and d actinide orbitals relative to the f. The former dominate the metal-ligand σ bonding in both calculations, as revealed by their contributions to the bonding molecular orbitals and the distinct ligand-induced destabilization of the metal-based levels.[24] In contrast, incorporation of relativistic corrections destroys any contribution of the metal 5f orbitals to metal-ligand bonding. There is practically no 5f character in the bonding levels, and the splitting in the 5f manifold is clearly caused by spin-orbit, rather than ligand field, interactions. There are two main sources of the distinction between these two orbital sets. The relativistic contraction of the 7s and 7p orbitals makes them better able to compete with the 5f and 6d orbitals for ligand charge density. Further, while the metal 6d and 5f orbitals are energetically similar, the spatial extent of the former is substantially greater than the latter and leads to better U 6d–H 1s overlap.[18,21,25]

The DFS results obtained for ThH$_3$ and PuH$_3$ (Figure 2) are analogous to those for UH$_3$. The nature of the interactions does not vary with metal, but the relative energetic orderings are modified. As Z increases, the 5f block drops in energy while the 6d and 7s remain relatively constant, in accord with atomic results.[6] As the 5f orbital energies decrease across the series, the 5f contribution to the An–H bonds does increase, although it remains a fraction of the d and s contributions.[25] More significant, however, is a change in the character of the HOMO, from a d/s hybrid in ThH$_3$ to f^3 in UH$_3$ and f^5 in PuH$_3$. Recently, Edelstein et al. used epr spectroscopy to show the ground state configuration of Cp"$_3$Th (Cp" = 1,3-bis(trimethylsilyl)cyclopentadienyl) to be 6d^1.[26] Our present and earlier results clearly support this assignment.[18] The d^1 ground configuration appears to be a result not of some ligand interactions unique to the cyclopentadienyl ligand but of the +3 oxidation state and three-coordination in general.

We conclude that, at least for the early actinides, σ bonding in a three-fold ligand environment occurs through s, p, and d metal orbitals. These results are not dependent on our choice of ligand: calculations on other models of a σ-only system, e.g. U(CH$_3$)$_3$ or UF$_3$, lead to the same qualitative conclusions.

Actinide-Ligand π Effects

The introduction of π-donor ligands into the actinide coordination sphere has interesting electronic consequences. The symmetry-allowed metal-ligand orbital matches for An(NH$_2$)$_3$ in the two possible D$_{3h}$ orientations, amide hydrogens

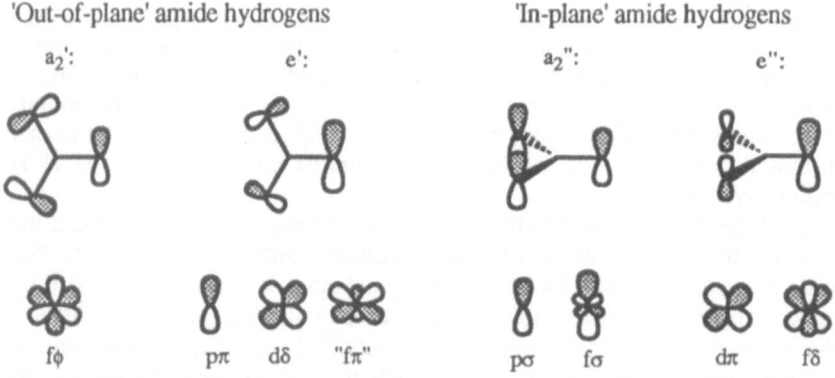

Figure 3. Symmetry matches between ligand π orbitals and metal based orbitals. Note that the fπ is rotated 90° relative to the pπ and dδ.

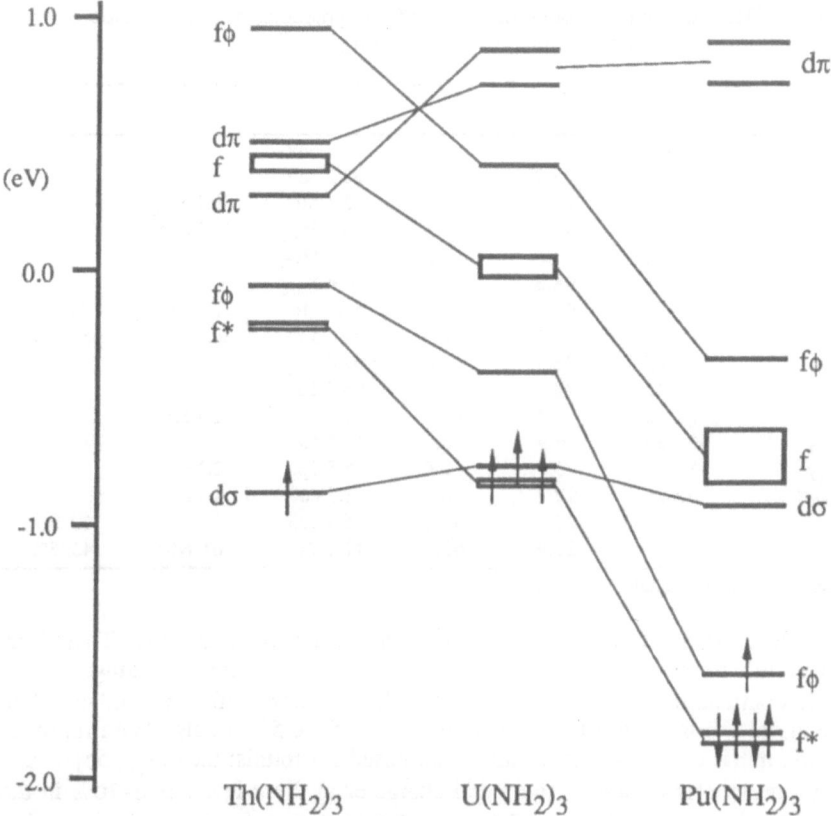

Figure 4. Comparison of the DFS metal-based-orbital energies for the perpendicular actinide triamides.

"out-of-plane" (perpendicular) and "in-plane" (parallel), are presented schematically in Figure 3. In the parallel case the ligand π orbitals are positioned normal to the molecular plane and can interact with similarly oriented metal p, d, and f orbitals. In the perpendicular case symmetry matches must be drawn from within the molecular plane and hence compete with metal-ligand σ bonds for a limited number of orbitals. The a_2' representation is particularly interesting, because the fϕ orbital is the only metal orbital of the proper symmetry to accept electron density from the ligand π orbital combination.

A comparison of the metal-based-orbital energies for perpendicular An(NH$_2$)$_3$, An = Th, U, Pu, is presented in Figure 4, and Table III contains the results specifically for U(NH$_2$)$_3$.[27] We shall consider first the U–N σ interactions, which are qualitatively similar to those in the AnH$_3$ complexes. Comparison of the results in Tables II and III reveals that the metal contribution to the σ bonds is diminished in the amide case relative to the hydride, a consequence of the greater electronegativity of nitrogen than hydrogen. The *nature* of the metal contribution is essentially unchanged—the metal 6d and 7s

Table III. Molecular orbital energies and compositions from relativistic calculations on perpendicular $U(NH_2)_3$.

Orbital	Energy (eV)	%U	%N	Primary Metal Contributions		
$8e_{3/2}$	0.891	95.4	1.2	94.8 $6d\pi$		
$10e_{1/2}$	0.733	95.2	1.4	88.2 $6d\pi$		
$9e_{1/2}$	0.426	92.9	7.0	83.5 $5f\phi$	8.2 $6\bar{d}_{1/2}$	
$8e_{1/2}$	0.055	99.2	0.3	76.6 $5f_{5/2}$	13.3 $6d\sigma$	
$7e_{3/2}$	0.032	97.6	2.2	94.8 $5f_{3/2}$		
$8e_{5/2}$	-0.037	98.3	1.2	96.4 $5f_{1/2}$		
$7e_{1/2}$	-0.401	95.9	3.4	55.3 $5f\phi$	32.6 $6d\sigma$	7.6 $7s$
$6e_{1/2}$ [a,b]	-0.774	90.1	9.3	43.0 $5f\phi$	30.5 $7s$	15.1 $6d\sigma$
$6e_{3/2}$ [a]	-0.805	98.7	1.0	97.0 $5\bar{f}_{3/2}$		
$7e_{5/2}$ [a]	-0.841	98.3	1.4	96.0 $5\bar{f}_{1/2}$		
$5e_{3/2}$	-3.762	8.4	90.5	5.2 $7p\pi$	2.9 $5f\pi$	
$5e_{1/2}$	-3.930	9.9	90.2	10.3 $5f\phi$		
$6e_{5/2}$	-3.956	8.9	90.6	5.5 $7p\pi$	2.2 $6d\delta$	
$4e_{3/2}$	-5.621	27.1	66.7	21.5 $6d\delta$	1.9 $5f\pi$	
$5e_{5/2}$	-5.837	23.5	69.2	16.5 $6d\delta$	3.6 $7p\pi$	2.1 $5f\pi$
$4e_{1/2}$	-6.639	23.6	67.8	12.6 $7s$	6.0 $6d\sigma$	4.2 $5f\phi$

[a] Singly occupied orbital. [b] HOMO.

orbitals dominate σ bond formation. Similar results are obtained for Th and Pu. Thus, little new is to be gained from an analysis of An–NH_2 σ bonding.

In contrast, the presence of the π-donor ligands has a rather marked effect on the metal-orbital manifolds, in particular that of the $5f$ orbitals. We first focus on the metal contributions to the ligand-based π orbitals: the $6e_{5/2}$, $5e_{1/2}$, and $5e_{3/2}$ molecular orbitals. The metal character in these levels is 8–10% in the uranium case. The $6e_{5/2}$ and $5e_{3/2}$ orbitals derive from the e' interaction diagrammed in Figure 3; their main metal contribution is $7p\pi$ with smaller amounts of $6d\delta$ and $5f\pi$. The $6d\delta$ orbital is involved in σ bond formation and thus cannot participate to a great extent in π bonding, and the $5f\pi$ orbital is poorly oriented for π bond participation. The $7p\pi$ orbital is relatively free of these hindrances, but it is too high in energy to form a strong π bond. As a result, the overall π interaction in these two levels is weak. The $5e_{1/2}$ orbital derives from the a_2' interaction shown in Figure 3 and does in fact incorporate a sizable (10%) metal $5f$ contribution. The importance of this interaction is reflected in the destabilization of the $5f\phi$ orbitals above the other members of the $5\bar{f}$ and $5f$ manifolds.

This $5f$ manifold splitting in perpendicular $An(NH_2)_3$ complexes can be analyzed in terms of a combination of strong ligand field and spin-orbit effects (Figure 5).[23] In a spherical potential all the $5f$ orbitals are degenerate. Introduction of the amide ligand field destabilizes one combination of the $f\phi$ orbitals, effectively removing it from consideration, but does not perturb the rest of the f levels, including the other $f\phi$ function. Application of the spin-orbit operator ($\xi l \cdot s$) is best considered in two steps. The "diagonal" part ($l_z \cdot s_z$) couples spatial and spin functions and uniformly splits the combinations by an amount proportional to their m_l value (7). This operator does not produce a

$$\langle \, Y_1^{m} \, (\theta,\phi) \; s(\tfrac{\pm 1}{2}) \; | \, l_z \cdot s_z \, | \; Y_1^{m} \, (\theta,\phi) \; s(\tfrac{\pm 1}{2}) \, \rangle = \pm \, m_l/2 \qquad (7)$$

diagonal perturbation of the lower-energy fϕ orbital. It does couple it to the higher-energy fϕ, but this interaction is small and can be ignored to first order. The "off-diagonal" spin-orbit operator $[(l_+ s_- + l_- s_+)/2]$ mixes the spin-orbitals with the same m_j value and splits them by an amount directly related to their energetic proximity. The $m_j = 1/2$ and $m_j = 3/2$ pairs split into degenerate sets of $\bar{f}$ ($j = 5/2$) and f ($j = 7/2$) functions. The lower-energy fϕ orbital contains an admixture of $m_j = 5/2$ and $7/2$ components and thus mixes weakly with the f$\delta_{5/2}$. The net splitting pattern appears at the right of Figure 5. In a spherical field with spin-orbit coupling the f block separates into $\bar{f}$ and f manifolds according to j value; the π ligand field perturbation in effect decouples both the fϕ orbitals, and they appear at energies above their respective manifolds.

The qualitative coupling scheme presented above is in excellent agreement with the results in Figure 4 and Table III. The molecular orbital diagrams for AnH_3 (Figure 2) and perpendicular $An(NH_2)_3$ (Figure 4) are qualitatively the same, except for the splitting off of the 5fϕ orbitals from the 5$\bar{f}$ and 5f manifolds in the latter molecules. The 5fϕ contributions conform to those expected for non-spin-orbit-coupled orbitals, as confirmed by analysis of the

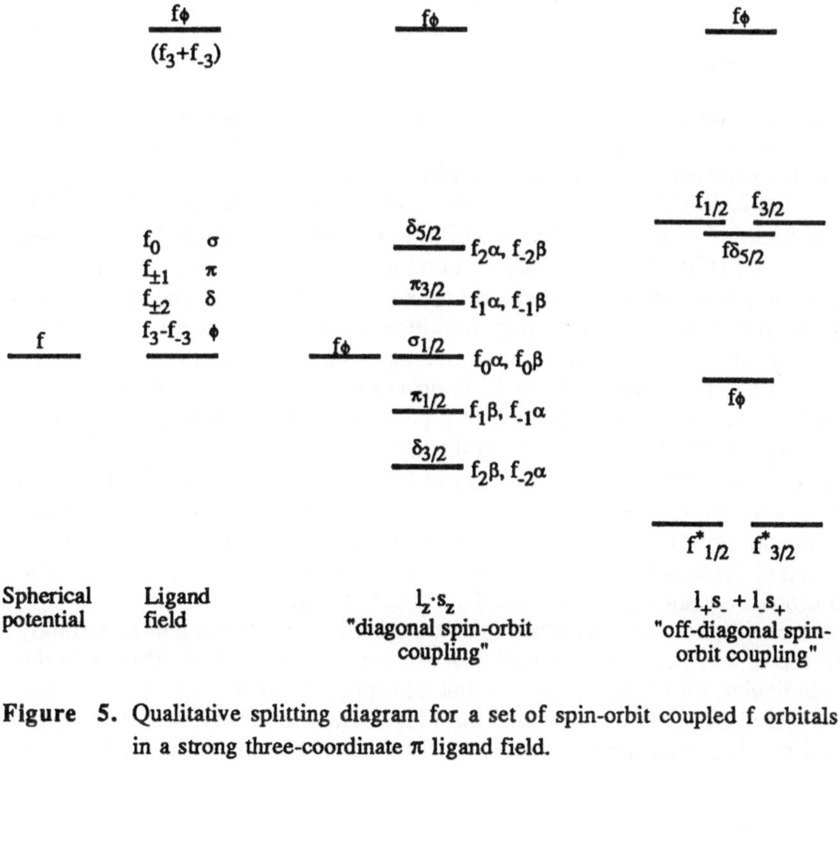

Figure 5. Qualitative splitting diagram for a set of spin-orbit coupled f orbitals in a strong three-coordinate π ligand field.

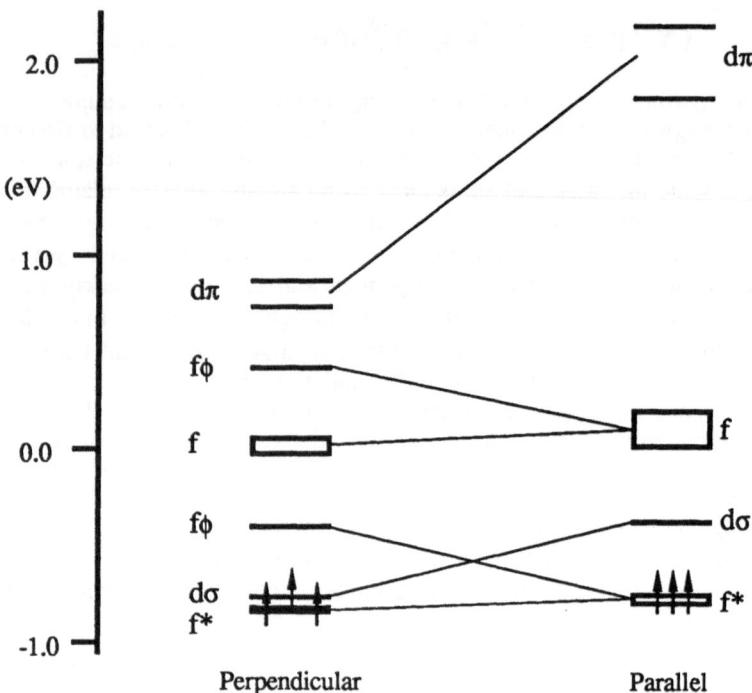

Figure 6. Comparison of metal-based orbital energies of U(NH$_2$)$_3$ in perpendicular and parallel ligand orientations.

orbital electron densities. This picture is somewhat clouded by higher order mixing that is not included in our simple analysis. For instance, in the uranium case, the low-energy destabilized fφ orbital is close in energy to the d/s hybrid that is the LUMO in UH$_3$. As a consequence, the two mix strongly to yield two orbitals of nearly equal 6d/7s and 5f composition (6e$_{1/2}$ and 7e$_{1/2}$, Table III), and the HOMO of U(NH$_2$)$_3$ is found to contain substantial d/s character. Similar mixings induced by energetic proximity are present in the thorium and plutonium calculations. The most important conclusion, however, is that π–donor ligands and actinide 5f orbitals can interact to a significant extent and that the interaction is manifested in 5f contributions to the ligand donor levels, modifications in the 5f orbital splitting patterns, and a quenching of the spin-orbit coupling of orbitals involved in the interaction.

Finally, a comparison of U(NH$_2$)$_3$ in the perpendicular and parallel ligand orientations is presented in Figure 6. The results are representative of the other metals and clearly reveal an increase in 6dπ participation in metal-ligand π bonding at the expense of the 5f contribution in the parallel conformation. The 6dπ orbitals are strongly destabilized upon ligand rotation, while the 5fφ orbitals coalesce with the other f orbitals. The 6d orbitals interact much more strongly with the ligand field in the parallel orientation than do the 5f orbitals in the perpendicular, for the same reasons that 6d orbitals form stronger σ bonds than do 5f orbitals. Therefore the parallel form is preferable electronically. The metal-ligand overlap population increases by 0.1 electrons in the parallel relative to perpendicular molecules. The one structurally characterized uranium triamide contains "propellered" amide ligands rotated 40.65° from the perpendicular form.[28] It is tempting to conclude that π electronic effects and steric repulsions compete to produce the observed equilibrium geometry.

Conclusions

The fully-relativistic implementation of the DV-Xα method has proved valuable in elucidating the interplay between ligand field and spin-orbit effects on the electronic structure of three-coordinate actinide complexes. The orbital interactions important in describing metal-ligand bonding have been delineated. Actinide-ligand σ bonding primarily involves the metal-based 7s, 7p, and 6d orbitals and has very little influence on the metal 5f manifold, which experiences instead a distinct spin-orbit splitting. In contrast, the 5f orbitals are able to participate to some extent in actinide-ligand π bonding under the proper symmetry conditions, with commensurate quenching of the spin-orbit interaction and destabilization of the interacting orbitals above their non-interacting partners. These results form a basis upon which a qualitative picture of actinide-ligand interactions in many different systems can be built.

References

1. (a) Katz, J. J.; Morss, L. R.; Seaborg, G. T., ed., *The Chemistry of the Actinide Elements*, Chapman and Hall, New York 1986. (b) Marks, T. J.; Fragalà, I. L., ed., *Fundamental and Technological Aspects of Organo-f-Element Chemistry*, Reidel, Dordrecht 1985. (c) Marks, T. J.; Ernst, R. D. in *Comprehensive Organometallic Chemistry*, Ch. 21, Wilkinson, G.; Stone, F. G. A.; Abel, E. W., ed., Pergamon Press, Oxford 1982.
2. (a) Pyykkö, P. *Chem. Rev.* **1988**, *88*, 563–594. (b) Pitzer, K. S. *Acc. Chem. Res.* **1979**, *12*, 271–276. (c) Pyykkö, P.; Desclaux, J.-P. *Acc. Chem. Res.* **1979**, *12*, 276–281.
3. Schiff, L. I. *Quantum Mechanics*, 3rd. ed., McGraw-Hill, New York 1968.
4. (a) Grant, I. P. *Adv. Phys.* **1970**, *19*, 747–811. (b) Grant, I. P. in *Methods in Computational Chemistry, Vol. 2: Relativistic Effects in Atoms and Molecules*, Ch. 1, Wilson, S. ed., Plenum Press, New York 1988.
    (c) Wilson, S. in *Methods in Computational Chemistry, Vol. 2: Relativistic Effects in Atoms and Molecules*, Ch. 2, Wilson, S. ed., Plenum Press, New York 1988.
5. Szabo, A.; Ostlund, N. *Modern Quantum Chemistry*, McGraw-Hill, New York 1989.
6. Desclaux, J.-P. *Atom. Data Nucl. Data. Tables* **1973**, *12*, 312–406.
7. Slater, J. C. *Adv. Quantum Chem.* **1972**, *6*, 1–92.
8. (a) Rosen, A.; Ellis, D. E. *Chem. Phys. Lett.* **1974**, *27*, 595–599.
    (b) Rosen, A.; Ellis, D. E. *J. Chem. Phys.* **1975**, *62*, 3039–3049.
9. (a) Powell, R. E. *J. Chem. Ed.* **1968**, *45*, 558–563. (b) White, H. E. *Phys. Rev.* **1931**, *38*, 513–520.

10. (a) Averill, F. W.; Ellis, D. E. *J. Chem. Phys.* **1973**, *59*, 6412–6418.
    (b) Doris, K. A.; Delley, B.; Ratner, M. A.; Marks, T. J.; Ellis, D. E. *J. Phys. Chem.* **1984**, *88*, 3157–3159.
11. Ellis, D. E.; Goodman, G. L. *Int. J. Quantum Chem.* **1984**, *25*, 185–200.
12. Baerends, E. J.; Ellis, D. E.; Ros, P. *Chem. Phys.* **1973**, *2*, 41–51.
13. Delley, B.; Ellis, D. E. *J. Chem. Phys.* **1982**, *76*, 1949–1960.
14. Van Der Sluys, W. G.; Burns, C. J.; Sattelberger, A. P. *Organometallics* **1989**, *8*, 855–857.
15. Andersen, R. A. *Inorg. Chem.* **1979**, *18*, 1507–1509.
16. Van Der Sluys, W. G.; Burns, C. J.; Huffman, J. C.; Sattelberger, A. P. *J. Am. Chem. Soc.* **1988**, *110*, 5924–5925.
17. Brennan, J. G. Ph. D. Dissertation, University of California, Berkeley, 1985.
18. Quasi-relativistic Xα-SW calculations have been used extensively in the investigation of Cp_3An and its adducts:
    (a) Bursten, B. E.; Rhodes, L. F.; Strittmatter, R. J. *J. Am. Chem. Soc.* **1989**, *111*, 2758–2766. (b) Bursten, B. E.; Rhodes, L. F.; Strittmatter, R. J. *J. Am. Chem. Soc.* **1989**, *111*, 2756–2758. (c) Bursten, B. E.; Rhodes, L. F.; Strittmatter, R. J. *J. Less Com. Met.* **1989**, *149*, 207–211. (d) Bursten, B. E.; Strittmatter, R. J. *J. Am. Chem. Soc.* **1987**, *109*, 6606–6608. (e) Bursten, B. E.; Strittmatter, R. J., submitted for publication.
19. Ziegler, T. *J. Am. Chem. Soc.* **1983**, *105*, 7543–7549, and references therein.
20. In this paper, metal-based orbitals are denoted σ, π, δ, and ϕ in correspondence with their symmetry about the principle molecular axis. Further, in dealing with jj-coupled orbitals, we adhere to the notation $\bar{l}$ (or l^*) and l for the lower and higher j-valued functions, respectively, and append a subscript referring to m_j if necessary. Refer to reference 9a for a clear introduction to these and related concepts.
21. Bursten, B. E.; Casarin, M.; DiBella, S.; Fang, A.; Fragalà, I. L. *Inorg. Chem.* **1985**, *24*, 2169–2173.
22. An An–H bond length of 2.0 Å was used in all the reported hydride results. Investigations of other bond lengths led to analogous conclusions.
23. Boerrigter, P. M.; Baerends, E. J.; Snijders, J. G. *Chem. Phys.* **1988**, *122*, 357–374.
24. The separation into metal manifolds of distinct orbital character is not entirely clean, due to higher-order mixing between orbitals of the same symmetry type. This mixing only clouds the overall picture and is not of any qualitative import.
25. (a) Rösch, N. *Inorg. Chim. Acta* **1984**, *94*, 297–299. (b) Rösch, N.; Streitwieser, A. *J. Am. Chem. Soc.* **1983**, *105*, 7237–7240. (c) Rösch, N.; Streitwieser, A. *J. Organomet. Chem.* **1978**, *145*, 195–200.
26. Kot, W. K.; Shalimoff, G. V.; Edelstein, N. M.; Edelman, M. A.; Lappert, M. F. *J. Am. Chem. Soc.* **1988**, *110*, 986–987.
27. Bond lengths and angles used in the amide calculations: An–N = 2.32 Å, N–H = 1.01 Å, An–N–H = 120.0°.
28. Stewart, J. L. Ph. D. Dissertation, University of California, Berkeley, 1988.

17
Local Density Functional Calculations on Metathesis Reaction Precursors

Dennis J. Caldwell and Patrick K. Redington

Abstract

Calculations were carried out to assess the role of Local Density Functional (LDF) programs in following the course of metathesis reactions in large molecular systems. This study was designed to provide preliminary information on: (1) the nature of weakly bound ligands in cocatalyst structures, (2) the influence of phenyl ring substitution on catalytic activity, (3) the role of $AlCl_3$ as a cocatalyst, and (4) the influence of macroenvironment on the basic electronic structure of the reactive site. For this purpose, calculations were carried out on an $MoCl_4O$/tetrahydrofuran (THF) complex, ring substituted derivatives of $MoCl_5O$-phenyl, the basic metallacyclobutane ring prototype, $Mo(CH_2)_3CH_3ClO$/$AlCl_3$, and a large fluorinated molybdenum complex.

The data indicate that atomic charges and Mayer bond orders (based on the second order density matrix) show considerable promise for making quantitative correlations (QSAR) between structure and catalytic activity. The results for the THF complex and the metallacyclobutane prototype suggest that increased catalytic activity may be associated with lowered bond orders at the active metal site. It was found that the orbital populations were 80-90% d in all cases. This is in contrast to the strong bonding associated with sp^3d^2 hybridization. There is evidently a certain amount of delocalization, which promotes the facile bonding changes necessary for low barrier catalysis.

Introduction

There is considerable interest in gaining new insights into the realm of transition metal catalysis using the power of quantum computational chemistry. Existing techniques have demonstrated their ability to provide unique data not available from experiment. For a given budget there are severe constraints on the size of the system which can be studied computationally. The most economical route is via a semiempirical formalism, where there has been considerable success in the calculation of ground state thermodynamic properties. In the calculation of transition state properties needed for reaction rate analysis, the situation is not so bright.

To carry out reasonably relevant investigations of reactivity in transition metal complexes, the computational standards are set by the array of ab initio methods. If the required data could be accurately generated at the SCF level, systems approaching the size of commercially viable catalysts could be routinely studied. Under these conditions, it is likely that the parametrization of companion semiempirical methods would be facilitated. Since it is well known that the SCF level of calculation is inadequate for general geometry optimized thermodynamic calculations, a severe limitation is placed on the allowed molecular size.

In the last decade, considerable advances have been made in the implementation of density functional theory. With a properly calibrated and monitored program, horizons are opened into areas of direct interest to industrial catalysis. In this approach, unlike the semiempirical, there is one central approximation, which is universal for all calculations. The method is based upon the existence of a universal functional of the total electron density, which gives the total energy of the system (Hohenberg and Kohn, 1964). As with solutions to the Schrödinger Equation, some suitable approximation must be made for practical calculations. Most approaches to this problem are based upon the Local Spin Density (LSD) assumption (Kohn and Sham, 1965). The unique advantages of density functional theory stem from the simplification which occurs when such approximations are implemented. See (Andzelm and Wimmer) and references therein.

Typically in the theoretical investigation of a particular catalytic process, a survey of the literature will reveal that ab initio calculations have been made on a number of small molecule model systems. Invariably the results are educational and relevant in a general sense, but the actual molecules treated are far from optimum in a practical sense, and in some cases would either be unusable or would exhibit very limited catalytic activity. For example, in the treatment of spectator ligands, heavy atoms or large organic groups are replaced by H as in substituting phosphine for triphenyl phosphine. This immediately removes the possibility of systematically studying the process of fine tuning catalysts to satisfy certain desired properties.

It has been fairly well established that SCF wave functions at the Hartree-Fock level are inadequate for studying transition metal complexes. Consequently, the HF calculations, which scale as N^4, must be supplemented with CI, where N is the number of basis functions. One of the simplest, MP2, scales as N^5. On the other hand, the LSD methods scale as N^3, and if one takes advantage of the basis set's sparsity, an N^2 dependence can be approached (Andzelm et al., 1989).

The implementation of computational studies with the faster and more efficient density functional methods allows us to pursue a computationally guided experimental research program which employs direct comparisons between calculated and measured values for the same transition metal system. In this way unique quantitative information can be gathered to supplement a data set for the application of predictive chemometric techniques.

Metathesis Reactions

A very efficient route to the synthesis of polymers is provided by the metathesis reaction. This is catalysed by generally hexavalent transition metal complexes with elements such as molybdenum or tungsten.

The rates of transition metal catalysed metathesis reactions are strongly dependent upon the ligands surrounding the metal. This comprises reactions of the type:

$$
\begin{array}{ccc}
\mathrm{R_1{-}C{=}C{-}R_2} & \mathrm{Metal} & \mathrm{R_1{-}C \quad C{-}R_2} \\
+ & \Longrightarrow & \| \ + \ \| \\
\mathrm{R_3{-}C{=}C{-}R_4} & \mathrm{Complex} & \mathrm{R_3{-}C \quad C{-}R_4}
\end{array}
$$

The propagator for this process is the complex:

$$
\begin{array}{c}
\mathrm{CH_2} \\
\| \\
\mathrm{X{-}M{-}Z} \\
| \\
\mathrm{Y}
\end{array}
$$

The key intermediate, a metallacyclobutane ring, is illustrated in the polymerization process:

```
    H   H                          H
    |   |                          |
R₁-C=[C-R-C]ₙ=CH-R-C-H            C
                  ||                \           ⟹
                X-M-Z    +   ||  R
                  |                 /
                  Y                C
                  |
                  H
```

```
    H   H           H  H
    |   |           |  |
R₁-C=[C-R-C]ₙ=CH-R-C—C             ⟹
                        \
                    |   | R
                        /
            X— M—C
              /\  |
              Y Z H
```

```
    H     H
    |     |
R₁-C=[C-R-C]ₙ₊₁=CH-R-CH
                  ||
                X-M-Z
                  |
                  Y
```

Here, the ligands X, Y, Z can be -Cl, -O-R, or =N-R.

The capacity of density functional methods for treating large systems can be utilized in a study of potential descriptors for ligands, which focuses on the following points:

1. Use of the total energy calculation to obtain barrier heights for the critical steps in the catalytic process.

2. Establishing the conditions under which the ring intermediate is a true thermodynamic species or a transition state.

3. Identifying a matrix of relevant computational parameters suitable for a statistical (QSAR) correlation of ligand effects on reactivity and selectivity.

In practice the actual structure of the intermediates is not well characterized. This gives computational chemistry a potentially large role in delin-
eating the most probable reaction paths. Enough is known about these
systems to narrow the field down to a manageable number of candidates
to provide a plausible program of investigation.

Calculations were carried out on four series of transition metal catalyst
prototype systems:

1. $MoCl_4O$/Tetrahydrofuran (THF) complex

2. Ring substituted derivatives of $MoCl_5O$-phenyl

3. $Mo(CH_2)_3$ CH_3ClO/$AlCl_3$

4. $Mo(CH_2CH_2CH_2)$ (N-aromatic) $[OC(CF_3)_2(CF_2CF_2CF_3)]_2$

These calculations were designed to explore the following areas:

1. The nature of weakly bound ligands in co-catalyst structures

2. The influence of phenyl ring substitution on catalytic activity

3. The influence of $AlCl_3$ as a co-catalyst

4. The electronic structure of a known stable metallacyclobutane

The calculations were carried out on the Cray XMP and YMP computers
using the DGAUSS Local Density Functional program developed by Jan
Andzelm (Andzelm and Wimmer). Molecules of up to 100 atoms can be
calculated in a few hours. This has accordingly provided us with a unique
opportunity to delve into the structure of the intermediates and catalyst
precursors in metathesis chemistry.

Calculations on the fluorinated molybdenum complex, for which crys-
tallogaphic data were available from the literature (Schrock et al., 1988),
were particularly demanding. This molecule has 80 atoms, and symmetry
cannot be utilized. An all electron calculation was performed, using a double zeta + polarization basis set (Andzelm et al., 1989). This amounts to
one thousand orbitals and over two thousand fitting basis functions. The
local spin density calculations took less than an hour and a half on a single processor YMP, which could be reduced to fifteen minutes if the eight
processor option were used.

These calculations provided the following information.

1. Atomic charges

2. Bond orders

3. Molecular geometry

4. Molecular energies

The last two were available on a limited basis in the first version of the program. This confined us either to the complete optimization of small molecules or calculations on fixed geometry conformations of large ones. The bond order matrix (Mayer, 1986) was calculated in all cases and has proved to be invaluable in elucidating many relevant concepts.

Computational Results

Below is a summary of the salient points for each series of calculations.

Calibration Compounds

In order to test the accuracy of the program on the types of compounds under consideration, several calculations were made on some representative systems. This included: $Mo(CO)_6$, $MoCl_6$, $MoCl_4O$, $TiCl_4$, and $AlCl_3$. The appropriate geometrical data are displayed in Table 1, along with the corresponding experimental quantities. In most cases bond distances agree to within 1 or 2 hundredths of an Angstrom, thus reenforcing our confidence in the ability of DGAUSS to treat transition metal complexes. A resume of the experimental data is given in (Cotton and Wilkinson, 1988).

Table 1: CALIBRATION COMPOUND GEOMETRIES

Compound	Parameter	Calc.	Exp.
$Mo(CO)_6$	R(Mo-C)	2.02	2.06
	R(C-O)	1.155	1.45
$MoCl_6$	R(Mo-Cl)	2.29	—
$MoCl_4O$	R(Mo-O)	1.66	1.66
	R(Mo-Cl)	2.29	2.28
	<O-Mo-Cl	103.5	103.0
$TiCl_4$	R(Ti-Cl)	2.16	2.17
$AlCl_3$	R(Al-Cl)	2.066	2.06

$MoCl_4O$/Tetrahydrofuran Complex

This series was studied in order to assess the role of weakly bound ligands on catalyst prototypes. The activity of transition metal complexes is intimately bound up with the nature of the ligands as well as the active site itself. Catalysts may be fine tuned either by substitution on strongly bound ligands or introducing weakly bonded structures into the coordination sphere.

First the $MoCl_4O$ complex was optimized by finding the minimum energy for its geometrical degrees of freedom. This corresponds to the distances

Mo - O $= 1.66$Å and Mo - Cl $= 2.29$Å with an O-Mo-Cl angle of $103.5°$. The Cl atoms lie in a plane and are folded back from the double bonded oxygen atom.

Next the tetrahydrofuran (THF) ring was optimized with MOPAC and the oxygen on the ring bonded to the Mo atom as shown.

```
Cl Cl          CH2—CH2
 | /            /
                |
OI=Mo- - - - -OII
 | \            \
Cl Cl          CH2—CH2
```

Two variables were studied: the Mo-furan oxygen distance R and the torsion angle ϕ. A rather shallow minimum was found at $R = 2.308$Åwith $\phi = 45°$ corresponding to the staggered position with respect to the Cl-Mo-Cl and C-O-C moieties. The Cl-Mo-O_I angle folded back to $97°$ to reach the minimum energy. This corresponds to an extra lowering of 6.4 kcal, emphasizing the importance of full geometry optimization in assessing reaction path energies.

The charges and bond orders for this system are shown in Tables 2, 3, and 4. The $90°$ angle is the eclipsed position and the $45°$ is the staggered. Initially the $MoCl_4O$ complex is electrically neutral. Proximity of the THF ring imparts an overall negative charge of about .23 units on the six original atoms of the complex. All atoms become about .04 units more negative. Rotations about the O_I-Mo-O_{II} axis involves small changes in individual atomic charges but the total is essentially unchanged.

Evidently the increase in negative charge on the $MoCl_4O$ complex due to the addition of the THF is associated with a net reduction in binding energy, since the total bond order decreases from 6.43 to 6.19. There does not appear to be any correlation between bond order changes and rotational barriers. The value of 2.70 kcal for the differences between staggered and eclipsed conformations is reasonable, but the bond order change is in the opposite direction. This is consistent with the rule of thumb that bond orders are a potential measure of energies associated with bond angles and bond distances but not dihedral angles.

The bond order between Mo and the THF oxygen is only .32, indicating a relatively weak linkage consistent with a valence of six rather than seven. Also one notes appreciable bond orders between the Cl and oxygen atoms, as well as between the Cl atoms, indicating a certain amount of delocalization. This is probably somewhat more than is encountered in typical compounds of nonmetallic elements such as hydrocarbons. An examination of the geometry indicates that bond orders between atoms on opposite ends of a line passing through the central metal atom are twice as large as those for adjacent next nearest neighbors. This is consistent with the

Table 2: CHARGES ON FURAN COMPLEXES

	90°	45°	MoCl₄O
Mo	.513	.543	.613
O(1)	- .377	- .376	- .330
Cl(1)	- .079	- .100	- .071
Cl(2)	- .106	- .100	- .071
Cl(3)	- .080	- .100	- .071
Cl(4)	- .107	- .100	- .071
O(2)	- .429	- .426	—
$CH_2(1)$	.316	.311	—
$CH_2(2)$	.017	.018	—
$CH_2(3)$	.016	.018	—
$CH_2(4)$	.317	.311	—
Total on Complex	- .236	- .233	0.000

Table 3:   NEAREST NEIGHBOR BOND ORDERS FOR FURAN SYSTEM

	MoCl₄O	Furan (90°)	Furan (45°)
Mo-O(1)	1.11	1.87	1.86
Mo-Cl(1)	1.13	1.14	1.09
Mo-Cl(2)	1.13	1.06	1.08
Mo-Cl(3)	1.13	1.14	1.08
Mo-Cl(4)	1.13	1.06	1.08
Mo-O(2)	—	0.316	0.322
O(2)-C(1)	—	0.87	0.87
O(2)-C(4)	—	0.87	0.87
C(1)-C(2)	—	0.99	0.99
C(2)-C(3)	—	0.99	0.99
C(3)-C(4)	—	0.99	0.99
Total Mo Bond Order		6.59	6.51

Barrier = 2.70 kcal (45 deg lower)

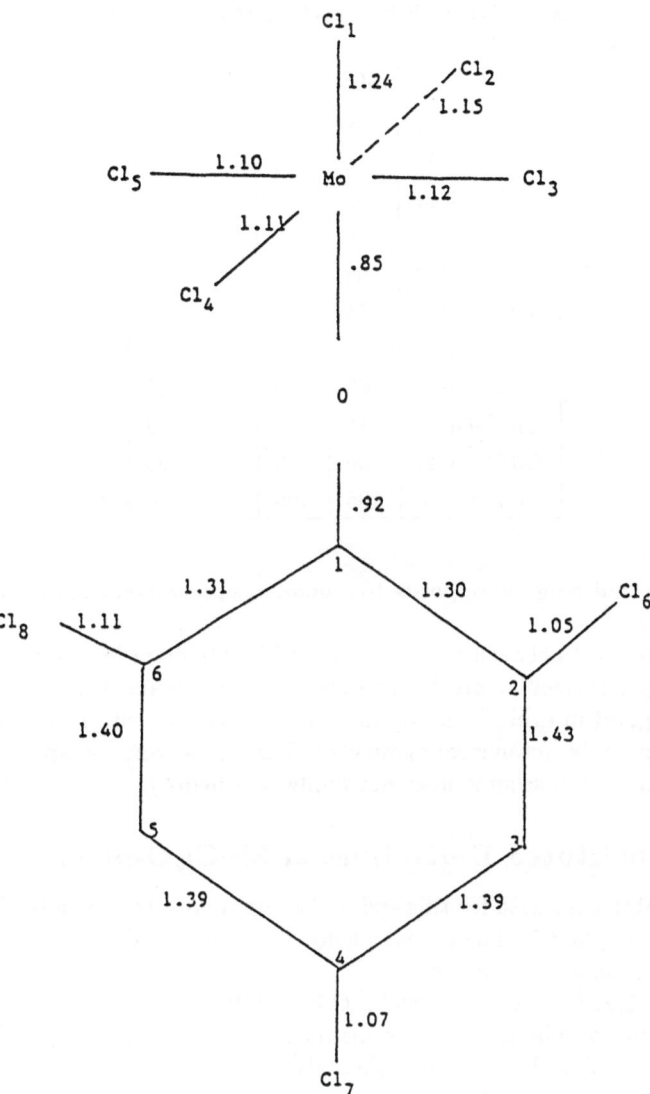

Figure 1: BOND ORDERS FOR STAGGERED TRICHLORO-PHENYL COMPLEX

Table   4:   NON-NEAREST   NEIGHBOR   BOND   ORDERS
FOR FURAN SYSTEM

	90°	45°	MoCl$_4$O
O(1)-Cl(1)	.09	.09	.12
O(1)-Cl(2)	.10	.09	.12
O(1)-Cl(3)	.09	.09	.12
O(1)-Cl(4)	.10	.09	.12
Cl(1)-Cl(2)	.05	.05	.06
Cl(1)-Cl(3)	.14	.14	.13
Cl(1)-Cl(4)	.05	.05	.06
Cl(2)-Cl(3)	.05	.05	.06
Cl(2)-Cl(4)	.14	.14	.13
Cl(3)-Cl(4)	.05	.05	.06
O(1)-O(2)	.05	.05	—

known fact that a ligand opposite to a bond has more effect than one which
is adjacent.

The ratio of 2:1 between the two types of Cl-Cl bond orders is reminiscent
of the simple molecular orbital picture in which vacant T_{2g} orbitals mix
with the ligand in orbitals giving the "back donation" effect in a complex
with approximate octahedral symmetry. Here, however, the sp^3d^2 sigma-
bond hybridization scheme does not apply (see below).

Ring Substituted Derivatives of MoCl$_5$O-phenyl

The computational data for this series of compounds are shown in Figure 1
and Tables 5,6, and 7. The parent phenyl complex was partially optimized
to give the following distances:

Mo-Cl = 2.29Å, Mo-O = 1.86Å, C-O = 1.40Å,

with standard defaults for the remaining distances in the ring. The Mo-
O-C angle was found to be approximately 145°.

Calculations were made on a dozen compounds listed in Tables 6 and
7 in both staggered (45°) and eclipsed (90°) conformations. A detailed
breakdown of the data for one of the complexes is shown in Table 5 and
Figure 1.

The most promising parameter for a ranking were total charge and total
bond order on the basic Mo complex MoCl$_5$O-. The ranking for these two
alternatives is shown in Tables 6 and 7. In both cases the extremes are
spanned by the 2,6 diamino and p-nitro derivatives, consistent with electron donating and withdrawing considerations. Although there are some
parallel trends in both cases, there are also considerable differences. Only

Table 5: CHARGE AND BOND ORDER DATA FOR 2,4,6 TRICHLORO PHENYL COMPLEX

Atom	Charges 45°	Charges 90°	Bond	Bond Orders 45°	Bond Orders 90°
Mo	.110	.081	Mo-Cl(1)	1.24	1.21
Cl(1)	- .013	- .017	Mo-Cl(2)	1.15	1.14
Cl(2)	0.000	- .005	Mo-Cl(3)	1.12	1.13
Cl(3)	- .021	- .016	Mo-Cl(4)	1.11	1.17
Cl(4)	- .021	- .004	Mo-Cl(5)	1.10	1.13
Cl(5)	- .027	- .016	Mo-O	.85	.84
O	- .513	- .510	O-C(1)	.92	.92
C(1)	.365	.352	C(1)-C(2)	1.30	1.31
C(2)	- .088	- .107	C(2)-C(3)	1.43	1.41
C(3)	- .108	- .104	C(3)-C(4)	1.39	1.39
C(4)	- .085	- .089	C(4)-C(5)	1.39	1.39
C(5)	- .109	- .102	C(5)-C(6)	1.40	1.41
C(6)	- .118	- .098	C(6)-C(1)	1.31	1.31
Cl(6)	.102	.110	C(2)-C(16)	1.05	1.09
Cl(7)	.061	.058	C(4)-C(17)	1.07	1.06
Cl(8)	.105	.108	C(6)-C(18)	1.11	1.09

Non-Adjacent Bond Orders	45°	90°
Mo-C(1)	.06	.06
Cl(1)-O	.12	.12
Cl(2)-Cl(4)	.20	.20
Cl(3)-Cl(5)	.20	.20

Rotational barrier = 16.0 kcal

Total bond order: 45°= 18.94

90°= 19.00

Net = - .06

Table 6: PHENYL SERIES: 45°

Complex	Σq	ΣBond Orders		Ranking	
		Mo	All	Charge	Bond
p-methyl	- .566	6.621	16.93	6	9
2,4,6 trichloro	- .485	6.586	18.94	11	5
2,6 dichloro p-methyl	- .505	6.582	18.88	9	4
2,6 dimethyl	- .554	6.538	17.78	7	2
2,6 diamino	- .734	6.330	17.48	1	1
p-ethyl	- .567	6.624	17.87	5	10
unsubstituted	- .537	6.638	16.03	8	11
p-amino	- .652	6.552	16.87	2	3
p-nitro	- .473	6.663	15.89	12	12
2,6 dichloro	- .493	6.589	—	10	6
p-isobutyl	- .575	6.616	—	4	8
p-n-butyl	- .634	6.614	15.92	3	7

Table 7: PHENYL SERIES: 90°

Complex	Σq	ΣBond Orders		ΔE(kcal)
		Mo	All	
p-methyl	- .559	6.63	16.96	- 1.534
2,4,6 trichloro	- .487	6.62	19.00	16.00
2,6 dichloro p-methyl	- .506	6.61	18.94	12.39
2,6 dimethyl	- .545	6.56	17.82	19.63
2,6 diamino	- .734	6.33	17.52	20.79
p-ethyl	- .562	6.65	17.91	- 1.373
unsubstituted	- .529	6.67	16.05	3.34
p-amino	- .650	6.59	16.89	4.85
p-nitro	- .466	6.70	15.94	1.82
2,6 dichloro	- .493	—	—	15.38
p-isobutyl	- .569	—	—	- 1.65
p-n-butyl	- .599	6.63	15.95	- 1.802

in the simple case of a diatomic model will charges and bond orders convey the same information. The spread and general behavior of the data indicates that this could prove to be a useful parameter in correlating catalytic activity.

As with the MoCl$_4$O-furan example there is no direct correlation between bond orders and rotational barriers. Total bond order changes ranged from .03 to .06 while barrier energies went from 1.4 to 20.8 kcal. If one considers a model for rotational barriers of the form

$$\Delta E = \Delta(\Sigma \rho_{ij} \beta_{ij}), \tag{1}$$

where ρ_{ij} is the bond order between the pair of atoms and β_{ij} is the exchange energy parameter, this would require substantial changes in the beta's as the geometry is changed. Higher bond orders could be realized by a combination of the inherent delocalization of the MoCl$_5$O- system and the phenyl system.

The diffuse nature of the transition metal orbitals may be playing a significant role in the enhancement of direct interactions between ortho substituents and the complex. It would be premature at this stage to put such predictions on the same firm footing as the more global structural parameters such as bond energies. It does raise the question of significant interactions between non-nearest neighbors, which is worth pursuing in subsequent investigations.

Again an examination of nonadjacent bond orders indicated a substantial amount of delocalization on the MoCl$_5$-O- moiety. As implied by the above relation for ΔE, large rotational barriers are associated with delocalization and large overlap between appropriate non-nearest neighbors such as ortho substituents and the metal or its ligands. It should be emphasized that in this context both delocalization and direct interaction of adjacent atoms are necessary to achieve the effect in question. This bears out the notion that rotational barriers essentially arise from a combination of through bond and direct orbital overlap effects.

More work needs to be done in this area. If these anomalous rotational barriers can be corroborated as real, it is tempting to associate them with delocalization in the Mo coordination sphere and the phenyl ring. This would also require a substantial overlap between ortho substituents and either the ligands or the Mo atom itself.

In this view, a reduced valence electron overlap between next nearest neighbors in the coordination sphere fosters electron delocalization and significant bond orders between non-nearest neighbors. This is coupled with enhanced overlap between atoms which are adjacent in a true geometrical, though not topological sense. It is also possible that the diffuse electrons on the metal can have a direct interaction with peripheral atoms in the coordination sphere.

The simple model for non-bonded interactions which arise from exchange

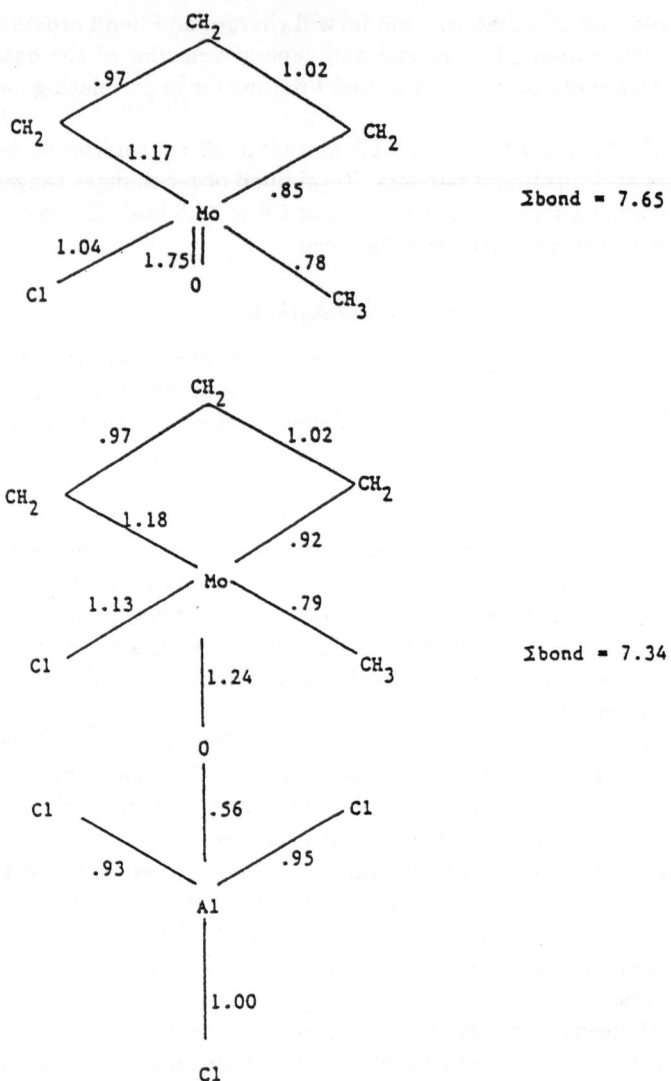

Figure 2: BOND ORDERS FOR METALLACYCLOBUTANE COM-
PLEX WITH AND WITHOUT AL

effects appropriate to SCF calculations assumes that the effect of correlation is essentially confined to nearest neighbor bonded interactions. This covers situations where the interactions between non-bonded atoms are primarily repulsive, arising from exchange interaction between closed shells. The situation is similar to the interaction between rare gas atoms, which is described to a first approximation by the "6-12" potential. Here the difference is that the distance between the atoms in the condensed phase is governed by the balance between attractive Van der Waals (correlation) and repulsive exchange forces.

The intramolecular forces in question depend upon another route for optimizing total energy. To a first approximation, the energy is minimized by adjusting the distribution of charge between pairs of bonded atoms via localized orbitals, which are orthogonal. Since there are large differences between, say s and p orbital energies, this initial value is lowered by increasing populations in the lower energy orbitals. In consequence, a certain amount of delocalization occurs, which would be quite substantial, were it not for an important compensating factor. In a purely localized picture, at the SCF level there are only electrostatic interactions between non-nearest neighbors. When the orbitals are partially delocalized, negative bond orders tend to occur between next-nearest neighbors, leading to repulsions. This greatly reduces the tendency to delocalize in order to satisfy the reduction in local atomic energies.

In hydrocarbons the repulsive energy is large enough to bring about a high degree of localization. Evidently in transition metal complexes overlap between next nearest neighbors is such that the repulsions are somewhat lower, and more delocalization can occur to reduce local atomic energies. This may be one mechanism giving rise to the remarkable properties of this compound class.

We note that the Mayer bond order, (Mayer, 1986), (Mayer, 1987), and (Lendvay, 1989), has been used in all of our analysis. This ensures that the best possible correspondence has been made between all electron computational data and intuitive chemical bonding concepts, which have been so useful over the past several decades.

$Mo(CH_2)_3CH_3ClO/AlCl_3$

In order to study the structure of active intermediates in a reaction cycle, a series of six structures were calculated based on the $Mo(CH_2)_3(CH_3)ClO$ parent complex. These were:

A The metallacyclobutane intermediate shown at the top of figure 2.

A + AlCl$_3$ The same complex with AlCl$_3$ attached to the oxygen atom, shown at the bottom of Figure 2.

B The metallacyclobutane with an elongated C-C bond to simulate a transition state

B + AlCl$_3$

C An ethylene complex with C_2H_4 weakly bound to the $Mo(CH_2)CH_3ClO$ complex.

C + $AlCl_3$

This approach is patterned after ab initio investigations of the basic metallacyclobutane intermediate, (Rappe et al., 1982), (Anslyn et al., 1988), and (Anslyn et al., 1989). These studies have provided considerable insight into the mechanism of the metathesis process, and we believe that the density functional approach is ideally suited to carry the work forward into more complex systems.

The results are summarized in Tables 8, 9, and Figure 2.

Table 8: ATOMIC CHARGES FOR REACTION CYCLE

	Structure A		Structure B		Structure C	
	A	A+Al	B	B+Al	C	C+Al
Mo	.659	.525	.376	.190	.421	.241
$CH_2(1)$	.116	.131	.288	.356	.240	.351
$CH_2(2)$	.026	.094	.243	.121	.213	.260
$CH_2(3)$	-.104	- .020	- .061	-.012	- .025	.014
Cl	-.190	- .113	- .229	- .162	- .242	-.174
O	-.467	- .567	- .486	- .607	- .460	-.608
CH_3	.043	.016	- .132	- .023	- .143	-.052
Al	—	.639	—	.706	—	.719
Cl(1)	—	- .192	—	- .181	—	-.197
Cl(2)	—	- .258	—	- .274	—	-.277
Cl(3)	—	- .257	—	- .272	—	-.278

First bond distances were optimized in the parent compound, assuming tetrahedral geometry. The results are:

$R(Mo-Cl) = 2.29$Å, $R(Mo-O) = 1.71$Å, $R(Mo-CH_3) = 2.05$Å, $R(Mo-CH_2) = 1.85$Å.

In all cases the positive charge on the Mo atom was reduced by more than 0.1 in the presence of the $AlCl_3$. In contrast with the phenyl series, the total charge on the Mo coordination sphere did not change significantly. The decrease in the Mo charge was offset by an increased positive or decreased negative charge on the ligands.

An alternative way of displaying the charge data is via bond polarity, obtained by subtracting the ligand atomic charge from that of the central metal atom. In all cases the bond polarity was decreased in the presence of $AlCl_3$. Only the Mo-O bond did not seem to be strongly affected with respect to polarity. On the other hand a larger change in bond order was observed for this bond.

In forming the $AlCl_3$ complex the bond order of the Mo-O linkage is

Table 9: BOND ORDERS FOR REACTION CYCLE

	Structure A		Structure B		Structure C	
	A	A+Al	B	B+Al	C	C+Al
Mo-C(1)	1.17	1.18	*	1.41	1.51	1.40
Mo-C(2)	.07	.09	*	.28	.22	.35
Mo-C(3)	.85	.92	*	.22	.10	.25
Mo-Cl	1.04	1.13	*	1.15	1.05	1.09
Mo-O	1.75	1.24	*	.84	1.44	.84
Mo-C(4)	.78	.79	*	.84	.81	.96
C(1)-C(2)	.97	.97	*	.03	.03	.06
C(2)-C(3)	1.02	1.02	*	1.80	1.80	1.68
Mo-Al	–	.065	–	.038	–	.06
Mo-Cl(1)	–	.24	–	.24	–	.22
Mo-Cl(2)	–	.11	–	.09	–	.09
Mo-Cl(3)	–	.038	–	.11	–	.09
O-Al	–	.56	–	.44	–	.42
Al-Cl(1)	–	.93	–	.96	–	.97
Al-Cl(2)	–	1.00	–	.99	–	1.00
Al-Cl(3)	–	.95	–	.99	–	.99
*Data not yet available						

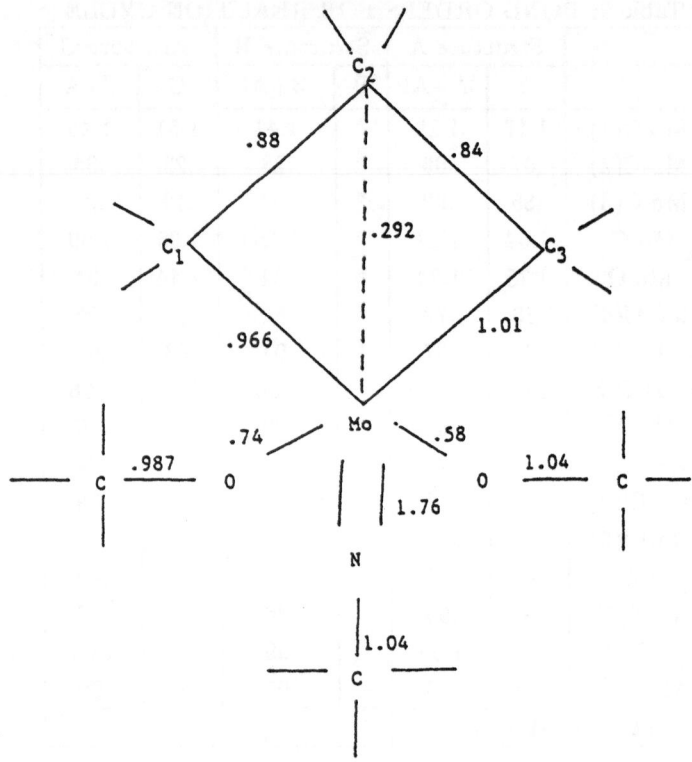

Figure 3: BOND ORDERS FOR METALLACYCLOBUTANE SKELE-
TON OF SCHROCK COMPOUND

decreased to a value more appropriate to that of a single bond. The ratio between Mo-O and Al-O bond orders suggests that the latter may be only half as strong as the former.

Bond orders in the ring are indicative of normal single bonds between the carbon atoms and at least formally single bonds from Mo to the carbon atoms. Appreciable bond orders exist between Mo and the Al along with its three Cl atoms. Particularly significant is the .24 bond order between Mo and Cl(1). This suggests that a more favorable structure for the resulting complex may involve a four membered Mo-O-Al-Cl ring, in analogy with other such rings involving Al and Cl.

The sum of bond orders for the three $AlCl_3$ substituted structures A, B, C comes out to be 7.34, 6.57 and 6.63, consistent with the evidence from other workers that A is the most stable, lying in energy below B and C with B and C nearly equivalent in energy. In these calculations our inability to carry out full geometry optimization prevented us from making a definitive assessment of this series.

We plan to repeat the calculations with the recently completed geometry optimization package available in DGAUSS. We feel that the charge distributions from the current results have some relevance in this preliminary investigation; but the energy differences are clearly misleading, being on the order of normal covalent bond energies. The calculations do indicate that bond order appears to be a promising criterion for formulating structure/activity relationships in the future.

$Mo(CH_2CH_2CH_2)(N-Ar) [OC(CF_3)_2(CF_2CF_2CF_3)]_2$ (Ar = 2,6-C_6H_3-di-isopropyl)

This compound was investigated because it is an example of a stable metal-lacyclobutane complex for which X-ray data were available in the literature (Schrock et al., 1988). The salient features of the compound are shown in Figure 3. In this case we may regard the geometry as essentially optimized, assuming reliable X-ray data. (Strictly speaking the data are for the W complex, but for bond order analysis the difference between W and Mo radii should not matter.)

The results of the calculations on this compound are summarized in Table 10.

Table 10: SCHROCK COMPOUND DATA SUMMARY

	Charges
Mo	.647
$CH_2(1)$	- .032
$CH_2(2)$	+ .075
$CH_2(3)$	- .025
N	- .523
O(1)	- .547
O(2)	- .536
Total	- .941

Bond Orders			
Nearest		Non-Nearest	
Mo-C(1)	.966	Mo-C(3)	.292
Mo-C(3)	1.010	Mo-C(4)	.072
Mo-N	1.764	Mo-C(5)	.076
Mo-O(1)	.585	Mo-C(6)	.068
Mo-O(2)	.73		
Total	5.06	Mo-F(1)	.058
		Mo-F(2)	.050
		Mo-F(3)	.078
C(1)-C(2)	.88	Mo-F(4)	.103
C(2)-C(3)	.84	Mo-F(5)	.085
		Mo-F(6)	.075
		Mo-F(7)	.085
		Total	1.042
		C(1)-C(3)	0.0

Several points may be noted:

1. Bond orders between the ring carbon atoms are 10-15% lower than normal.

2. A substantial bond order exists between Mo and the opposite carbon atom.

3. The five nearest neighbors of Mo account for a total bond order of 5, while about a dozen next nearest neighbors give a total contribution of 1.

4. The net charge on the Mo coordination sphere including the four membered ring is -.94.

There is considerable delocalization even out to second next nearest neighbors, namely the fluorine atoms. This suggests that the electronic structure of the active site may be influenced by a larger portion of the ligand structure than would appear at first consideration.

The reduced bond orders for $C(1)$-$C(3)$ and the increased Mo-$C(2)$ bond order evidently set the stage for a ring cleavage for which bonding power is essentially conserved, conducive to a low barrier.

Summary

This preliminary study indicates that atomic charges and bond orders are a promising tool for correlating chemical behavior with molecular structure. It appears that basic ligand structure is crucial for providing the environment that ensures shallow wells and low barriers for a particular conversion process. The Schrock compound, which is known to have catalytic activity, provides a template for establishing certain gross features of electronic environment needed to catalyze metathesis.

If one compares charges on the metal, this compound as well as the $MoOCl_4$ and the $Mo(O)(Cl)(CH_2)(CH_3)$ precursor has a large positive charge on the metal atom (approximately 0.6). On the other hand, the $MoCl_5O$-phenyl compounds displayed 0.1 units or less. In line with conclusions expressed in the literature (Rappe et al., 1982), a double bonded O or N seems to be needed to provide this desirable environment.

The study has examined two separate routes to ligand fine tuning, either by substituting onto strongly bound groups as in the phenyl series or by introducing a new weakly bound ligand as in the THF study.

The calculations of the THF and $Mo(CH_2)_3(CH_3)ClO/Al$ series indicate that improvement (increase) in activity is associated with lowered bond orders at the active metal site. Large migrations of charge may ultimately be correlated with changes in hybridization of the reaction intermediates and, where appropriate, with secondary solvent effects.

It should be noted that the s,p,d orbital populations in all cases favored the d to the extent of 85-90%, thus ruling out conventional strong bonding schemes such as sp^3d^2 sigma-bonds and d^3 pi-bonds in an octahedral complex. A substantial amount of delocalization occurs, which sets the stage for facile changes in bonding between atoms; this is the hallmark of an effective catalyst.

This investigation demonstrates that Local Density Functional methodology has the capability of extending all electron calculations into the realm of practical interest for transition metal catalysis. In subsequent work it should be possible to provide insights into the energetics of the catalytic cycle and provide data for equilibria and rate assessment.

Acknowledgement

The authors wish to thank Cray Research for making their facilities available to perform these calculations. In particular we thank Jan Andzelm and Erich Wimmer for their efforts, without which this project would not have been possible.

References

Andzelm, J. and Wimmer, E. and Salahub, D. R., 1989. *The Challenge of d and f Electrons. Theory and Computation*, volume 394. ACS Symposium Series, Ed. D. Salahub and M. Zerner.

Andzelm, J. and Wimmer, E., To Be Published.

Anslyn, Eric V. and Brusich, Mark J. and Goddard, William A.,III, 1988. *Organometallics*, 7:98–105.

Anslyn, Eric V. and Goddard, William A.,III, 1989. *Organometallics*, 8:1550–1558.

Cotton, F. A. and Wilkinson, G., 1988. *Advanced Inorganic Chemistry*. Wiley Interscience, New York, fifth edition.

Hohenberg, H. and Kohn, W., 1964. *Phys. Rev. B*, 136:864.

Kohn, W. and Sham, L. J., 1965. *Phys. Rev. A*, 140:1133.

Lendvay, G., 1989. *J. Phys. Chem.*, 93:4422–4429.

Mayer, I., 1986. *Int. J. Quant. Chem.*, 29:477–483.

Mayer, I., 1987. *J. Molec. Struct.*, 149:81–89.

Rappe, Anthony K. and Goddard, William A.,III, 1982. *J. Am. Chem. Soc.*, 104:448–456.

Schrock, R.R. and DePue, R.T. and Feldman, J. and Schaverien, C.J. and Dewan, J.C. and Liu, A.H., 1988. *J. Am. Chem. Soc.*, 110:1423–1435.

18
An Algorithm in Direct Space for the Local Electronic Structure of Ferromagnetic Phases: Co(bcc) and Ni(fcc)

MIGUEL CASTRO, FERNANDO ESTRADA, AND VICENTE SORIA

Abstract. We apply the multiple scattering cluster-in-condensed-matter technique to study the electronic structure of a single atom in a condensed matter like boundary potential. This central atom may be considered as an impurity embedded in its own material. From the results it is possible to analyzed the trends of the magnetizations, exchange splittings, band widths, etc. in the closed packed Co(bcc) and Ni(fcc) ferromagnetic materials. Ignoring geometric effects it is obtained an increase of the *sd* hybridization from Co to Ni. In fact, the results for nickel predicts a zero magnetic moment for copper (in a rigid band model scheme).

I. Introduction

A basic theoretical problem is the study of the existence of a ferromagnetic state in terms of modern electronic structure. Here, the ab-initio self-consistent band structure techniques have been successfully applied and provided a quantitative understanding of many ground states properties in iron, cobalt and nickel (Connolly, 1967; Wang et al., 1977; Anderson et al., 1979). However, magnetic properties of a highly local character also may be analyzed in a real space formalism by means of cluster calculations, both in free space (Yang et al., 1981) or embedded in the bulk of the material (Keller et al., 1982); in this scheme the concerning information for those properties is more readily extracted than in a band theory calculation, and the theoretical explanation is given in terms of quantum chemical concepts which provides complementary information, and in some cases a new interpretation, to that commonly given for the solid state phenomena by means of the reciprocal space methodology.

For finite small clusters in free space (molecular boundary conditions) the self-consistent field results shows an anisotropic magnetic distribution and a strong dependence of this property on cluster size and geometry (Yang et al., 1981; Keller et al., 1982; Fournier et al., 1986).

In this contribution we present a simple algorithm, designed in the real space formalism, that allows easily the computation of the local electronic structure in ferromagnetic materials. For this purpose an atom from the crystal is studied self-consistently in a spin polarized environment representing the ferromagnetic bulk. The calculations are performed with multiple scattering techniques and local spin density functionals for the exchange-correlation effects in the way explained below.

II. Computational Procedure

A) Isolated atoms were computed self-consistently in the $3d^x 4s^1$ atomic configurations, where $x = 8$ and 9 and with initial d-spin polarizations of 2 and 1 for the cobalt and nickel atoms, respectively. In all cases the $4s$ orbitals remains unpolarized and there are not $4p$ electrons at this initial step. Such atomic configurations are widely employed in band structure calculations. However, in our case they are only the initial trial electronic population parameters. These atomic calculations are performed with a modified version of a relativistic self-consistent field method (Liberman et al., 1965). The modifications include the use of the α-β exchange-correlation potential (Herman et al., 1969) and the possibility of minimizing the total energy as a function of the fractional occupation of the valence levels. The electronic spin densities are the main products of this atomic step.

B) The environment to a given atom in the ferromagnetic bulk of the material was represented by a potential which takes into account the contribution of the first six layers of atoms. This one-electron crystalline potential was built by superposition of atomic charge densities, only the spherically symmetric part is kept, then the Poisson equation is solved for the coulombic part and the exchange-correlation contribution is approximated in the α-β scheme.

In the real space formalism the atomic cells radii, R_c, corresponds to the volume per atom. In Co(bcc) and Ni(fcc) the radius of the atomic spheres are of 2.63 bohrs and of 2.60 bohrs, respectively. The total number of electrons in the central cell is equal to the atomic number Z. In the single site approximation for the crystalline potential, a search for the set of resonances was carried out by means of the Cellular Multiple Scattering $X_{\alpha\beta}$ technique. The potential outside this central region was kept frozen. The purpose of our study is to obtain more understanding of the way the condensed matter boundary conditions for the one electron wave functions act on a group of atoms to promote a ferromagnetic state in it.

C) A charge density for the central cell was constructed in a band-like picture fashion: all s, p, d, f and g spin-orbitals with eigenvalues less than a E_{max} are occupied with a local density of states, E_{max} is that energy for which the total number of electrons in the cell, with eigenvalues $\epsilon_i \leq E_{max}$, is equal to Z.

In the multiple scattering technique, the local density of states for the atomic species i, at energy E, is given by integration over cell i of the diagonal elements of the imaginary part of Green's function (Keller et al., 1982; Castro et al., 1982)

$$-\frac{1}{\pi} \int_0^{R_c} dr \langle \mathbf{r} | Im \; \mathbf{G}^+(E) | \mathbf{r} \rangle = \int_0^{R_c} \rho(\mathbf{r}; E) d\mathbf{r} \; . \tag{1}$$

The G^+ matrix of the system is given in terms of the free electron Green's function G_0^+ and the array of the single site scattering matrices K, for stationary wave boundary conditions. G^+ can be expressed in a supermatrix angular momentum representation ($\tilde{G}^+$) as

$$\tilde{G}^+_{iL,jL'} = \sum_{kL''} \tilde{G}^+_{0iL,kL''} \left[(\tilde{1} - \tilde{G}_0^+ \tilde{K})^{-1} \right]_{kL'',jL'} , \tag{2}$$

L is the ordered pair (l, m) and i, j, k denote scatterer atoms. The radial local density of states can be written in terms of the supermatrix as

$$\rho_i(r'; E) = -4 \; \langle\langle \mathbf{r}' | Im \; \mathbf{G}^+(E) | \mathbf{r}' \rangle\rangle_{sp} ,$$

$$= -4 \sum_L R_l^2(r'; E) \; Im \; G^+_{Li,Li} , \tag{3}$$

where sp = spherical average; $r' = |\mathbf{r}'| = |\mathbf{r} - \mathbf{r}'|$; $R_l(r'; E)$ are the solutions for energy E of the radial Schrödinger equation for scatterer i in the self-consistent atomic potential with the boundary condition:

$$R_l(r_c'; E) = j_l(\kappa r_c') \; \cos \eta_l - n_l(\kappa r_c') \; \sin \eta_l , \tag{4}$$

where $E = \kappa^2$.

The potentials are computed from the charge densities using the exchange $X_{\alpha\beta}$ technique (Herman et al., 1969). Correlation was also included with a local density functional (von Barth et al., 1972). In the self consistent procedure for the central (cluster) region the eigenstates were divided into core and conduction bands. The conduction band was divided itself into s, p, d, f, and g cluster resonances for which the integration in Eq (1) gives fractional occupation numbers. All orbitals with eigenvalues less than E_{max} are occupied with that local density of states. The crystalline potential outside the central cell was kept frozen during the SCF process and at the end of this step the central cell has acquired a different electronic arrangement from that of the atoms in the frozen layers. Then with these new spin-orbital occupations the atomic step is repeated, a new crystalline potential is constructed and the electronic structure at the central cell is recalculated in the above fashion. After three $A \rightarrow B \rightarrow C$ cycles the electronic populations of the central atom is not quite different from that of the atoms in the frozen environment. At this point E_{max} defines the Fermi level, ϵ_f, of the system.

In Co(bcc) and Ni(fcc), the orbital energies of the conduction band begins to appear at deeper values, however the set of resonances in the range $-4.00 \, \text{Ry} \le \epsilon_i \le -1.50 \, \text{Ry}$ contains a negligible electronic charge contribution, less than 0.001 electrons, in the cell of the central atom, such wavefunctions belong to the atoms of the frozen layers. For this reason we report , for cobalt and nickel , only those eigenvalues which contains a significant electronic charge contribution in the central atom; all the physics of the problem is given by this set of eigenvalues.

III. Results Co(bcc)

In table 1 are reported the self-consistent eigenvalues, ϵ_i, together with the occupation numbers, n_i (given by Eq. 1), for Co(bcc) . This electronic structure gives a magnetization of 1.78 spins per atom (spa), which is greater than the estimated experimental value, 1.53 spa (Prinz, 1985), for this meta-stable phase of cobalt. The band theory techniques also report a high magnetic moment, 1.68 spa (Moruzzi et al., 1986) and 1.65 spa (Schwarz et al., 1984). We obtain an average d-exchange splitting of $1.56\,\mathrm{eV}(\Delta\epsilon_{ex} = 2.58\,\mathrm{eV}$ for the d-levels at the Fermi level and $\Delta\epsilon_{ex} = 0.60\,\mathrm{eV}$ for the couple of d-levels at $\approx -13.5\,\mathrm{eV}$, the e_g and t_{2g} levels of a cubic symmetry), this value falls in the range of the band theory results, $1.49\,\mathrm{eV}$ and $1.72\,\mathrm{eV}$ (Jansen et al., 1980). Our d-band width is of $5.12\,\mathrm{eV}$ ($4.08\,\mathrm{eV}$ for the majority spin and $6.16\,\mathrm{eV}$ for the minority spin), while Slater (Slater, 1929) have reported $4.84\,\mathrm{eV}$ for that property.

Table 1.- Eigenvalues, ϵ_i, and occupation numbers, n_i, for Co(bcc). The sum of the accumulated charge for each spin at the Fermi energy, $\epsilon_i = \epsilon_f$, is equal to the number of valence electrons per atom; the difference gives a magnetization of 1.78 spa.

Orbital	$\epsilon_i(Ry)$	n_i	Ocup.	Total ↑	Total ↓	↑ + ↓
$f\!\uparrow$	−1.4624	0.00571	0.00571	0.11026		0.21468
$f\!\downarrow$	−1.4315	0.00579	0.00579		0.11021	0.22047
$g\!\uparrow$	−1.3473	0.00217	0.00217	0.11243		0.22260
$g\!\downarrow$	−1.3172	0.00219	0.00219		0.11240	0.22482
$s\!\uparrow$	−1.2710	0.08684	0.08684	0.19927		0.31164
$s\!\downarrow$	−1.2420	0.08836	0.08836		0.20076	0.40031
$p\!\uparrow$	−1.1543	0.17264	0.17264	0.37191		0.57268
$p\!\downarrow$	−1.1240	0.17764	0.17764		0.37840	0.75036
$d\!\uparrow$	−0.9896	0.79089	0.79089	1.16280		1.54121
$d\!\downarrow$	−0.9457	0.46530	0.46530		0.84370	2.00657
$f\!\uparrow$	−0.6957	0.09210	0.09210	1.25490		2.09862
$d\!\uparrow$	−0.6884	4.13752	4.13752	5.39242		6.23821
$f\!\downarrow$	−0.6626	0.09210	0.09210		0.93580	6.32821
$d\!\downarrow$	−0.5001	4.26791	2.67178		3.60758	9.00000
$g\!\uparrow$	−0.4178	0.03367				
$s\!\uparrow$	−0.4172	0.15076				
$s\!\downarrow$	−0.3928	0.13912				
$g\!\downarrow$	−0.3868	0.03342				
$p\!\uparrow$	−0.3220	0.16096				
$p\!\downarrow$	−0.3050	0.14595				
$d\!\uparrow$	−0.2322	0.10746				
$d\!\downarrow$	−0.2164	0.20344				

IV. Results Ni(bcc)

The table 2 contains the SCF electronic structure for Ni(fcc), our computed magnetization of 0.54 spa is very close to the experimental result of 0.56 spa (Dannan et al., 1968). In this system the highest occupied cluster orbitals are of d-character, the e_g and t_{2g} orbitals of a cubic symmetry, with exchange splittings of 0.023 Ry and 0.019 Ry, respectively; Eastman (Eastman et al., 1980) have reported an splitting of 0.022 Ry for Ni(fcc). Our total population numbers, in electrons, for the majority and minority d-bands are 4.71 and 4.16, respectively, these values are close to those obtained by Langlais and Callaway (Langlais et al., 1972), 4.78 and 4.22.

Table 2.- Eigenvalues, ϵ_i, and occupation numbers, n_i, for Ni(bcc). The sum of the accumulated charge for each spin at the Fermi energy, $\epsilon_i = \epsilon_f$, is equal to the number of valence electrons per atom; the difference gives a magnetization of 0.54 spa.

Orbital	$\epsilon_i(Ry)$	n_i	Ocup.	Total ↑	Total ↓	↑ + ↓
$g\uparrow$	−1.1853	0.00489	0.00489	0.13692		0.27050
$g\downarrow$	−1.1732	0.00492	0.00492		0.13850	0.27542
$s\uparrow$	−0.9870	0.17812	0.17812	0.31504		0.45354
$s\downarrow$	−0.9790	0.18037	0.18037		0.31887	0.63391
$p\uparrow$	−0.8253	0.28987	0.28987	0.60491		0.92378
$p\downarrow$	−0.8160	0.29418	0.29418		0.61305	1.21796
$d\uparrow$	−0.7496	3.10445	3.10445	3.70936		4.32241
$d\downarrow$	−0.7267	2.80486	2.80486		3.41791	7.12727
$d\uparrow$	−0.5729	1.56035	1.56035	5.26971		8.68762
$d\downarrow$	−0.5534	1.80220	1.31238		4.73029	10.00000
$s\uparrow$	−0.4996	0.07929				
$f\uparrow$	−0.4991	0.03147				
$s\downarrow$	−0.4928	0.07831				
$f\downarrow$	−0.4902	0.03124				
$p\uparrow$	−0.4138	0.22955				
$p\downarrow$	−0.4078	0.22880				
$g\downarrow$	−0.3860	0.00000				
$g\uparrow$	−0.3780	0.00710				

In table 3 is presented the d-exchange splitting computed with different model potentials: X_α ($\alpha = 2/3$), $X_{\alpha\beta}$ ($\alpha = 2/3$, $\beta = 0.0025$), and the von Barth-Hedin (vBH) scheme (von Barth et al., 1972). We compare our results with those obtained by Wang and Callaway (Wang et al., 1977) who employed the Kohn-Sham (Kohn et al., 1965) and the vBH exchange correlation potentials.

Table 3.- Exchange splittings for Ni(fcc), in band theory and in this work, with different exchange-correlation potentials. Energies in rydbergs.

Wang and Callaway	Exchange-correlation potential		
Band	Kohn-Sham		vBH
$X \uparrow$	−0.4123		−0.5561
$X \downarrow$	−0.3478		−0.5098
$X \uparrow - X \downarrow$	0.0645		0.0463
This Work			
Orbital	X_α	$X_{\alpha-\beta}$	vBH
$d \uparrow$	−0.486	−0.479	−0.488
$d \downarrow$	−0.457	−0.446	−0.444
$d \uparrow - d \downarrow$	0.029	0.032	0.044

V. Conclusions

Combining the multiple scattering techniques together with the density functional exchange correlation potentials, we have designed an algorithm that allows easily the computation of the local electronic structure of ferromagnetic materials. Our results are not far away of those obtained with the more sophisticated band theory techniques. The algorithm can be applied to the study of meta-stable phases, e. g. Co(bcc), also it is possible to follow the changes of the electronic structure that occurs in the transition magnetic phases, e. g. Pd(fcc) (Castro et al., 1990).

Acknowledgments

Partial financial support was provided by DGAPA-UNAM under project IN-I0-40-89. We want to thank Dr. Carlos Amador for enthusiastic discussions.

References

Anderson J. R., et al., *Phys. Rev.* **B20**, 3172 (1979).
Castro M., Keller J., and Rius P., *Hyperfine Interact.*, **12**, 21 (1982).
Castro M. and Soria V., to be published.
Connolly J. W. D., *Phys. Rev.* **159**, 298 (1967).
Dannan H., et al., *J. Appl. Phys.* **39**, 669 (1968).
Eastman D. E., Himpsel F. J., and Knapp J. A., *Phys. Rev. Lett.* **44**, 95 (1980).
Fournier R. and Salahub D. R., *Int. J. of Quantum Chemistry* **XXXIX**, 1077 (1986).
Herman F., Van Dyke J. P., and Ortenburger I. B., *Phys. Rev. Lett.* **22**, 807 (1969).
Jansen H., et al., *Physics of Transition Metals 1980*, P. Rhodes. London (1980).
Keller J., Castro M., and de Paoli A. L., *J. Appl. Phys.* **53**, 8850 (1982).
Kohn W. and Sham L. J., *Phys. Rev.* **140**, A1133 (1965).

Langlais J. and Callaway J., *Phys. Rev.* **B5**, 124 (1972).

Liberman D., Waber J. T., and Cromer D. T., *Phys. Rev.* **137**, 27 (1965)

Moruzzi V. L., et al., *Phys. Rev.* **B34**, 1784 (1986).

Prinz G. A., *Phys. Rev. Lett.* **54**, 1051 (1985).

Schwarz K., et al., *J. Phys. F: Metal Phys.* **14**, 2659 (1984).

Slater J. C., *Phys. Rev.* **34**, 1293 (1929).

von Barth U. and Hedin L., *J. Phys.* **C5**, 1629 (1972).

Wang C. and Callaway J., *Phys. Rev.* **B15**, 298 (1977).

Yang C. Y., et al., *Phys. Rev.* **B24**, 5673 (1981).

19
Structural Phase Transitions in Cesium Halides

Andrés Cedillo, Alberto Vela, and José Gázquez

Abstract

In the last few years it has been established that the pressure induced phase transitions from bcc to tetragonal in Cesium halides are of second order, and show an almost constant volume ratio. In this work we present a first principles approach based on density functional theory to calculate the interaction potential and the p-V curve at zero temperature. The results thus obtained are in good agreement with the experimental values, and show a well defined second order transition.

Introduction

The development of experimental techniques for high pressures has lead to the discovery of new interesting phenomena in material science.   The insulator-metal transition was studied experimentally in Xe and its isoelectronic systems, such as CsI and BaTe, and their analogs CsBr, CsCl, RbI, KI and BaSe.   Using the diamond anvil cell it is possible to determine the optical absorption, reflectivity, Raman and Brillouin scattering, and also X-ray scattering.   These studies show the pressure induced insulator-conductor phase transition in some systems, (Vohra, 1985; Ruoff, 1986; Vohra, 1986) and other phenomena.   For example the alkaline earth chalcogenides show a fcc to bcc phase transition at about 10 *GPa* (Syassen, 1986).   The alkaline halides show a very special behavior; fifteen have a fcc stable structure and only three (CsCl, CsBr and CsI) have bcc structure.   Those with fcc structure show a fcc to bcc phase transition at 1 *GPa*, while the bcc systems have a bcc to tetragonal transition at 10 *GPa*.   Also, the former systems have a first order transition and in the latter case it is second order (no volume change) (Huang, 1984; Brister, 1985). This paper presents a density functional approach to study the phase stability in ionic crystals and it is applied to the Cesium halides phase transition.

Model

Working in the pair interaction approximation, it is necessary to determine the interaction energies in the crystal.   They are calculated as follows

$$V_{ij}(r_{ij}) = E[\rho_i^* + \rho_j^*] - E[\rho_i^0] - E[\rho_j^0] + \frac{Z_i Z_j}{r_{ij}} \qquad (1)$$

where ρ_i^* is the density of the ith ion in the crystal, ρ_i^0 is the density of the ith free ion, and the last term is the nuclear repulsion.

The densities in the crystal can be calculated with a mean field approach (Vela, 1986).   To determine the density of ith ion in the crystal, the other ions can be approximated by point charges and their potential can be spherically averaged, so the ρ_i^* can be calculated with a Watson sphere to simulate an ion shell.   Then, the interaction potential can be evaluated with ρ_i^0 and ρ_j^*.

For a specific crystal, with a given crystalline structure   and   lattice   parameters,   the   total interaction energy, E, is given by

$$E = \frac{1}{2} \sum_{i \neq j} \sum V_{ij}(r_{ij}) \qquad . \qquad (2)$$

When the solid is made by two types of ions, the energy can be grouped in two terms, and it will be given by

$$E = \frac{1}{2} (E_A + E_B) \qquad (3)$$

with

$$E_i = N_i \sum_s n_s V_{iX_s}(r_{iX_s}) \qquad (i = A, B) \qquad (4)$$

where N_1 is the number of ions "i" in the crystal. The sum is made over the ions shells, X_s is the type of ion in the "s" shell, and n_s is the number of ions in the "s" shell.

For a 1:1 binary system, the lattice energy is given by

$$E = \frac{N}{2} \sum_s n_s \left(V_{AX_s}(r_{AX_s}) + V_{BX_s}(r_{BX_s}) \right) \quad (5)$$

where 2N is the total number of ions in the crystal (n_s and r_{iXs} depend on the crystalline structure). In the interaction potential it is possible to separate the point charge interaction, so

$$V_{ij}(r_{ij}) = \left[V_{ij}(r_{ij}) - \frac{Q_i Q_j}{r_{ij}} \right] + \frac{Q_i Q_j}{r_{ij}}$$

$$V_{ij}(r_{ij}) = W_{ij}(r_{ij}) + \frac{Q_i Q_j}{r_{ij}} \quad (6)$$

At large r, $V(r) \approx \frac{Q_i Q_j}{r}$, so $W(r)$ decays very rapidly and the sums over the shells can be separated as

$$\sum_s n_s V_{iX_s}(r_{iX_s}) = \sum_s n_s W_{iX_s}(r_{iX_s}) + \sum_s n_s \frac{Q_i Q_{X_s}}{r_{iX_s}} \quad (7)$$

where the second sum corresponds to the Madelung term. Substituting the last equation for A and B in Eq. (5), one obtains that

$$E = \frac{N}{2} \left\{ \sum_s n_s \left(W_{AX_s}(r_{AX_s}) + W_{BX_s}(r_{BX_s}) \right) \right.$$

$$\left. + 2 \frac{\mathcal{A}}{R} \right\} \quad (8)$$

where A is the Madelung constant of the crystal, and R is the nearest neighbors distance. Expanding the sum in the last equation, one obtains that

$$E = \frac{N}{2} \left\{ 2 \frac{A}{R} + n_1 \left(W_{AB}(R) + W_{BA}(R) \right) \right.$$
$$\left. + n_2 \left(W_{AA}(r_2) + W_{BB}(r_2) \right) + \ldots \right\} \qquad (9)$$

where r_2 is the second nearest neighboors distance. By noting that from the definition of $W(r)$, the sum converges very rapidly, one can neglect the terms corresponding to shells with $s \geq 2$, then the lattice energy per particle is

$$E(R) = \frac{1}{2} \left[n_1 W_{AB}(R) + \frac{A}{R} \right] . \qquad (10)$$

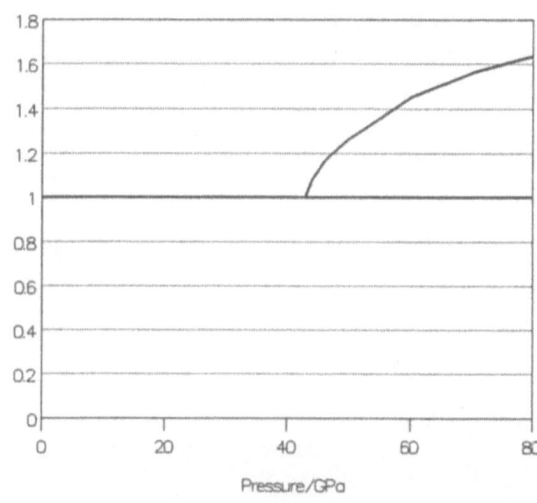

Figure 1.   Lattice parameter ratio for CsCl

The free energy is given by

$$G = E + pV - TS \qquad (11)$$

and at zero Kelvin, the last term vanishes so

$$G(R) = E(R) + p \ V(R) \qquad (12)$$

where p is the independent variable and V depends on the crystalline structure.

At each pressure it is possible to calculate the free energy of each crystal, with its own structure and its own lattice parameters.   Constructing a set of crystals with different structures and volumes, the free energy allows one to determine the most stable phase at each pressure.

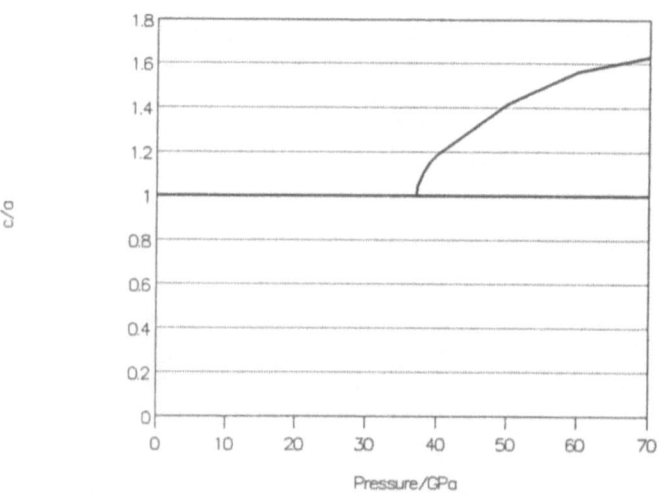

Figure 2.   Lattice parameter ratio for CsBr

Cesium Halides

The Cesium chloride, bromide and iodide show a second order pressure induced phase transition from a bcc structure to the tetragonal one (Huang, 1984; Brister, 1985).   In the second order transitions there is no change in the volume, and these Cesium halides have an almost constant transition volume ratio V/V_0, where V is the volume at the transition and V_0 is the zero pressure volume.

A body centered structure has eight first

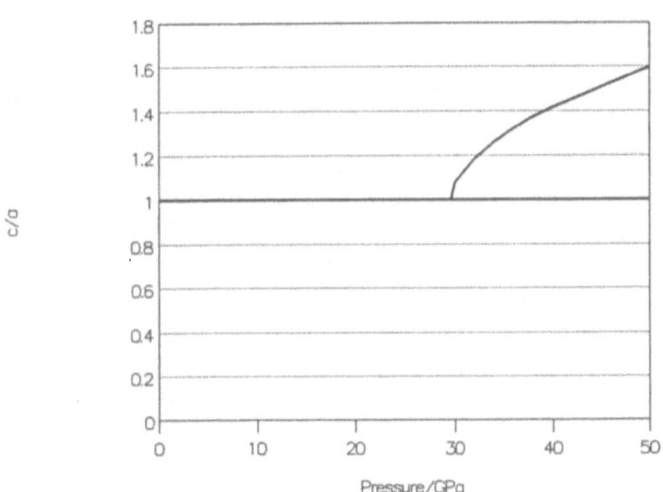

Figure 3.   Lattice parameter ratio for CsI

neighbors:

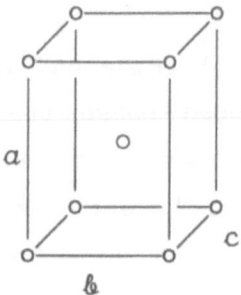

In the bcc, the three lattice parameters are equal, $a=b=c$, and it is characterized by one lattice parameter a or the first neighbors distance $R = \dfrac{\sqrt{3}}{2}a$; while the tetragonal has two equal and one different, $a=b\neq c$, and it is characterized by two parameters R and $x=c/a$ (The bcc structure is the tetragonal with x=1).

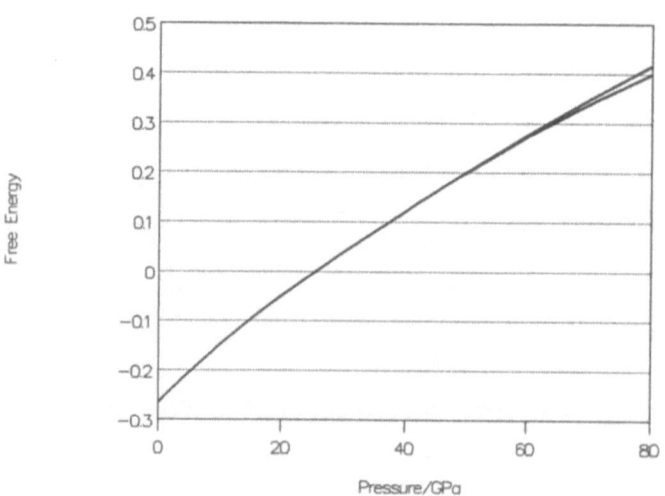

Figure 4.   Free energy per particle for CsCl

The atomic densities were calculated with a Watson sphere in a Kohn-Sham program (Vela, 1986), the interaction potentials were computed using Eq. (1) with a Gordon-Kim methodology (Vela, 1986), and the Madelung constant for the tetragonal crystal was evaluated with the Ewald method (Vohra, 1985).

In this case, the interaction energies and the Madelung constant can not be obtained analytically and it is necessary to make a set of crystals each one with different structure and different size (the bcc crystals are labeled with its nearest neighbors distance R and the tetragonal with R and x).   At each pressure the free energy is computed for every crystal in the set and that with the minimum free energy is

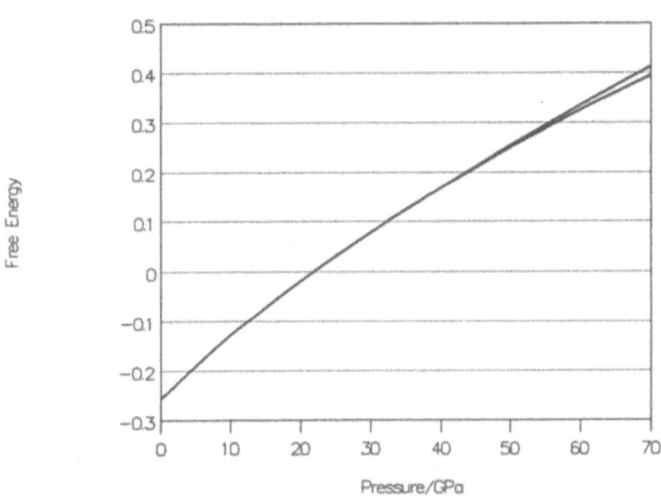

Figure 5.   Free energy per particle for CsBr

the stable and its volume corresponds to the volume of
the system.

Results

The calculations described above were performed for
CsCl, CsBr and CsI, taking into account only the first
shell potential in the computation of ρ_1^* and using Eq.
(10) for the lattice energy.   The free energy was
calculated for bcc and tetragonal structures, the last
with x between 1 and 1.6, and pressures in the range
from 0 to 70 *GPa*.

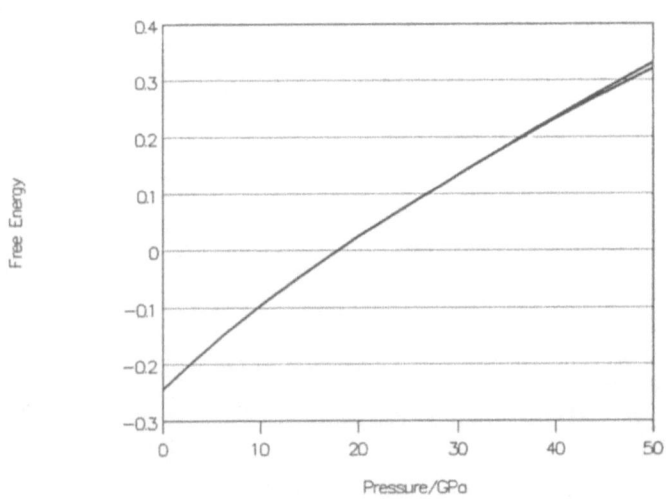

Figure 6.   Free energy per particle for CsI

The lattice parameters, free energy and volume of the stable phase are plotted for each pressure in Figs 1-9.   From Figs 1-3 it is very clear that the tetragonal structure becomes stable at pressures around 40 *GPa*.   At the phase transition the free energy is continuous (figures 4-6), as well as its derivative, that is, the volume (figures 7-9),   The experimental and calculated transition pressures are shown in Table I.   The values calculated in this work are smaller than the experimental results, but they have the same trend and in the three cases the transition is second order as the experiment shows, with a discontinuity in the slope of the p-V curve at the transition.   Also, the calculated volume ratios are in very good agreement with the experimental ones,

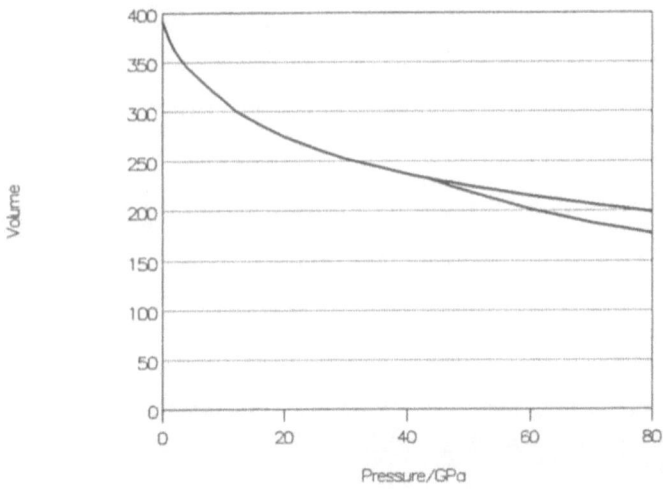

Figure 7.   Volume per particle for CsCl

and show an almost constant value.

	p/GPa		V/V_0	
	Experimental	This work	Experimental	This work
CsCl	65	43	0.530	0.592
CsBr	53	37	0.546	0.599
CsI	39	30	0.544	0.597

TABLE I.

Discussion

The methodology presented in this work allows the calculation of any kind of ionic crystal, and the

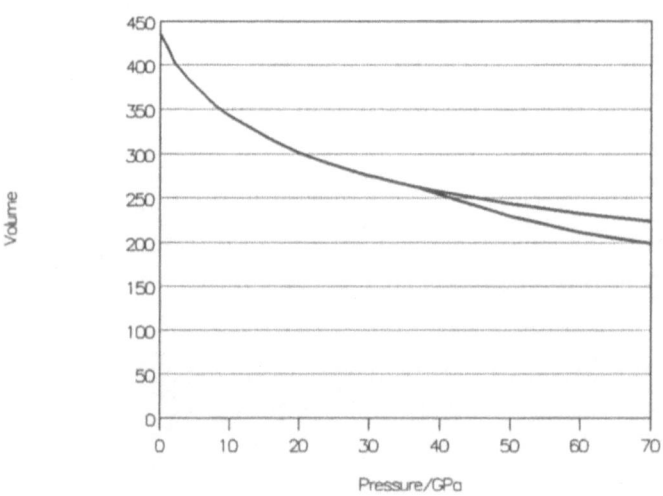

Figure 8.   Volume per particle for CsBr

accuracy of the model can be easily improved in the crystal densities and the lattice energies.   However the results above presented show a very good description of the macroscopic behavior of the systems studied.

This methodology is a complete density functional procedure, the computational time is relatively small, and in the simplest case provides a very good description of the transition order and the volume ratio.   We expect an improvement in the accuracy of the numerical results if more shells are included in the lattice energy sums and in the crystal potential used for the crystal densities.

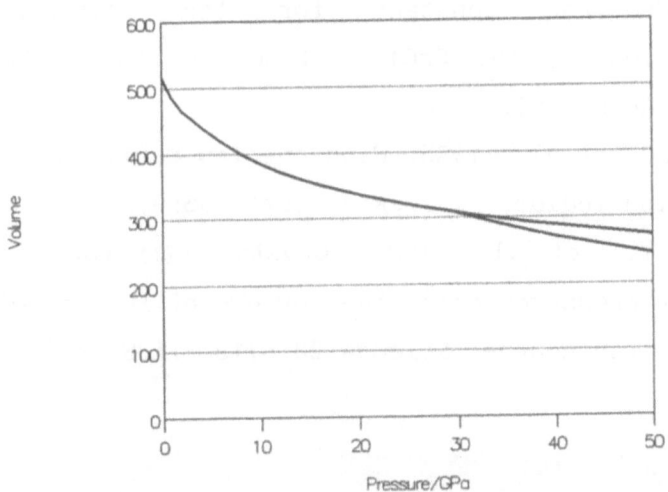

Figure 9.   Volume per particle for CsI

References

Brister,K. E., Vohra, Y. K., and Ruoff, A. L., 1985, High-pressure phase transition in CsCl at V/V_0=0.53, *Phys. Rev. B* **31**: 4657-4658.

Huang, T., Brister, K. E., and Ruoff, A. L., 1984, Pressure-induced structural transition of CsI and CsBr, *Phys. Rev. B* **30**: 2968-2969.

Ruoff, A. L., et al., 1986, Phase transitions near 1 Mbar, *Physica* **136 & 140B**: 209-214.

Syassen, K., 1986, Ionic monochalcogenides under pressure, *Physica* **136 & 140B**: 277-283.

Vela, A., Cedillo, A., and Gázquez, J. L., 1986, Interatomic interactions in density functional theory, *Int. J. Quantum Chem.* **29**: 937-948.

Vohra, Y. K., Duclos, S. J., and Ruoff, A. L., 1985, The Madelung constant for the tetragonal distortion of the CsCl lattice, *J. Phys. Chem Solids* **46**: 515-517.

Vohra, Y. K., et al., 1985, Band-overlap metallization of Cesium iodide, *Phys. Rev. Lett.* **55**: 977-979.

Vohra, Y. K., et al., 1986, Crystal structures at megabar pressures determined by use of the Cornell synchrotron source, *Science* **231**: 1136-1138.

20
Overview of the Degeneracy-Dependent Self-Interaction Correction (D-SIC)

PIETRO CORTONA

Abstract

D-SIC is a method to correct the local-density approximation for the self-interaction effects which is strictly based on the homogeneous gas theory. In this chapter, the theoretical foundations of the method will be reviewed and a large set of tests and applications in atomic calculations will be presented. Furthermore, an extensive comparison with a variety of other approximations in use in the density-functional theory will be given.

Introduction

The density-functional theory (DFT) states that the knowledge of the ground state charge density of an electronic system is sufficient to determine (at least in principle) all the properties of that system (Hohenberg et al., 1964). In particular, the total energy (and the different pieces in which it is usually separated) are functionals of the ground state charge density. The usual way of determining the latter consists in solving the Kohn and Sham (1965) one-electron equation

$$\left(-\frac{1}{2} \nabla^2 + V^{en} + V^{eff} \right) \psi_i = \varepsilon_i \psi_i \tag{1}$$

and then to calculate the density as if the electrons were noninteracting. In Eq.(1) V^{en} is the electron-nucleus potential and V^{eff} is the effective electron-electron potential. The latter is generally divided in two parts, V^c and V^{xc}, where V^c is the coulombic potential generated by the total charge density of the system

$$V^c = \int \frac{\rho(r')}{|r - r'|} d^3 r', \tag{2}$$

and V^{xc} is given in terms of the functional derivative of the exchange-correlation ground state energy with respect to the charge density ρ:

$$V^{xc} = \frac{\delta E^{xc}}{\delta \rho}. \tag{3}$$

307

The scheme sketched above gives an exact description of the system, in the sense that it would produce the exact charge density, if the exact exchange-correlation potential was used in Eq. (1). Of course, the problem is that one needs the knowledge of the explicit functional dependence of E^{xc} from ρ and this dependence is unknown. Thus, a central problem in the DFT is to find accurate approximate expressions of this functional.

The approximation which has been used in most applications of the theory is the local-density approximation (LDA), which consists in assuming that the functional dependence of E^{xc} from ρ is the same as for the homogeneous electronic gas. Several different methods have been used to find more accurate approximations of $E^{xc}[\rho]$, and some very recent examples are illustrated in other chapters of this book. Here we will be interested in the so called self-interaction corrections.

These methods are based on the remark that the coulombic potential (Eq. 2) contains a self-interaction contribution, which reasonably has to be compensated by the exchange-correlation potential. This is confirmed by the exact formal theory, since it has been rigorously proved that V^{xc} has the following three properties: i) The "exchange-correlation hole" satisfies the sum rule (Gunnarsson et al., 1976); ii) V^{xc} reduces to $-V^c$ for a system containing only one electron; iii) at large distance from a neutral finite system, $V^{en}+V^{eff}$ decreases proportionally to $1/r$ (Almbladh et al., 1985). Most approximations in use in the DFT do not satisfy these three conditions. To find approximate functionals $E^{xc}[\rho]$, which give rise to potentials satisfying them, defines the self-interaction correction problem in the context of the standard Kohn and Sham theory.

In contrast to the use of Eqs. (1) - (3), another approach to the determination of the electronic charge density is possible. In fact, one can separate the exchange-correlation energy in an exchange and a correlation term and treat the first one as in the Hartree-Fock (HF) theory. Also this approach was proposed by Kohn and Sham (1965), and its main difference with respect to the preceding one is that the potential is no longer identical for all the electrons. More precisely, one has different local potentials for the various electrons or, equivalently, a unique, but nonlocal, potential. The advantages of this second approach are essentially the following: i) only approximations of the correlation energy functional are required; ii) the self-coulombic interaction is exactly cancelled out by a corresponding self-exchange term. On the other hand, one has to perform complex HF-like calculations and, furthermore, it is not clear that treating exchange and correlation on different footing one really obtains improved results (see, for example, Lagowski et al., 1989; Cortona et al., 1989).

In general, one calls self-interaction corrections approximations which cancel out explicitly the coulombic self-interaction of each electron. As they give rise to different potentials for the various electrons - the exchange-correlation potential for the i-th electron of the system has the following general form:

$$V_i^{xc} = - \int \frac{\rho_i(r')}{|r - r'|} d^3r' + V_i^{xc,ie} , \qquad (4)$$

where $V_i^{xc,ie}$ is some kind of approximation of the interelectron contribution - they are approximate versions of the second kind of methods rather than of the former.

In this chapter we will be concerned, in particular, with the so called degeneracy-dependent self-interaction correction (D-SIC), which is a method based on an expression of the exchange energy of a homogeneous gas containing N electrons given by Rae several years ago (1975). The discussion of that method will be organized as follows. We will start by reviewing the Rae theory; then, the D-SIC method will be introduced and his relations with the Perdew and Zunger (1981) self-interaction correction will be analyzed. Later on, we will report a few tests of the theory and we will apply it to the calculation of ionization potentials and electron affinities. Finally, some conclusions will be drawn.

Self-interaction correction for the electronic homogeneous gas (Rae, 1975)

Let us consider N_σ electrons of spin σ in a box of volume V, with periodic boundary conditions. The (interelectron) exchange energy-per-unit volume is given by:

$$E_\sigma^{x,ie} = - \frac{1}{2V^2} \sum_{\substack{i,j=1 \\ i \neq j}}^{N_\sigma} \frac{4\pi}{|k_i - k_j|^2} . \qquad (5)$$

The standard way of evaluating this expression consists in trasforming it into the integral

$$E_\sigma^x = - \frac{1}{2V^2} \left(\frac{V}{8\pi^3} \right)^2 \int_{k_2 < k_F} \int_{k_1 < k_F} \frac{4\pi}{|k_1 - k_2|^2} d^3k_1 d^3k_2 , \qquad (6)$$

where k_F is the Fermi momentum. Doing that, the $i \neq j$ condition is disregarded, and this implies the inclusion of the self-interaction terms. To take into account this condition, Rae's idea was to eliminate from the integration region in Eq.(6), a small volume corresponding to approximatively equal values of k_1 and k_2. This volume is defined by requiring that it contains N_σ pair of states:

$$N_\sigma = \left(\frac{V}{8\pi^3}\right)^2 \iint\limits_{|k_1 - k_2| \le \beta k_F} d^3k_1\, d^3k_2 \,. \tag{7}$$

Calculating the integral one obtains the equation:

$$N_\sigma^{-1} = \beta^3 - \frac{9}{16}\beta^4 + \frac{1}{32}\beta^6 \,, \tag{8}$$

which determines β as a function of N_σ.   Now, one can evaluate the integral in Eq. (6) under the condition $|k_1 - k_2| > \beta k_F$. The result is:

$$E_\sigma^{x,ie} = -\frac{3}{4}\left(\frac{6}{\pi}\right)^{1/3}\left(\frac{N_\sigma}{V}\right)^{4/3}\gamma(N_\sigma) = \gamma(N_\sigma)\, E_\sigma^x \,, \tag{9}$$

where

$$\gamma(N_\sigma) = 1 - \frac{4}{3}\beta + \frac{1}{2}\beta^2 - \frac{1}{48}\beta^4 \,. \tag{10}$$

Thus, the self-interaction-corrected exchange energy of a homogeneous gas of N_σ electrons of spin σ is given by the classical expression for the infinite gas multiplied for a factor γ which is a function of N_σ. As $\gamma=0$ ($\beta=2$) when $N_\sigma=1$, and $\gamma=1$ ($\beta=0$) when $N_\sigma\to\infty$, the limiting cases are correctly obtained.

Self-interaction correction for inhomogeneous systems (Cortona, 1986)

In order to correct the LDA for the self-interaction effects, it is natural to use Eq. (9) to derive an approximate expression for the interelectron exchange. However, the more direct way of doing this - to use the total charge density ρ_σ and the total number of electrons N_σ in Eq. (9) - gives rise to an objection of principle.

In fact, it is sufficient to add one electron at very large distance from the system, to change the value of γ and to produce unphysical changes in the properties of the system. Thus, a more careful analysis is required.

   In general, it is possible to partition an inhomogeneous system in parts which are "more similar" to a homogeneous gas than the overall system. An example of such a partition is given by two groups of electrons of charge density ρ_1 and ρ_2 localized very far apart; another well known example is given by the electronic shells of an atomic system. Suppose then to have partitioned a system in n groups of $N_{a\sigma}$ electrons and let the charge density of the a-th group be $\rho_{a\sigma}$. $N_{a\sigma}$ and $\rho_{a\sigma}$ are such that

$$N_\sigma = \sum_{a=1}^{n} N_{a\sigma} \,, \tag{11}$$

$$\rho_\sigma = \sum_{a=1}^{n} \rho_{a\sigma} \,. \tag{12}$$

We assume that each subsystem is sufficiently similar to a homogeneous gas. Then, we can apply Eq. (9) separately to each group of electrons. We obtain the following expression for the *intragroups* exchange:

$$E_\sigma^{x,intra} = \sum_{a=1}^{n} \gamma(N_{a\sigma}) E_\sigma^x[\rho_{a\sigma}] \,. \tag{13}$$

In order to derive an expression for the interelectron exchange of the overall system, we have to add to Eq. (13) the contribution of the *intergroups* exchange. In the spirit of the LDA, this contribution is given by

$$E_\sigma^{x,inter} = E_\sigma^x[\rho_\sigma] - \sum_{a=1}^{n} E_\sigma^x[\rho_{a\sigma}] \,. \tag{14}$$

Thus, one obtains the following expression for the total *interelectron* exchange:

$$E_\sigma^{x,ie} = E_\sigma^x[\rho_\sigma] - \sum_{a=1}^{n} \{ 1 - \gamma(N_{a\sigma}) \} E_\sigma^x[\rho_{a\sigma}] \tag{15}$$

and, after having performed the variation, the following potential for each electron belonging to the a-th group:

$$V_{a\sigma}^{x,ie} = -\left(\frac{6}{\pi} \rho_\sigma \right)^{1/3} + \left(\frac{6}{\pi} \rho_{a\sigma} \right)^{1/3} \{ 1 - \gamma(N_{a\sigma}) \} \,. \tag{16}$$

We note that to use the LDA for the intergroups exchange does not contrast with the use of Eq. (9) for the intragroups contributions, because there is no self-interaction in the intergroups exchange. We note also that the LDA is certainly a bad approximation for the intergroups exchange, and thus one cannot expect a good description of the properties which crucially depend from it.

Focusing now on atomic systems and partitioning them in shells, Eq. (15) transforms in the following:

$$E_\sigma^{x,ie} = \frac{3}{4} \sum_{nl} N_{nl\sigma} \rho_{nl\sigma} V_{nl\sigma}^{x,ie} \,, \tag{17}$$

where $V^{x,ie}_{nl\sigma}$ is the interelectron exchange potential given by:

$$V^{x,ie}_{nl\sigma} = - \left(\frac{6}{\pi}\rho_\sigma\right)^{1/3} + \left(\frac{6}{\pi}\rho_{nl\sigma}\right)^{1/3} N^{1/3}_{nl\sigma} \{ 1 - \gamma(N_{nl\sigma}) \} . \quad (18)$$

In these latter two equations, $\rho_{nl\sigma}$ is the charge density normalized to 1 of the shell of quantum numbers $nl\sigma$ and $N_{nl\sigma}$ is the occupation number of that shell.

The coefficient $N^{1/3}_{nl\sigma}\{1-\gamma(N_{nl\sigma})\}$ characterizes the approximation and summarizes the informations coming from the homogeneous gas theory. To replace it with $D_l^{1/3}\{1-\gamma(D_l)\}$, where $D_l=(2l+1)$ is the degeneracy of the shell, makes no difference for a closed shell atom and has a minor effect on the total energy of an open shell atom. On the other hand, this (arbitrary) substitution permits to profit better of the compensation of errors in taking total energies differences. To use Eqs. (17) and (18) with the coefficient $N^{1/3}_{nl\sigma}\{1-\gamma(N_{nl\sigma})\}$ replaced by $D_l^{1/3}\{1-\gamma(D_l)\}$ defines the degeneracy-dependent self-interaction correction for atomic systems[1].

Other ways of performing the self-interaction correction

An analysis of the relations between D-SIC and some other forms of self-interaction correction has been recently done by Cortona et al. (1989). We refer to that paper for this kind of discussion, while a more extended review of various forms of self-interaction correction can be found in a paper by Perdew and Zunger (1981). Here, we only wish to point out the very strong analogies between D-SIC and the Perdew and Zunger (1981) method.

The form of self-interaction correction proposed by these two authors - usually referred to as SIC - is based on the following expression for the interelectron exchange-correlation energy functional:

$$E^{xc,ie} = E^{xc}[\rho\uparrow, \rho\downarrow] - \sum_{i\sigma} E^{xc}[\rho_{i\sigma}, 0], \quad (19)$$

where $\rho_{i\sigma}$ is the charge density of the i-th electron of the system and $E^{xc}[\rho\uparrow,\rho\downarrow]$ is a generic approximation of the exchange-correlation functional. This expression was originally proposed in order to obtain a functional which

---

[1] For practical applications to atomic calculations, one needs only few values of $D_l^{1/3}\{1-\gamma(D_l)\}$. It is equal to 1, 1.130153, 1.167141 for s, p, d electrons respectively.

vanishes for a system containing only one electron. Supposing now that $E^{xc}[\rho\uparrow,\rho\downarrow]$ is the local expression of the functional, the exchange part of Eq.(19) coincides with the D-SIC exchange [Eq. (15)] under the condition $N_{a\sigma}=1$. This implies to have chosen a partition with just one electron in each group: if one systematically uses this particular partition for any inhomogeneous system, D-SIC and SIC reduce to different theoretical points of view of the same approximation. However, Rae's theory suggests that this is not always the best thing to do: using other partitions, D-SIC takes more effectively into account the results of the homogeneous gas theory.

On the other hand, the strong analogies between the two methods make quite natural to use the Perdew and Zunger correlation together the D-SIC exchange. This is the choice we have systematically adopted in our calculations. Thus, the differences between the SIC and D-SIC results reported in this paper, are entirely due to the exchange part of the functional.

In the following, we will also discuss results obtained by two other forms of self-interaction correction. The first one, due to Stoll, Pavlidou and Preuss (SPP) (1978), is an approximation for the correlation energy based on the expression

$$E^{corr,ie} = E^{corr}[\rho\uparrow,\rho\downarrow] - \sum_{\sigma} E^{corr}[\rho_{\sigma},0] \ , \qquad (20)$$

which completely neglects the correlation between electrons having the same spin.

The second one - due to Vosko and Wilk (VW) (1983) - is based on the functional

$$E^{corr,ie} = E^{corr}[\rho\uparrow,\rho\downarrow] - \sum_{\sigma} N_{\sigma} E^{corr}\left[\frac{\rho_{\sigma}}{N_{\sigma}},0\right] \qquad (21)$$

and essentially is an average correction similar to the Fermi-Amaldi correction (1934) of the coulombic energy in the context of the Thomas-Fermi theory.

The SPP approximation has not an analogous expression for the exchange contribution and is generally used in HF calculations or together with some other approximation for the exchange energy; the VW correction is easily extended to the exchange term, but, in general, it is used in the context of HF-like calculations.

Tests of the D-SIC exchange

As D-SIC is an approximation for the exchange contribution to the exchange-correlation energy functional, it is quite natural to test it by performing exchange-only calculations and by comparing the results with the corresponding HF results. A number of these tests were done in the first paper on the D-SIC

TABLE I - Total energies (in Hartrees) calculated by various exchange-only approximations. HF: Hartree-Fock; LDA: local-density approximation; SIC: Perdew and Zunger (1981) self-interaction correction; D-SIC: degeneracy-dependent self-interaction correction.

	HF	LDA - HF	SIC - HF	D-SIC - HF
Sc	759.736	2.725	-1.032	0.263
Ti	848.406	2.872	-1.136	0.289
V	942.884	3.008	-1.261	0.302
Cr	1043.355	3.079	-1.460	0.271
Mn	1149.866	3.279	-1.533	0.328
Fe	1262.443	3.401	-1.689	0.330
Co	1381.414	3.549	-1.825	0.357
Ni	1506.871	3.678	-1.990	0.365
Cu	1638.963	3.719	-2.240	0.319
Zn	1777.848	3.933	-2.341	0.379
average % error		0.28	0.13	0.03

TABLE II - Exchange energies (in Hartrees) calculated by various exchange-only approximations. HF, LDA, SIC and D-SIC as in Table I.

	HF	LDA - HF	SIC - HF	D-SIC - HF
Sc	-38.03	2.83	-1.00	0.36
Ti	-41.04	3.00	-1.09	0.41
V	-44.20	3.16	-1.21	0.44
Cr	-47.76	3.29	-1.39	0.46
Mn	-50.98	3.46	-1.48	0.49
Fe	-54.38	3.60	-1.62	0.51
Co	-57.97	3.77	-1.74	0.56
Ni	-61.68	3.90	-1.92	0.56
Cu	-65.79	4.03	-2.13	0.58
Zn	-69.64	4.21	-2.24	0.62
average % error		6.7	2.9	0.9

TABLE III - Average % error of various exchange-only approximations with respect to HF. The average is performed on all the atoms from Li to Ar.

	LDA	SIC	D-SIC
total energies	1.11	0.16	0.02
exchange energies	10.5	1.9	0.6

method (Cortona, 1986), where atoms belonging to the first two rows of the Periodic Table were considered. We extend now this kind of analysis by reporting results for the third row transition metal atoms.

Total and exchange energies calculated by the LDA, SIC and D-SIC methods are compared with the corresponding HF quantities in Tables I and II respectively. For the sake of completeness, in Table III, we summarize the results obtained for the atoms from Li to Ar. The HF total energies reported in these Tables are directly taken from Clementi and Roetti (1974), while the exchange energies are the values recently calculated by Becke (1988) using the Clementi and Roetti wave functions.

The main comment about Tables I and II is that D-SIC *systematically* improves the total as well as the exchange energies calculated by both the LDA and the SIC methods. This is the same result found for the atoms of the first two rows of the Periodic Table. The improvements are quite large: for the total energies of the transition metal atoms, they amount to a factor 10 with respect to LDA and to a factor 4 with respect to SIC. Furthermore, the maximum error we have found in the total energies calculated by D-SIC is 0.04 % (in the case of Ar). We have also tested the accuracy of D-SIC for heavier atoms by performing calculations for the remaining rare gas atoms: the errors we have found are 0.03% on the total energies and 1.2%, 1.6%, 1.7% on the exchange energies of Kr, Xe and Rn respectively[2].

In the original paper, we also studied ionization potentials (IP's), electron affinities (EA's), charge densities and energy eigenvalues in the exchange-only approximation. Without entering in a detailed discussion of these results, they can be summarized as follows. The IP's and the EA's calculated by LDA, SIC and D-SIC compare with HF about in the same way as the corresponding quantities including the correlation contributions compare with the experimental results (see following sections). In general, there is a partial compensation of errors between the exchange and the correlation contributions and thus the agreement with the experimental values is better than the agreement of the exchange-only quantities with HF. The charge densities are not very sensitive to the approximation used. In fact, with the exception of the atoms of the first row of the Periodic Table, where D-SIC gives the best results, LDA, SIC and D-SIC agree very well with each other and with the HF results. Finally, the energy eigenvalues calculated by SIC and D-SIC are very similar, and quite different from the LDA ones. A more careful analysis, however, reveals that the difference of two levels having the same value of n is almost the same as the corresponding LDA difference: the main effect of the self-interaction correction is an almost rigid shift of each set of levels having the same n. This shift makes the eigenvalues more similar to the HF ones.

---

[2] The HF total energy of Rn is taken from McLean et al. (1981).

Applications: the ionization potentials

The IP's and the EA's are very interesting quantities to calculate not only for their important role in the chemical binding, but also because they are very sensitive tools to test the approximations used and their ability to compensate the errors. Furthermore, extensive calculations of these quantities have been recently performed by Lagowski and Vosko (LV) (1988), who used a variety of different approximations, and thus a detailed comparison of D-SIC with other methods is possible.

The set of results relative to the IP's is reported in Tables IVa and IVb. These results are obtained by three kinds of theoretical methods. First of all, methods using an approximate expression for the exchange as well as for the correlation functionals. These are LDA, SIC and D-SIC and the relative calculations were performed by ourselves. Second, methods using the HF exchange and a local correlation expression. The latter can be the full local correlation (KS) or, eventually, contain some kind of self-interaction correction (SPP, VW). The IP's calculated by these methods are taken from the paper by

TABLE IVa - Experimental ionization potentials (in eV) and errors of the values calculated by various theoretical methods. The relativistic contributions, reported in column three, are included in the calculated values. LDA, SIC and D-SIC as in Table I (but with the correlation contributions included).

	Exp.	Rel. Contr.	LDA - Exp.	SIC - Exp.	D-SIC - Exp.
s electrons					
Li	5.39	0.00	0.08	0.05	0.05
Be	9.32	0.00	-0.28	-0.21	-0.21
Na	5.14	0.01	0.23	0.17	0.21
Mg	7.65	0.01	0.10	0.10	0.13
K	4.34	0.01	0.20	0.15	0.19
Ca	6.11	0.02	0.12	0.11	0.13
p electrons					
B	8.30	0.00	0.26	0.57	-0.06
C	11.26	0.00	0.46	0.81	0.11
N	14.53	-0.01	0.40	0.76	0.02
O	13.62	-0.01	0.36	0.61	-0.09
F	17.42	-0.01	0.68	0.94	0.18
Ne	21.56	-0.02	0.61	0.86	0.06
Al	5.99	-0.01	-0.01	0.03	-0.37
Si	8.15	-0.01	0.09	0.17	-0.26
P	10.49	-0.01	0.00	0.12	-0.34
S	10.36	-0.01	0.25	0.41	-0.05
Cl	12.97	-0.02	0.30	0.51	0.00
Ar	15.76	-0.02	0.17	0.41	-0.13
average error			0.26	0.39	0.14

TABLE IVb - Errors of the ionization potentials calculated by various theoretical methods. The relativistic contributions (see Table IVa) are included in the calculated values. KS, SPP, VW: Hartree-Fock exchange plus local correlation, eventually self-interaction-corrected by the methods due to Stoll et al. (1978) and to Vosko and Wilk (1983); HL, P: Hartree-Fock exchange plus the nonlocal correlation functionals by Hu and Langreth (1985) and by Perdew (1986).

	KS - Exp.	SPP - Exp.	VW - Exp.	HL - Exp.	P - Exp.
s electrons					
Li	0.39	0.00	0.11	0.50	0.15
Be	0.12	-0.33	-0.20	0.19	-0.23
Na	0.29	-0.10	0.11	0.48	0.04
Mg	0.20	-0.21	0.00	0.41	-0.07
K	0.14	-0.22	0.01	0.36	-0.05
Ca	0.09	-0.28	-0.05	0.34	-0.10
p electrons					
B	0.67	0.09	0.26	0.80	0.35
C	0.55	-0.11	0.09	0.73	0.24
N	0.45	-0.27	-0.04	0.62	0.12
O	0.31	-0.34	-0.16	0.43	-0.19
F	0.13	-0.59	-0.37	0.29	-0.35
Ne	-0.01	-0.78	-0.54	0.14	-0.48
Al	0.34	-0.14	0.09	0.56	0.20
Si	0.34	-0.21	0.04	0.60	0.19
P	0.41	-0.20	0.07	0.67	0.24
S	0.37	-0.18	0.07	0.56	0.07
Cl	0.38	-0.23	0.03	0.61	0.10
Ar	0.48	-0.19	0.10	0.72	0.21
average error	0.32	0.25	0.13	0.50	0.19

LV (1988) as well as the last sets of IP's, labelled HL and P. These two latter methods are completely nonlocal: they include the correlation in HF-like calculations by the nonlocal functionals proposed by Hu and Langreth (1985) and by Perdew (1986) respectively. Finally, we note that all the results include the relativistic contributions. The latter were evaluated perturbatively by LV and are included in their IP's. Indeed, for the atoms considered in Table IV, their contribution is very small, but, in order to do a coherent comparison with the LV results, we have added them to our LDA, SIC and D-SIC values.

We begin the discussion of Table IV by comparing the LDA, SIC and D-SIC IP's. It is immediately verified that D-SIC is the method in better general agreement with experiment. This is quantified by the average error, which also shows that SIC does not improve the LDA. A more detailed analysis shows that, if the electron concerned in the ionization process is an *s* electron, SIC is slightly better of both D-SIC and LDA, while these two latter approximations are almost equivalent. For *p* electrons, D-SIC is the method which gives the most accurate results, while LDA is systematically better than SIC.

The comparison with the nonlocal methods gives rise to some surprises. At first, considering the two completely nonlocal methods, it can be seen that HL is the approximation which gives rise to the greatest average error (0.5 eV). On the other hand, P, which is certainly the most sophisticated of all these methods, turns out to be quite accurate, but no as accurate as D-SIC, which has a smaller average and maximum error and that gives better IP's for a greater number of atoms. Among the approximations using the HF exchange and a local correlation, the VW one is the most accurate. It also results to be slightly better than D-SIC, because the average error is a little smaller and because it gives best IP's in a greater number of cases. The maximum error, however, is greater than the maximum error of D-SIC.

Finally, as a last comment on the IP's, we note that the errors of the most accurate approximations - VW, D-SIC and P - do not show any particular trend, while LDA, SIC, KS and HL generally overestimate the IP's and SPP underestimates them.

Applications: the electron affinities

The EA's calculated by the methods described in the preceding section are reported in Table V and in Table VI. Of course, there are not LDA results because, as it is well known, this approximation does not give, in general, stable negative ions. As in the case of the IP's, the relativistic contributions and the KS, SPP, VW, HL and P values are taken from the LV work, while the LDA, SIC and D-SIC EA's are those calculated by Cortona et al. (1989).

The discussion of the EA's reported in Table V gives rise to considerations similar to those we have done for the IP's, with some few differences. First, D-SIC is now about equivalent to SIC for s electrons and systematically better for p electrons. Second, the approximations based on the HF exchange are less accurate than D-SIC: this is not systematically true, but D-SIC gives the smallest average and maximum errors and better EA's in a greater number of cases. Finally, the accuracy of the D-SIC results is quite good: the maximum error is 0.18 eV (for Cu).

In Table VI are reported the EA's of atoms for which the electron concerned is a d electron. All the methods considered meet serious problems in treating these electrons. In particular, the approximations using the HF exchange greatly underestimate the EA's and, in general, do not predict stable negative ions (only one exception: V in the HL approximation). As a consequence, the HL approximation, which generally overestimates the EA's as well as the IP's, results to be the HF-based method which gives the best results. One time more D-SIC improves the other methods. However, its errors are now large and the results for these EA's cannot be considered satisfactory.

TABLE V - Experimental electron affinities (in eV) and errors of the values calculated by various theoretical methods. The relativistic contributions, reported in column three, are included in the calculated values. SIC, D-SIC, KS, SPP, VW, HL and P as in Tables IVa and IVb.

	Exp.	Rel. Contr.	SIC - Exp.	D-SIC - Exp.
s electrons				
Li	0.62	0.00	-0.06	-0.06
Na	0.55	0.00	0.04	0.04
K	0.50	0.00	0.06	0.06
Cr	0.67	-0.02	-0.15	-0.09
Cu	1.23	0.03	0.08	0.18
p electrons				
B	0.28	-0.01	0.36	0.08
C	1.26	0.00	0.40	0.04
O	1.46	-0.01	0.34	-0.01
F	3.40	-0.01	0.34	-0.07
Al	0.44	-0.01	0.17	-0.02
Si	1.39	-0.01	0.17	-0.08
P	0.75	-0.01	0.34	0.09
S	2.08	-0.02	0.43	0.13
Cl	3.62	-0.02	0.37	0.03
average error			0.24	0.07

	KS - Exp.	SPP - Exp.	VW - Exp.	HL - Exp.	P - Exp.
s electrons					
Li	0.12	-0.12	-0.02	0.22	-0.01
Na	0.15	-0.09	0.08	0.29	0.03
K	0.11	-0.10	0.09	0.27	0.02
Cr	-0.11	-0.34	-0.15	0.01	-0.33
Cu	-0.36	-0.66	-0.43	-0.18	-0.52
p electrons					
B	0.10	-0.35	-0.17	0.37	-0.02
C	0.03	-0.52	-0.31	0.30	-0.14
O	-0.50	-1.06	-0.87	-0.28	-0.83
F	-0.62	-1.25	-1.03	-0.41	-0.96
Al	0.13	-0.26	-0.04	0.42	0.06
Si	0.20	-0.28	-0.03	0.51	0.11
P	0.12	-0.29	-0.07	0.33	-0.09
S	0.13	-0.36	-0.12	0.38	-0.07
Cl	0.21	-0.36	-0.09	0.47	0.01
average error	0.21	0.43	0.25	0.32	0.23

TABLE VI - Experimental electron affinities (in eV) and errors of the values calculated by various theoretical methods. The relativistic contributions, reported in column three, are included in the calculated values. SIC, D-SIC, KS, SPP, VW, HL and P as in Tables IVa and IVb.

	Exp.	Rel. Contr.	SIC - Exp.	D-SIC - Exp.
d electrons				
Ti	0.08	-0.15	0.84	0.29
V	0.53	-0.18	1.15	0.57
Fe	0.16	-0.26	0.32	-0.22
Co	0.66	-0.30	0.71	0.16
Ni	1.16	-0.35	1.06	0.49
average error			0.82	0.35

	KS - Exp.	SPP - Exp.	VW - Exp.	HL - Exp.	P - Exp.
d electrons					
Ti	-0.41	-0.89	-0.63	-0.21	-0.52
V	-0.58	-1.09	-0.82	-0.39	-0.70
Fe	-1.18	-1.65	-1.41	-0.97	-1.40
Co	-1.42	-1.91	-1.66	-1.21	-1.63
Ni	-1.67	-2.19	-1.92	-1.47	-1.87
average error	1.05	1.55	1.29	0.85	1.22

Conclusions

We believe that the results reported in this chapter make evident the importance of using the self-interaction correction, at least in atomic calculations. In fact, in contrast to SIC, which often gives results of quality similar or even inferior to the corresponding LDA results, D-SIC produces almost systematic improvements of them.

More difficult to analyze is the comparison with the HF-based methods, and, in particular, with the completely nonlocal methods HL and P. HL is the only method which makes no use of the Monte-Carlo results for the correlation energy and certainly suffers from that, but P, from the theoretical standpoint, should give better results than D-SIC. In practice, the opposite happens. There are two possible reasons of this surprising result. The first one is that D-SIC uses a self-interaction-corrected correlation energy, while this is not the case of P. An indication that this could be an important point is also given by the good results obtained by the VW method. The second one is that to use similar approximations for the exchange and for the correlation gives rise to a more effective compensation of errors. These two facts, added to the good quality of the D-SIC exchange, probably explain the good results obtained by D-SIC.

Finally,we would like to make just a comment about the interconfigurational energies and the IP's (for d electrons) of the transition metal atoms, that we have not reported here. Indeed, we have performed (Cortona, 1988) a nonrelativistic study of these quantities which has shown that D-SIC systematically improves both LDA and SIC. The remaining (large) errors were attributed to the nonlocal interactions, which certainly strongly affect these quantities. However, a very recent work by Lagowski and Vosko (1989) indicates that part of these errors should be attributed to unexpectedly large relativistic effects. Unfortunately the results reported by Lagowski and Vosko are not sufficient to completely reanalyze our calculations; however, it seems that to take into account the relativistic effects does not change the relative quality of the three approximations. This is also the result of some preliminary calculations we have performed by our fully-relativistic spin-polarized program (Cortona, 1989). We will return on this subject in a forthcoming paper.

References

Almbladh, C. O., and von Barth, U., 1985, *Phys. Rev. B* **31**: 3231.

Becke, A. D., 1988, *Phys. Rev. A* **38**: 3098.

Clementi, E., and Roetti, C., 1974, *At. Data and Nucl. Data Tables*, **14**: 177.

Cortona, P., 1986, *Phys. Rev. A* **34**: 769.

Cortona, P., 1988, *Phys. Rev. A* **38**: 3850.

Cortona, P., Böbel, G., and Fumi, F. G., 1989, *J. Phys. France* **50**: 2647.

Cortona, P., 1989, *Phys. Rev. B* **40**: 12105.

Fermi, E., and Amaldi, E., 1934, *Mem. Accad. d'Italia* **6**: 119.

Gunnarsson, O., and Lundqvist, B. I., 1976, *Phys. Rev. B* **13**: 4274.

Hotop, H., and Lineberger, W.C., 1985, *J. Phys. Chem. Ref. Data* **14**: 731.

Hohenberg, P., and Kohn, W., 1964, *Phys. Rev.* **136**: B864.

Hu, C. D., and Langreth, D. C., 1985, *Phys. Scr.* **32**: 391.

Kohn, W., and Sham, L. J., 1965, *Phys. Rev.* **140**: A1133.

Lagowski, J. B., and Vosko, S. H., 1988, *J. Phys. B: At. Mol. Opt. Phys.* **21**: 203.

Lagowski, J. B., and Vosko, S. H., 1989, *Phys. Rev. A* **39**: 4972.

McLean, A. D., and McLean, R.S., 1981, *At. Data and Nucl. Data Tables* **26**: 197.

Perdew, J. P., and Zunger, A., 1981, *Phys. Rev. B* **23**: 5048.

Perdew, J. P., 1986, *Phys. Rev. B* **33**: 8822.

Rae, A. I. M., 1975, *Mol. Phys.* **29**: 467.

Stoll, H., Pavlidou, C. M. E., and Preuss, H., 1978, *Theor. Chim. Acta* **149**: 143.

Vosko, S. H., and Wilk, L., 1983, *J. Phys. B: At. Mol. Phys.* **16**: 3687.

21

Correlation Effects on Ionization Energies. A Comparison of Ab Initio and LDA Results

P. Decleva, G. Fronzoni, and A. Lisini

Introduction

Since the advent of photoelectron spectroscopy (Turner et al., 1970; Siegbahn et al., 1971) the interest in atomic and molecular ionization energies has remained very high. At a first approximation the experimental data can be interpreted as a measurement of orbital energies, making therefore a direct connection with orbital theories of electronic structure (Siegbahn and Karlsson, 1982). Notably, in the ab-initio framework, the most immediate connection is with the eigenvalues of the Fock operator, through the Koopmans theorem (KT). Within a single determinant approach a refined description may be achieved by reoptimizing orbitals for each final ionic state, the so called ΔSCF calculation. The energy gained in the ionic states, termed relaxation energy, lowers the calculated IEs, and is particularly important for the deeper hole states, although it may be very significant also in the valence shell, typically in transition metal compounds. It has been soon realized however that important deviations, notably inversions in the ordering of the ionic states, are sometimes obtained both at the KT and ΔSCF levels. Also additional states become accessible, corresponding to multiple electron excitations, which are forbidden at the KT level, and can be only partly explained by the ΔSCF approach through the loss of orbital orthogonality.

Calculation of ionization energies by DF approaches also started quite early (Connolly and Johnson, 1971), stimulated by the analysis of orbital eigenvalues in the Xα approach, which led to Slater's transition state formalism (Slater, 1972), a practical and accurate approximation to ΔSCF calculations within the Xα approach. It appeared immediately that IEs so obtained were often better than ab-initio KT or ΔSCF values (Baerends and Ros, 1973; Sambe and Felton, 1974; Salahub et al., 1976), notably in cases like F_2, N_2, O_3, where the latter gave an incorrect ordering, and this behaviour was found quite general (De Alti et al., 1982).

As a matter of fact, even the definition of correlation effects is not unambiguous, both in the HF and DF schemes, and not easily comparable in the two approaches. Nonetheless the study of these effects on IEs has added a great deal to their understanding and the design of refined models in the HF scheme. It may be hoped that a similar advancement may be achieved in the DF context, and that knowledge gained in the ab-initio scheme may be transferred to the DF framework, and contribute to the generation of improved functionals, despite the diversity of the two approaches.

Last, as the interpretation of photoelectron spectra provides important information on the electronic structure of the system, it appears important to assess the accuracy of DF schemes, especially in areas, typically transition metal complexes, where the ab-initio treatment meets severe difficulties.

Let us briefly examine correlation in the HF scheme (Sinanoğlu, 1969; Beck and Nicolaides, 1979). It is well known that the mixing of determinants caused by the interelectronic interaction is associated with two types of physically distinct effects. The first is tied to the presence of degeneracy, actual or related to a zero order description of the system considered. Typically it is the hydrogenic degeneracy of atomic orbitals of the same principal quantum number, or that of the bonding-antibonding molecular orbitals at infinite internuclear separation. Also when exact degeneracy is lost because of additional interactions, the mixing of these quasi-degenerate (QD) configurations is very strong, because of the strong spatial localization, which maximizes the interelectronic interaction, and the relative closeness in energy. Well known examples are the H_2 dissociation problem, the $2s^2-2p^2$ interaction in Be, and the $3s3p^6-3s^23p^43d$ one in Ar. The latter is an example of a typical strong correlation present in ionic states, between a single hole and a two holes-one particle (2h-1p) hole filling configuration.

The other type of mixing (dynamical correlation) is caused by the instantaneous interelectronic repulsion that is left when quasi degeneracy is taken into account, and is present alone in truly closed shell systems, like the ground state of He or Ne atoms. Despite converging very slowly in a CI treatment, it is a weak effect, with a regular trend.

In the DF schemes, correlation is defined instead with reference to a model system, which is usually taken as the electron gas. Typically the exchange-correlation functional is approximated locally by the value of an electron gas with the same density. With respect to this system exchange and correlation can be defined exactly like in HF, and therefore the two contributions separated. In this respect $X\alpha$ with $\alpha=2/3$ is a pure exchange approximation (Jones and Gunnarsson, 1989). The separation is however not intrinsic to DF formalism, and given the diversity between the free electron gas and an actual molecular system it may be not particularly illuminating. A better understanding may be achieved by a comparison with ab-initio results for the same system. An investigation of atomic multiplet structure gave evidence of the absence of correlation in $X\alpha$ approaches (Wood, 1980), but there is now ample evidence on the contrary from both ground and excited state properties (Baerends and Ros, 1978;   Dunlap et al., 1979; Salahub et al., 1982; Cook and Karplus, 1987).

Results and Discussion

Selected data for atomic ionizations are reported in table 1. The Ne atom is the prototype of a closed shell system. The ΔSCF values for the 2p and 1s IEs are slightly too low for the neglect of the ground state correlation lost, while the 2s IE is above the experimental value because of the 2h-1p correlations active in the ionic state. Cancellation with relaxation makes KT a fair approximation to the outermost ionization.

While a similar behaviour is apparent in Pd, a marked difference is shown by ΔSCF for the 3d^{10} ionization in Ni, giving a very low IE, indicating the presence of large correlation and relaxation associated with the 3d^{10} shell. The large interelectronic repulsion makes this shell similar to loosely bound anion states, dominated by a characteristic in-out correlation (Froese-Fischer, 1977; Botch et al.,1981).

In transition metal compounds, where the molecular environment pushes the metal towards the high d occupation, the large electron-electron repulsion is relieved by depletion of the d orbitals in favour of those of the ligands, forcing the former in the virtual space, causing therefore an unbalanced description of the ground state MO composition and the appearance of huge relaxation associated with d ionizations (Coutière et al., 1972; Demuynck and Veillard, 1973). As a result KT becomes often totally inadequate, ΔSCF underestimates significantly metal d IEs, and in general ab-initio methods find

	KT	ΔSCF	Xα R	Xα U	VWN	VWN/S	Exp
Ne							
1s	891.8	868.6	874.8	862.4	862.8	861.9(.26)	870.2
2s	52.5	49.3	44.7	44.0	45.0	44.2(.79)	48.5
2p	23.1	19.8	21.9	21.3	22.3	21.5(-.11)	21.6
Ni							
3d	7.23	2.14	5.76	5.49	6.45	5.75(-.15)	5.88
Pd							
4d	9.15	6.67	8.68	8.49	9.37	8.74(-.04)	8.51

Table 1. Ionization Energies for Ne, Ni and Pd.
Exp.: Moore, 1949, 1952, 1957; Siegbahn and Karlsson, 1982.
VWN: Vosko et al., 1980. VWN/S: correction by Stoll et al., 1980.
Gradient correction (Becke, 1986) in parentheses.

	KT[a]	ΔSCF[a]	2h-1p[b]	GF[c]	R	Exp[d]
N₂						
3σ₉	17.28	15.99	13.30	15.70	0.91	15.6
1π₄	-0.48	-0.62	1.30	1.15	0.92	1.4
2σ₄	3.89	3.78	3.07	3.26	0.82	3.2
2σ₉	22.99	21.90	22.77	21.73	0.30	21.7
F₂						
1π₉	18.04	16.27	13.03	15.86		15.9
1π₄	3.87	4.92	3.04	3.13		2.9
3σ₉	2.26	3.33	4.52	5.17		5.2
2σ₄	22.63	22.52	23.45	20.31	0.71	
2σ₉	29.76		28.24	24.67	0.56	
O₃				CI[e]		
6a₁	15.37	13.85	10.10	13.41	0.79	12.73
4b₂	0.45	-0.03	0.18	0.25	0.74	0.27
1a₂	-1.89	-1.96	0.14	0.86	0.75	0.81
3b₂	6.35		6.46	7.84	0.44	6.3-
5a₁	7.46		6.50	6.33	0.24	8.8
1b₁	6.19		9.43	8.13	0.32	
4a₁	14.51		12.34	13.14	0.19	11.4
2b₂	24.16		20.77			
3a₁	33.08					

Table 2. Ab-initio Ionization Energies for N₂, F₂ and O₃.
Absolute energy for the first state. For all other states
relative energies are reported.
a) N₂: Cade et al., 1966; F₂: Cade et al., 1974; O₃: Decleva and
Lisini, 1985, Thunemann et al., 1978.
b) Decleva and Lisini, 1985.
c) N₂: Schirmer and Walter, 1983; F₂: Bieri et al., 1981.
d) N₂: Turner et al., 1970, Siegbahn et al., 1971; F₂: Bieri et
al., 1981; O₃: Katsumata et al., 1984.
e) Decleva et al., 1988.

difficulty in the correct relative placing of d and ligand levels.

The restricted (or non spin-polarized) Xα functional (Xα R) offers
in this case a very satisfactory agreement for the outer IEs, notably a
smooth behaviour for the three systems examined. The behaviour of the
other potentials is substantially similar. Xα unrestricted (Xα U) shows
a slight lowering of IEs, the VWN potential (Vosko et al., 1980) is
globally more attractive, resulting in a sensible worsening of the outer
IEs, while use of the Stoll correction (Stoll et al., 1980) brings back
to values similar to the Xα R ones. In all cases the agreement is good,
notably better than ab-initio KT or ΔSCF, and essentially the same in
all systems, including Ni 3d¹⁰.

The large deviations for the deeper levels 2s and 1s in Ne are
associated with inadequate treatment of the exchange, notably the self-

	Xα R	Xα U	VWN	VWN/S	VWN/SB	Exp
N₂						
$3\sigma_g$	15.16	14.57	15.63	14.95	15.25	15.6
$1\pi_u$	1.79	1.79	1.82	1.83	1.44	1.4
$2\sigma_u$	2.91	3.01	2.97	2.97	3.15	3.2
$2\sigma_g$	20.22	19.86	19.91	19.88	19.35	21.7
F₂						
$1\pi_g$	15.38	14.84	15.86	15.12	15.37	15.9
$1\pi_u$	3.17	3.25	3.25	3.26	3.18	2.9
$3\sigma_g$	6.24	6.13	6.19	6.20	6.75	5.2
$2\sigma_u$	17.58	17.78	17.71	17.68	18.08	-
$2\sigma_g$	24.44	24.50	24.49	24.49	25.53	-
O₃						
$6a_1$	12.28	11.89	12.89	12.19	12.41	12.73
$4b_2$	0.29	0.17	0.21	0.21	0.18	0.27
$1a_2$	1.38	1.17	1.26	1.27	1.02	0.81
$3b_2$	6.96	6.90	6.93	6.92	7.48	6.3-
$5a_1$	6.75	6.71	6.74	6.73	7.41	8.8
$1b_1$	7.52	7.20	7.32	7.33	6.96	
$4a_1$	11.27	11.36	11.32	11.29	11.50	11.4
$2b_2$	20.08	20.07	20.08	20.06	20.44	
$3a_1$	27.32	27.20	27.24	27.23	27.58	

Table 3. Ionization Energies in the Local Density Approximation for N_2, F_2 and O_3.
Absolute energy for the first state. For all other states relative energies are reported.
VWN potential by Vosko et al., 1980.
VWN/S: correction by Stoll et al., 1980.
VWN/SB: gradient correction by Becke, 1986.
Exp.: see table 2.

interaction (Perdew, 1984), for the inner orbitals, as is clear by comparison with ΔSCF results.

In the molecular case the presence of QD effects can give rise to large energy shifts. These effects are associated with the valence virtual orbitals, typically π^* or σ^*, like $1\pi_g$ or $3\sigma_u$ in N_2 and F_2. In table 2 are reported ab-initio results for the classic systems N_2, F_2 and O_3, and in table 3 the corresponding DF values. Although partly available in the literature, the latter have been recalculated to get an accurate and complete set of data.

The inversions present in the ab-initio KT and ΔSCF results for N_2 and F_2 are well known (Cade et al., 1966; Cornford et al., 1971). Correlation effects separate neatly in the QD and dynamical contributions. The first originates from the 2h-1p correlations in the ionic states, in fact a CI calculation limited to these configurations

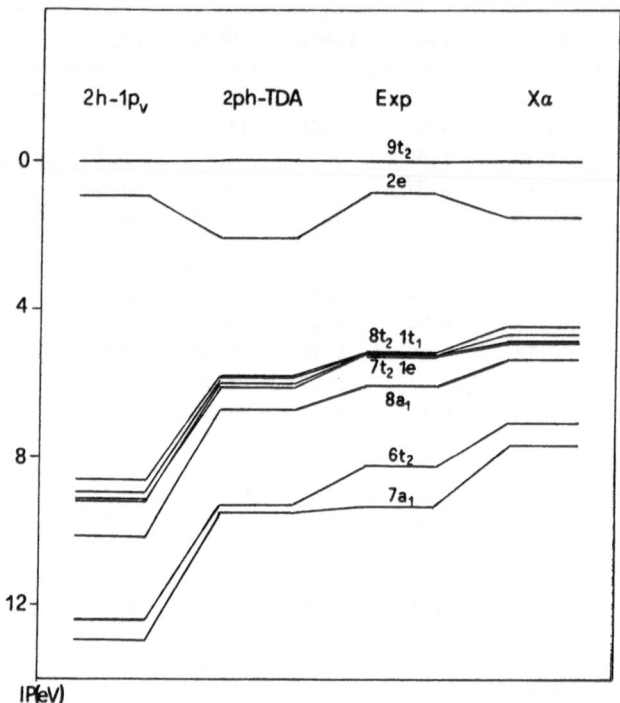

Figure 1. Ionization Energies for Ni(CO)₄.
9t₂ IE (eV): 2h-1p 5.32, 2ph-TDA 9.1, Xα 9.32, Exp. 8.90.
2h-1p, Xα: Decleva et al., 1990. 2ph-TDA: Smith et al., 1989.
Exp.: Hillier et al., 1974.

(2h-1p CI) reproduces correctly the sequence and spacings of ionic
states (Decleva and Lisini, 1985). The second gives the pair
correlations lost upon ionization, it is sensibly constant for the
various states, and mainly affects the absolute values of IEs. If both
contributions are accurately described, as in the ADC(3) (or 2ph-TDA)
Green's Function scheme (Schirmer et al., 1983), it is possible to get
excellent agreement with the experimental values. The bulk of
correlation contribution originates from the valence space. In fact
calculations limited to this space, e.g. by extracting valence virtual
levels by projection techniques, give very similar results at the 2h-1p
CI level, and a fair quantitative agreement at the full CI (FCI) level
(Decleva and Lisini, 1987).

In O₃ QD effects relative to the π orbitals   (b₁,a₂) are already
strong in the ground state (Hay et al., 1975), and make schemes based on
the single reference determinant to converge slowly. In fact 2h-1p CI,
like perturbative GF approaches, fails to correct the double inversion
present at the KT and ΔSCF levels. It is necessary to include higher
excitations within the π space to regain the correct sequence. Here also
a valence space FCI calculation gives a fair agreement with the
experimental data. The very strong correlation effects in this molecule

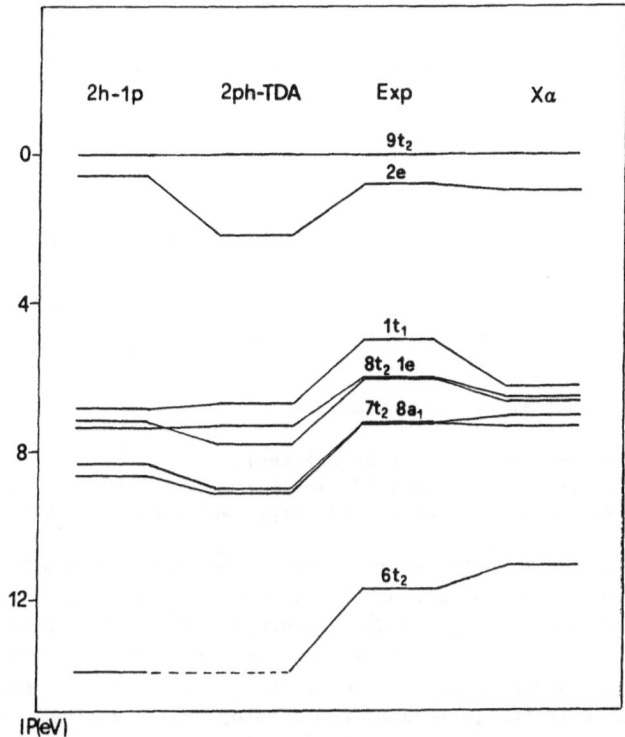

Figure 2. Ionization Energies for Cr(NO)₄.
9t₂ IE (eV): 2h-1p 5.87, 2ph-TDA 8.6, Xα 8.65, Exp. 8.7.
2h-1p, Xα: Decleva ·et al., 1990. 2ph-TDA: Smith et al., 1989.
Exp.: Plummer et al., 1980.

give rise to intense satellites (not reported here) already following
the three outermost ionizations. This is shown also by the very low
spectral intensities (R) of the following states, of which the single
most intense component is reported.

The absence of any noticeable quirk in the DF results is remarkable.
The major difference between the various potentials concerns the
absolute values, reflected in the first IE, which is reproduced
accurately by the VWN potential, while it is variously underestimated in
the other cases. Energy separations are very similar, with the exception
of the values incorporating the gradient correction (Becke, 1986).

In the case of N_2, and in general for the outermost ionizations, the
correction is in the right direction with respect to VWN/S, and brings
to a good agreement with the experimental data, while a worsening is
obtained for the $3\sigma_g$ IE of F_2, already significantly overestimated. The
lack of data for the inner states prevents a closer comparison, although
the underestimate of the $2\sigma_u$ IE in N_2 is expected to be fairly general,
because of the deficiency already noted.

On the whole the agreement is very satisfactory and, with the
exception of the $3\sigma_g$ IE in F_2, only slightly inferior to highly

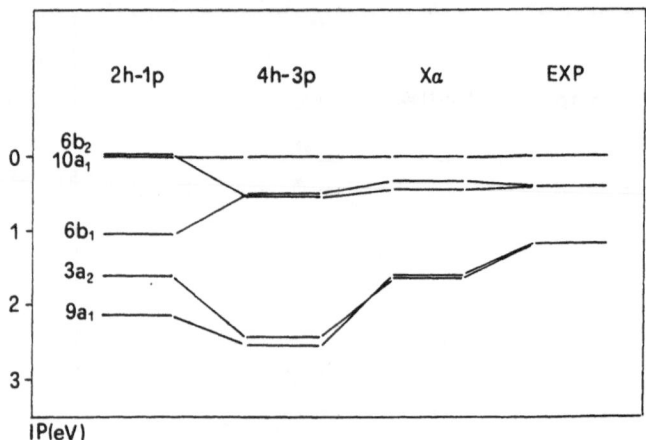

Figure 3. Ionization Energies for Fe(CO)$_2$(NO)$_2$.
10a$_1$ IE (eV): 2h-1p 5.07, 4h-3p 7.97, Xα 8.74, Exp. 8.56.
2h-1p, 4h-3p, Xα: Decleva et al., 1990. Exp.: Guest et al., 1980.

correlated ab-initio results, also in the particularly difficult case of
O$_3$. Results on the outer IEs of other small systems reinforce this
conclusion (Baerends et al., 1984, Tschinke, 1989). The more recent
potentials, notably the Becke correction, show promise of improvement,
although the difference with the simple Xα R results is not great. As
the differences with the experiment are already small, further progress
clearly requires a very accurate numerical comparison.

    Results for the isoelectronic carbonyls/nitrosyls Ni(CO)$_4$, Cr(NO)$_4$,
Fe(CO)$_2$(NO)$_2$ are presented in figures 1, 2 and 3. The common relaxation
effects associated with d ionizations are less important in the Ni and
Cr molecules, but strong in the Fe compound, despite the similarity in
their electronic structure, well apparent in the experimental spectra
(Hillier et al., 1974).

    In the 2h-1p CI results for Ni(CO)$_4$ (Decleva et al., 1990) the d IEs
are underestimated, giving a too large separation with the ligand
levels, while both structures are separately described rather
satisfactorily. The effect is less apparent in Cr(NO)$_4$, presumably
because of the smaller number of d electrons. The 2ph-TDA level (Smith
et al., 1989) gives a great improvement in the d-L separation, but
notably overestimates the splitting of the d band. In any case the
agreement with the experiment is definitely less satisfactory than in
the case of non transition molecules. The Fe compound is the most
difficult at the ab-initio level and the structure of the d band is not
properly reproduced at the 2h-1p CI level (figure 3). It is necessary to
include selected 4h-3p excitations in the CI scheme to regain the
correct structure, still with a grossly overestimated separation.

    A generally satisfactory agreement, of comparable quality to that
obtained in the non transition compounds, is given by the Xα results,
which again are hardly affected by the correlations important in the ab-

SYM	2h-1p	3h-2p	4h-3p	Xα	VWN	Exp
7Aᵤ	5.91	6.42	7.15	6.71	7.8	7.76
13Aₑ	-0.82	1.06	0.33	0.47	0.2	0.43
6Bₑ	-0.65	1.27	0.46	0.87	0.6	
12Aₑ	-0.33	1.61	0.84	1.34	0.4	0.82
11Aₑ	0.28	2.26	1.37	1.73	1.4	
5Bₑ	0.30	1.90	1.40	1.89	1.7	1.64
11Bᵤ	3.80	3.99	3.89	2.85	2.9	2.62
10Aₑ	4.70	5.34	4.64	4.14	4.1	3.79
6Aᵤ	6.39			4.43	4.3-	
9Aₑ	6.64			4.49	4.7	4.94
10Bₑ	6.66			4.29		

Table 4 . Outer Valence Ionization Energies for bis(π-allyl) Nickel.
Absolute energy for the 7Aᵤ state. For all other states relative
energies are reported.
2h-1p, 3h-2p, 4h-3p, Xα: Decleva et al., 1989.
VWN: Andzelm et al., 1989.
Exp.: Böhm et al., 1980.

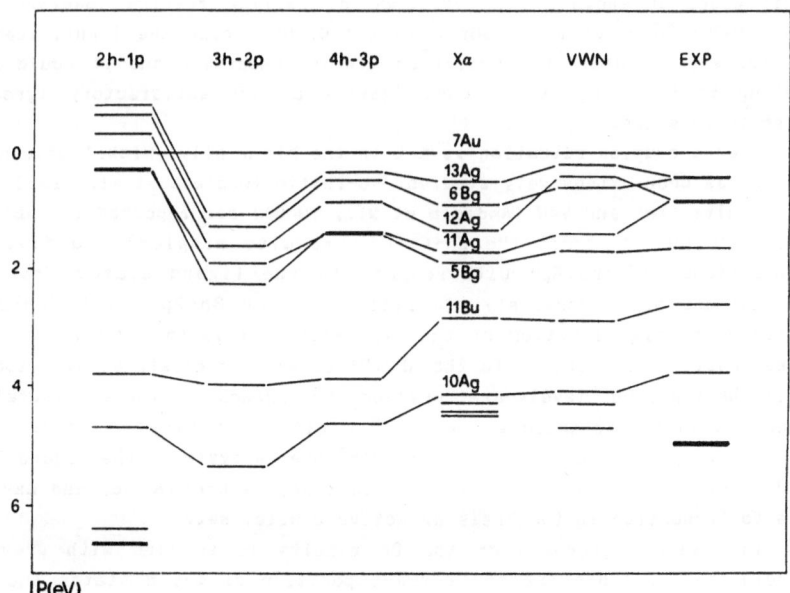

IP(eV)

Figure 4. Ionization Energies for bis(π-allyl) Nickel.
2h-1p, 3h-2p, 4h-3p: Decleva et al., 1989.
VWN: Andzelm et al., 1989. Exp.: Böhm et al., 1980.

SYM	2h-1p	4h-3p	Xα	VWN	Exp
7A$_u$	5.27	6.73	6.48	7.77	7.56
13A$_g$	1.69	2.13	1.06	1.10	1.16
6B$_g$	2.09	2.85	1.78	1.80	1.95
12A$_g$	3.26	3.48	2.84	2.83	
5B$_g$	3.27	3.46	3.21	3.32	2.22
11A$_g$	3.52	3.88	3.31	3.37	1.69
11B$_u$	3.90	3.74	3.13	3.12	2.89
10A$_g$	5.54	5.46	4.39	4.42	4.01
10B$_u$	6.70		4.56	4.50	4.94
6A$_u$	6.75		4.70	4.63	5.44
4B$_g$	6.91		5.46	5.47	6.69
9A$_g$	7.51		5.39	5.45	7.94
5A$_u$	8.17		6.26	6.23	

Table 5. Outer Valence Ionization Energies for bis(π-allyl) Palladium. Absolute energy for the 7A$_u$ state. For all other states relative energies are reported.
Exp.: Böhm et al., 1980.

initio context. Also the absolute values of the IEs are very satisfactory. A slight underestimate of the ligand IEs, which increases for the deeper states, can be detected. The overestimate of the d bands splitting found in Ni(CO)$_4$ may be due to an inadequate basis set for this case, as shown by the variance of earlier results (Baerends and Ros, 1975; Rösch et al., 1986). In the Cr(NO)$_4$ case the identification of the experimental 1t$_1$ ionization appears uncertain and it could also belong to the next 8t$_2$,1e band, giving a very satisfactory agreement also in this case.

A more complex situation is met in the bis(π-allyl)Nickel molecule, which has been extensively studied. Ab-initio (Decleva et al., 1989) and DF results (Xα and VWN (Andzelm et al., 1989)) are reported in table 4 and in figure 4. Again the 2h-1p CI results misplace the five d ionizations (A$_g$ and B$_g$) with respect to the ligand states. Metal ionizations are excessively stabilized at the 3h-2p level, which recovers a large fraction of the dynamical correlation, giving in this case results very similar to the ADC(3) scheme (Moncrieff et al., 1985). Also the 11A$_g$, 5B$_g$ levels are reversed. Enlargement of the configuration space to 4h-3p excitations within QD orbitals eventually leads to satisfactory agreement for these levels. Overestimate of the ligand 7A$_u$-11B$_u$ separation, present in all calculations, is unexpected, and may be due to truncation in the basis or active orbital set.

The overal agreement of the DF results is in line with previous conclusions. Notable are the correct position of the d states and the good 7A$_u$-11B$_u$ separation. The structure of the d band is somewhat less satisfactory, although the errors are small, and the ligand levels following the 10A$_g$ are too close. Xα underestimates the first IE by

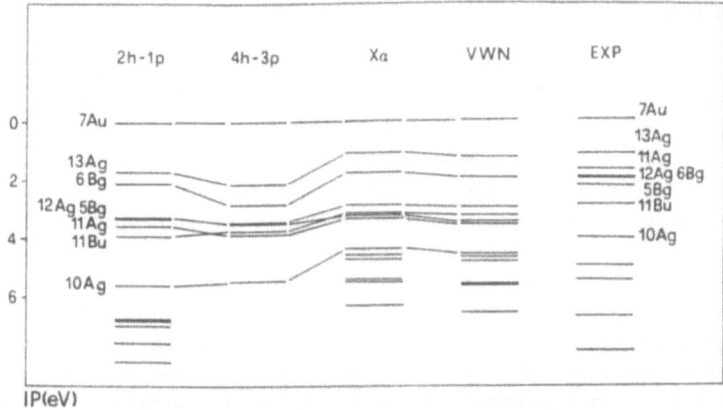

Figure 5. Ionization Energies for bis(π-allyl) Palladium.
Exp.: Böhm et al., 1980.

about 1 eV. The value is close to the experiment in the VWN case, which additionally shows a noticeable difference in the structure of the d band, destabilizing the 12A$_g$ level by about 1 eV. Again resolution of these smaller scale discrepancies requires very accurate examination of this problem.

The large increase in the Pd 4d IE reflects also in the Pd(C_3H_5)$_2$ molecule, with metal bands shifting to higher energies, and gives a major clue to the assignment of the spectra in these molecules (Bohm et al., 1980). Ab-initio and DF results are reported in table 5 and figure 5. As anticipated, the 2h-1p CI results are now more in line with the experimental pattern, although nowhere as accurate as in non transition compounds. 4h-3p CI analogous to the Ni case closes up the six IEs following the 7A$_u$ level, giving a resonable structure, while the separations with the 7A$_u$ and the following 10A$_g$ levels are significantly overestimated. The DF results are in close agreement with each other, apart from the absolute IEs. VWN gives again excellent agreement for the outermost IE, which is underestimated by about 1 eV in Xα . A preliminary more extended calculation increases however the 7A$_u$ Xα IE by about 0.5 eV, so that the VWN value could be significantly overestimated. The rest of the structure is almost indistiguishable, and generally in good agreement with the experimental pattern, also allowing for uncertainties in the experimental assignment. For instance the consistent placing of 6B$_g$ as the third outermost IE strongly suggests the revision of the original 11A$_g$ attribution. Like in the Ni compound the most noticeable deficiencies concern a possibly unsatisfactory d band structure, and the spacing of the levels following the 10A$_g$, with the same considerations applying also here.

Conclusions

Ionization of closed shell systems provides much richer information than total energy alone, as it tests removal of electrons from individual orbitals, yet avoiding the multiplet problem, still not neatly solved in the DF scheme.

From the results considered a few trends are apparent.

The first is the success of the DF scheme in situations plagued by complex correlation effects at the HF level. As discussed by Cook and Karplus (1987) this must be primarily due to the fact that QD correlations, which involve orbitals of similar spatial localization, although profoundly affecting the two particle density, have a relatively modest influence on the density. It is therefore just a consequence of the dependence of the energy on the density alone, and is common to all DF schemes.

A general inadequacy of the description of inner orbitals is clearly apparent, which is known to be associated with poor treatment of the self-interaction. This may be an important issue as it affects already 2s electrons, which are significantly involved in bonding.

The major difference between the potentials examined concerns the absolute values of the IEs, reflecting a difference in the treatment of the dynamical correlation. In the atomic case the VWN potential gives significant overbinding, which is effectively removed by the Stoll correction, giving good agreement with the experiment and the Xα R values. The same difference appears in molecules, but now VWN gives excellent agreement, while VWN/S and Xα R significantly underestimate ionization energies. Apparently there is a compensation between atomic overbinding and underestimate of orbital stabilization upon bond formation, which may be connected to the overestimate of bond energies usually given by the LDA schemes. In the few cases examined the results of the gradient correction (Becke, 1986), which has proven very effective in the correction of bond energies, does not appear to give a consistent improvement, apart for the outermost IEs, although more extensive analysis is certainly needed.

The last point concerns the consistently good reproduction of the experimental IEs in the few but significantly complex systems examined. The agreement obtainable in small systems is only slightly inferior to that given by highly correlated ab-initio treatments, and improvement in the potentials shows promise of further refinement. Most important the results do not appear to deteriorate in going to more complex systems, like small transition metal complexes, where they are already competitive with HF based treatments. Because of the much lower computational demands of the DF schemes, they appear to be the preferred choice whenever a highly correlated ab-initio treatment cannot be afforded.

Acknowlegdments

We are deeply indebted to Prof. E. J. Baerends and Prof. E. R. Davidson
for making available the DF Amsterdam codes and the ab-initio MELDF set
of programs which have been employed in the calculations reported.

References

Andzelm, J., Wimmer,- E., and Salahub, D. R., 1989, The Challenge of d
    and f Electrons (Edited by Salahub, D. R., and Zerner, M. C.), p. 228,
    ACS, Washington, DC.
Baerends, E. J., and Ros, P., 1973, Chem. Phys. 2:52.
Baerends, E. J., and Ros, P., 1975, Mol. Phys. 30:1735.
Baerends, E. J., and Ros, P., 1978, Int. J. Quantum Chem. 12S:169.
Baerends, E. J., Snijders, I. G., de Lange, C. A., and Jonkers, G.,
    1984, Local Density Approximations in Quantum Chemistry and Solid
    State Physics (Edited by Dahl, J. P., and Avery, J.), p. 415,
    Plenum, New York.
Beck, D. R., and Nicolaides, C. A., 1979, Excited States in Quantum
    Chemistry (Edited by Nicolaides, C. A., and Beck, D. R.),
    p. 105,329, Reidel, Dordrecht.
Becke, A. D., 1986, J. Chem. Phys. 84:4524.
Bieri, G., Asbrink, L., and von Niessen, W., 1981, J. Electron Spectry.
    23:281.
Bohm, M. C., Gleiter, R., and Batich, C. D., 1980, Helv. Chim. Acta
    63:990.
Botch, B. H., Dunning, T. H., Jr., and Harrison, J. F., 1981, J. Chem.
    Phys. 75:3466.
Cade, P. E., Sales, K. D., and Wahl, A. C., 1966, J. Chem. Phys. 44:1973.
Cade, P. E., and Wahl, A. C., 1974, At. Data Nucl. Data Tables 13:339.
Conford, A. B., Frost, D. C., Mc Dowell, C. A., Ragle, I. L., and
    Stenhouse, F. A., 1971, J. Chem. Phys. 54:2651.
Connolly, J. W. D., and Johnson, K. H., 1971, Chem. Phys. Letters 10:616.
Cook, M., and Karplus, M., 1987, J. Phys. Chem. 91:31.
Coutiere, M. M., Demuynck, J., and Veillard, A., 1972, Theor. Chim. Acta
    27:281.
De Alti, G., Decleva, P., and Lisini, A., 1982, Chem. Phys. 66:425.
Decleva, P., De Alti, G., and Lisini, A., 1988, J. Chem. Phys. 89:367.
Decleva, P., Fronzoni, G., and Lisini, A., 1989, Chem. Phys. 134:307.
Decleva, P., Fronzoni, G., De Alti, G., and Lisini, A., 1990, J. Mol.
    Struct., Theochem, in press.
Decleva, P., and Lisini, A., 1985, Chem. Phys. 97:95.
Decleva, P., and Lisini, A., 1987, Chem. Phys. 112:339.
Demuynck, J., and Veillard, A., 1973, Theor. Chim. Acta 28:241.
Dunlap, B. I., Connolly, J. W. D., and Sabin, J. R., 1979, J. Chem.
    Phys. 67:3970.
Froese-Fischer, C., 1977, J. Phys. B 10:1241.
Guest, M. F., Hillier, I. H., Mc Dowell, A. A., and Berry, M., 1980, Mol.
    Phys. 41:519.
Hay, P. J., Dunning, T. H., Jr., and Goddard III, W. A., 1975, J. Chem.
    Phys. 62:3912.
Hillier, I.H., Guest, M.F., Higginson, B.R., and Lloyd, D.R., 1974, Mol.
    Phys. 27:215.
Jones, R. O., and Gunnarsson, O., 1989, Rev. Mod. Phys. 61:689.
Katsumata, S., Shiromaru, H., and Kimura, T., 1984, Bull. Chem. Soc.

Jpn. 57:1784.

Moncrieff, D., Hillier, I. H., Saunders, V. R., and von Niessen, W., 1985, Inorg. Chem. 24:4247.

Moore, C. E., 1949, 1952, 1957, Atomic Energy Levels, NBS Circular No 467, Washington, DC.

Perdew, J. P., 1984, Local Density Approximations in Quantum Chemistry and Solid State Physics (Edited by Dahl, J. P., and Avery, J.), p. 173, Plenum, New York.

Plummer, E. W., Loubriel, G., Rajoria, D., Albert, M. R., Seddon, L. G., and Salaneck, W. R., 1980, J. Electron Spectry. 19:35.

Rosch, N., Jorg, H., and Dunlap, B. I., 1986, Quantum Chemistry: the Challenge of Transition Metals and Coordination Chemistry (Edited by Veillard, A.) p. 179, Reidel, Dordrecht.

Salahub, D. R., Lamson, S. H., and Messmer, R. P., 1982, Chem. Phys. Letters 85:430.

Salahub, D. R., Messmer, R. P., and Johnson, K. H., 1976, Mol. Phys. 31:529.

Sambe, H., and Felton, R. H., 1974, J. Chem. Phys. 61:3862.

Schirmer, J., Cederbaum, L. S., and Walter, O., 1983, Phys. Rev. A 28:1237.

Schirmer, J., and Walter, O., 1983, Chem. Phys. 78:201.

Siegbahn, H., and Karlsson, L., 1982, Handbuch der Physik, Vol. 31, (Edited by Melhorn, W.), p. 215, Springer, Berlin.

Siegbahn, K., Nordling, C., Johansson, G., Hedmak, J., Heden, P. F., Hamrin, K., Gelius, U., Bergmark, T., Werme, L. O., Manne, R., and Baer, Y., 1971, ESCA Applied to Free Molecules, North-Holland, Amsterdam.

Sinanoglu, O., 1969, Atomic Physics, Vol. 1 (Edited by Bederson, B., Cohen, V. W., and Pichanik, F. M. S.), p. 131, Plenum, New York.

Slater, J. C., 1972, Adv. Quantum Chem. 6:1.

Smith, S., Hillier, I. H., von Niessen, W., and Guest, M. F., 1989, Chem. Phys. 135:357.

Stoll, H., Golka, E., and Preuss, H., 1980, Theor. Chim. Acta 55:29.

Thunemann, K.-H., Peyerimhoff, S. D., and Buencker, R. J., 1978, J. Mol. Spectry. 70:432.

Tschinke, V., 1989, Thesis, University of Calgary.

Turner, D.W., Baker, C., Baker, A.D., and Brundle, C.R., 1970, Molecular Photoelectron Spectroscopy, Wiley, New York.

Vosko, S.H., Wilk, L., and Nusair, M., 1980, Can. J. Phys. 58:1200.

Wahl, A. C., 1964, J. Chem. Phys. 41:2600.

Wood, J. H., 1980, J. Phys. B: At. Mol. Phys. 13:1.

22
Improved Variational Calculations with Atomic Energy Functionals Using an Additional Restriction on the Density

M. DANIEL GLOSSMAN AND EDUARDO A. CASTRO

Introduction

The main aim of density functional theory (Hohenberg et al., 1964; R.G. Parr, 1983; Parr et al, 1989) is to use the density as the fundamental variable instead of the many-particle wave function and then to explore the ground-state properties of a system of interacting electrons. The fact that the ground-state properties of a system of interacting electrons are functionals of the electron density provides the basic framework of such formalism.

According to the Hohenberg and Kohn theorems, both the ground-state density ρ and the energy $E(\rho)$ can be variationally determined by minimizing the energy functional $E(\rho)$ with respect to a trial density subject to the normalization condition or by means of the use of the stationary condition $\delta(E(\rho) - \mu\int\rho dr) = 0$, where μ is a Lagrange multiplier that has been interpreted as a chemical potential. The exact form of the functional $E(\rho)$ is unknown. The design of such functional would lead to a single differential equation useful for an *ab initio* calculation of the electron density. So, it would be highly desirable to obtain good approximate forms of the functional $E(\rho)$ and then, by solving the Euler equation for ρ, to obtain both E and ρ.

It has been shown (Kim et al., 1974; Shih et al., 1976; Wang et al., 1976) that for atoms, energy density functionals including gradient terms can give energy values that have excellent agreement with the experimental ones when Hartree-Fock charge densities are used (Ghosh et al., 1982). But it would be very important to minimize the energy functionals in a variational way without making use of any other method.

Of course, the differential equations to be solved are non-linear and a variation of them represents a very complicated task. Although accurate numerical solutions of the variational equations are available for atoms and ions (Tomishima et al., 1966; Stich et al., 1982; Yang, 1986), analytical approximations for the ground-state density are of interest. An alternative way of solving these models is by means of the use of a trial density function with several variational parameters whose values are determined by minimization of the corresponding energy functional. This technique has proved to be useful in obtaining approximate variational solutions for the Thomas - Fermi (TF), Thomas - Fermi - Dirac (TFD) and Thomas - Fermi - Dirac - Weizsäcker (TFDW) equations, not only for atoms but for positive ions (Csavinszky, 1968, 1969, 1973). In particular, we note the work by Csavinszky (1968), who solved the TF equation by replacing it by an equivalent variational principle first postulated by Wesselow (1937) and was able to introduce the shell structure into the energy functionals by means of a density constructed using H-like wave-functions (Csavinszky, 1981, 1983, 1986; Csavinszky et al., 1983ab). This technique has also been applied in another cases (Glossman et al., 1983, 1984, 1985, 1987abc, 1989ab, 1990; Donnamaría et al, 1983, 1986), such as in the study of atoms in superstrong magnetic fields (Glossman et al., 1986, 1987, 1988abc). But the choice of the trial density function is somewhat arbitrary. For most purposes, a decaying exponential is a natural choice, because it has been shown that the electronic density possesses a monotonically decreasing nature (Weinstein et al., 1975). In this way, different combinations of exponentials have been proposed (Pathak et al., 1983; Battacharya et al., 1985; Scheidemann et al., 1986) and applied to the calculation of energies and several related atomic properties. Indeed, the results of these calculations are greatly dependent on the particular election of the trial density function. At this point, one should ask if there is any way to obtain a trial density that when applied kind to these calculations leads to the same results as those obtained by solving the differential equations. If this is possible, then one can apply the method to solve variationally the equations associated with different models for which solution are unknown or very difficult to obtain, as we are going to show below.

It is also known that the Thomas-Fermi theory of the electrons gives an electron density that diverges at an atomic nucleus. Not only is the density discontinuous, but also $\nabla^2 \rho$ is not integrable. Of course, for a wave-mechanical density, it is true that the Laplacian of the density must integrate to zero, and then, this is a natural condition to impose for Coulomb systems. An attempt has been made recently (Parr et al., 1986) to remedy this, by solving the TF equation imposing some constraints. In particular, it should be required that the exact electronic density function computed for any state of an atom must obey the cusp condition in the form established through Kato's theorem (Kato, 1957) and its extension by Steiner (1963). The results gave improved calculations of ground-state electron densities and energies, and a rigorous mathematical foundation for their work was later provided by the work of Goldstein and Ruiz-Rieder (1987). But no general solutions are known for the TF, TFD and TFDW equations with the cusp condition enforced. The major purpose of this work is to show that approximate solutions to those equations can be found by using a trial density function constructed as a summation of exponentials, restricted to obey the cusp condition and the requirement of integrable Laplacian of the density, and to investigate the effect of this improved density distribution on several physical quantities.

Theory

The ground-state energy of an atomic system in the TF model is given by the functional

$$E_{TF} = C_F \int \rho^{5/3} \, dr - \int \frac{Z}{r} \rho \, dr + \frac{1}{2} \int\int \frac{\rho(r) \; \rho(r')}{|r - r'|} \, dr \; dr' \quad (1)$$

where the different terms represent the kinetic energy of the electrons, the electron-nuclear attraction and the repulsion between electrons, respectively, and the constant $C_F = 3/10 \, (3\pi^2)^{2/3}$.

Inclusion of the exchange term due to Dirac (1930) leads to the TFD functional

$$E_{TFD} = E_{TF} - C_X \int \rho^{4/3} \, dr \quad (2)$$

with $C_x = 3/4 \ (3/\pi)^{1/3}$.

The second order correction to the kinetic energy was first given by von Weizsäcker (1935) and the corresponding model is

$$E_{TFDW} = E_{TFD} + \frac{\lambda}{8} \int \frac{|\nabla\rho|^2}{\rho} \, dr \qquad (3)$$

The importance of the coefficient λ has been deeply studied. There is a good revision in the work by Yang (1986). We have chosen $\lambda = 1/9$ because it has been shown (Wang et al., 1976) that it is the one that well reproduces Hartree-Fock (HF) energies when HF densities are used.

Variation of any of these functionals with respect to the density under the auxiliary condition of fixed particle number yields the Euler-Lagrange equations

$$5/3 \ C_F \ \rho^{2/3} + v(r) + \mu = 0 \qquad (4)$$

$$5/3 \ C_F \ \rho^{2/3} - 4/3 \ C_x \ \rho^{1/3} + v(r) + \mu = 0 \qquad (5)$$

$$5/3 \ C_F \ \rho^{2/3} - 4/3 \ C_x \ \rho^{1/3} + \frac{\lambda}{8} \left[(\nabla\rho/\rho)^2 - 2 \ (\nabla^2\rho/\rho) \right]$$

$$+ \ v(r) + \mu = 0 \qquad (6)$$

for the TF, TFD and TFDW cases respectively. In the above equations

$$v(r) = -\frac{Z}{r} + \int \frac{\rho(r')}{|r - r'|} \, dr' \qquad (7)$$

and μ is a Lagrange multiplier that has been identified as the negative of the electronegativity (Parr et al., 1978). Although numerical solutions for these equations are known, it is generally necessary to resort to complicated methods to solve them.

An alternative technique involves use of a trial density function with several variational parameters in one or more or the functionals given above. This gives the energy as a function of the parameters, and these can be determined by minimization of the energy with respect to them, provided the normalization condition $\int \rho \, dr = N$ has been taken into account.

In this work, we choose trial density functions of the type

$$\rho^{1/3} = \sum_i c_i \exp (-\alpha_i \, r) \tag{8}$$

where c_i, α_i are the variational parameters. This kind of trial density function with different values of N has been proven useful (Pathak et al., 1983; Battacharya et al., 1985). This election is not arbitrary: it has been shown that the electron density can be written as a summation of piecewise decaying exponentials (Wang et al., 1977) and this density is finite at the nucleus unlike the Thomas–Fermi density that diverges near the origin. One should also mention the work by Gázquez and Parr (1978), who used a somewhat different trial density function, but with the desired characteristics: monotonically decreasing, finite at the nucleus, and behaving as a single decaying exponential. The use of the third power greatly simplifies the calculation of the kinetic and exchange energy terms.

The nuclear cusp condition, namely

$$\frac{d\rho}{dr}_{r=0} = - \, 2 \, Z \, \rho(0) \tag{9}$$

is not always satisfied if one solves the self-consistent Eqs. (4) – (6). With the present technique, the cusp condition imposed to the trial density function leads to a very simple relationship between the variational coefficients

$$\sum_i c_i \, (\alpha_i - \frac{2}{3} \, Z) = 0 \tag{10}$$

From a practical point of view, this means that one of the coefficients is already fixed during the minimization procedure. With this election for the trial density function, the condition of integrable Laplacian $\int \nabla^2 \rho \, dr = 0$ is automatically fulfilled.

To check the accuracy of the optimized densities obtained we make plots of the corresponding radial densities and we calculate the density at the nucleus, given in the present model by

$$\rho(0) = \left[\sum_i c_i \right]^3 \qquad (11)$$

We are now going to apply these findings in the calculation of total atomic energies and related properties for several elements.

Calculation of atomic energies

The computation of total atomic energies for the TF, TFD, TFDW models has been done (Glossman et al., 1989) by means of the trial density function of Eq. (8), using a different number of terms in the expansion, and with and without consideration of the restriction of the density by application of the cusp condition. The results are shown in Table 1 for the case of the Ne atom. For all three cases of TF, TFD and TFDW, as the number of terms in the summation for the model density is increased, the values of total energies approach the exact self-consistent results (SCR). But with the corrected cusp constraint on the trial density, after N =3, the total energies do not change even with a further increase of the number of terms in the expansion. The cusp constraint gives most improvement for the TF case with a value very close to the HF result, which in the present case is 0.5967 (in units of $-E/Z^{7/3}$).

Table 2 shows a similar test of convergence for the case of a heavy atom like Xe, where the HF value is 0.6562 (in the same units). The reason for the fast convergence of

Table 1 - Ground-state total energies of Ne $(-E/Z^{7/3})$ for the TF, TFD and TFDW models, with and without cusp condition restriction, as a function of the number of terms in the summation expansion.

Model N	TF	TF/cusp	TFD	TFD/cusp	TFDW	TFDW/cusp
1	0.4087	——	0.4650	——	0.4465	——
2	0.6410	0.5935	0.6960	0.6421	0.6216	0.6014
3	0.7179	0.5945	0.7706	0.6422	0.6467	0.6019
4	0.7467	0.5945	0.7979	0.6422	0.6468	0.6019
5	0.7585	0.5945	0.8090	0.6422	0.6491	0.6019
6	0.7638	0.5945	0.8139	0.6422	0.6493	0.6019
SCR	0.7687	——	0.8183	——	0.6494	——

SCR: Self-consistent results

the cusp-constrained variational calculation might come from the following (Glossman et al., 1989): without the imposition of the cusp constraint, an increase in the number of terms in the expansion of the trial density gives better corrections for the density at the place near the nucleus than the place far away from the nucleus. This explains that the TFDW variation needs less terms in the expansion for the trial density than in the TF case to approach the self-consistent calculation results. It can be seen in Tables 1 and 2 for neon and xenon atoms. As we know , the density far away from the nucleus can be expressed as an exponential function. In this way, by

Table 2 - Ground-state total energies of Xe $(-E/Z^{7/3})$ for the TF, TFD and TFDW models, with and without cusp condition restriction, as a function of the number of terms in the summation expansion.

Model N	TF	TF/cusp	TFD	TFD/cusp	TFDW	TFDW/cusp
1	0.4087	——	0.4266	——	0.4209	——
2	0.6410	0.6410	0.6585	0.6584	0.6310	0.6291
3	0.7179	0.6692	0.7346	0.6848	0.6773	0.6587
4	0.7467	0.6694	0.7630	0.6849	0.6846	0.6589
5	0.7585	0.6694	0.7746	0.6849	0.6856	0.6589
SCR	0.7687	——	0.7845	——	0.6862	——

SCR: Self-consistent   results

means of the imposition of the cusp constraint at the site of the nucleus, convergence can be obtained just by using a few terms of expansion for the model density.

Ground-state total energies of noble gas atoms have been calculated (Glossman et al., 1989) for the TF, TFD and TFDW models with the cusp condition restriction and the results are shown in Table 3. The number of terms in the expansion is three because, as we can see from Tables 1 and 2, this is enough to achieve convergence. The results are very close to those obtained through a HF calculation. The interesting point is that the results belong to models whose solutions are unknown or are very difficult and bothering to calculate. By inspection of Tables 1 and 2, we can understand that provided the number of terms in the expansion is great enough, then the solution of the variational calculation tends to the solution of the differential equation associated with the corresponding models. As far as this is true, the same should be for the differential equations associated with different models and this has been validated by recent calculations (Glossman et al., 1989). A particular case is that of a model including corrections like the ones discussed here. The improvement is then twofold: first of all, we can solve the models including the cusp condition restriction for the TF (Parr et al., 1986), the TFD (Zhou et al., 1988), TFDW and any other model we are interested in. Second, due to the inclusion of the cusp condition restriction, the convergence is achieved with a small number of terms, calculations are shorter and very easy to perform.

Table 3 - Ground-state total energies of noble gas atoms $(-E/Z^{7/3})$ for the TF, TFD and TFDW models with cusp condition restriction (N=3) as compared with HF results

Atom Model	He	Ne	Ar	Kr	Xe
TF/cusp	0.4734	0.5945	0.6254	0.6550	0.6694
TFD/cusp	0.6129	0.6422	0.6576	0.6753	0.6848
TFDW/cusp	0.5550	0.6019	0.6228	0.6462	0.6589
HF	0.5678	0.5967	0.6204	0.6431	0.6562

Table 4 - Electron densities at the site of the nucleus $\rho(0)$ of noble gas atoms for the TF, TFD and TFDW models with cusp condition restriction (N=3) as compared with Hartree-Fock results (in units of $\rho(0)/Z^3$)

Atom Model	He	Ne	Ar	Kr	Xe
TF/cusp	0.3774	0.5358	0.5989	0.6605	0.6853
TFD/cusp	0.4631	0.5596	0.5856	0.6109	0.6221
TFDW/cusp	0.4031	0.5004	0.5266	0.5516	0.5650
HF[a]	0.4495	0.6199	0.6584	0.6909	0.7063

[a] HF values from Parr et al. (1986)

A simple test of the validity of the proposed technique consists in the calculation of the atomic densities at the site of the nucleus. The results are given in Table 4 for different noble gas atoms together with the values obtained from Hartree-Fock calculations. From Table 4 one can see that the values are a little low when compared with the HF ones, and the increasing trend of them down the Periodic Table is well reproduced.

If we call the total energy obtained through this method E_0, then $E_0 = T_0 + V_0$, where the terms represent the total variational kinetic energy and the total variational potential energy. It is observed that these components do not satisfy the virial theorem. In order to remedy this, it necessary to apply a scaling procedure. Following Parr et al. (1986), the scaled density

$$\rho_s(r) = s^3 \rho_0(sr) \qquad (12)$$

is defined, where the density in the right hand of the equation is that obtained after the variational procedure has been applied. For the density of the last equation, it should be

$$E = s^2 T_0 + s V_0 \qquad (13)$$

The best scale factor s is then obtained by minimization of the energy, that is $s = - V_0/2 T_0$, and the

Table 5 - Scaled ground-state total energies of noble gas atoms $(-E/Z^{7/3})$ for the TF, TFD and TFDW models with cusp condition restriction (N=3) as compared with HF results

Atom Model	He	Ne	Ar	Kr	Xe
TF/cusp	0.4480	0.5982	0.6278	0.6562	0.6702
TFD/cusp	0.6250	0.6457	0.6599	0.6766	0.6857
TFDW/cusp	0.5613	0.6037	0.6239	0.6469	0.6592
HF	0.5678	0.5967	0.6204	0.6431	0.6562

corresponding scaled energy is then given by

$$E = - V_0^2 / 4 \, T_0 \tag{14}$$

The results of the application of this scaling procedure is given in Table 5 for the case of the total atomic energies of noble gas atoms, and in Table 6 for the densities at the nucleus for the same neutral atoms. Not only is the previous trend well reproduced, but the values of scaled atomic energies and densities at the nucleus are closer to the HF results than the unscaled values.

Table 6 - Scaled electron densities at the site of the nucleus $\rho(0)$ of noble gas atoms for the TF, TFD and TFDW models with cusp condition restriction (N=3) as compared with Hartree-Fock results (in units of $\rho(0)/Z^3$)

Atom Model	He	Ne	Ar	Kr	Xe
TF/cusp	0.6669	0.6856	0.6898	0.6910	0.6938
TFD/cusp	0.7223	0.7045	0.7012	0.6971	0.6974
TFDW/cusp	0.5642	0.5933	0.6015	0.6096	0.6136
HF[a]	0.4495	0.6199	0.6584	0.6909	0.7063

[a] HF values from Parr et al. (1986)

Calculation of atomic properties

We now proceed to calculate various atomic properties using the solution of Eq. (8).

Diamagnetic susceptibility

The diamagnetic susceptibility of an atom is given by the well-known formula (Gombás, 1946)

$$\chi = \frac{- Na\ e^2}{6\ m\ c^2} <r^2> \tag{15}$$

where Na is the Avogadro's number, m the electron-mass and

$$<r^2> = \int r^2\ \rho\ d\tau = 4\ \pi \int \rho\ r^4\ dr \tag{16}$$

Using Eq. (8) and Eq. (16) one gets

$$<r^2> = 6\ Z \sum_i (c_i/p_i^2) \tag{17}$$

From Eq. (15), one gets

$$\chi = 2.8283 \times 10^{10} cm\ mole^{-1} <r^2> \tag{18}$$

The values of χ for noble gas atoms are listed in Table 7, together with those obtained through the original TF model (Schey et al, 1965), the Lenz - Jensen modification of this model (Lenz, 1932; Jensen, 1936; Gombás, 1946) and the experimental values (Dehn et al., 1968). As we can see from the table, the diamagnetic susceptibilities obtained with the densities determined with the particular technique explained before are closer to the experimental values than those obtained on the basis of the usual TF model and they are better than the other available analytic approximation based on the modification of the TF model due to Lenz and Jensen. This is due to the fact that the TF density at large distances vanishes too slowly (as explained before) whereas the density used in this work has an exponential decaying. Notwithstanding, they are not good enough to be considered useful and different models should be tried in order to improve the results.

Table 7 - Comparison of calculated and experimental
        diamagnetic susceptibilities $(-\chi \times 10^6 \text{ cm}^3/\text{mole})$

Atom Model	He	Ne	Ar	Kr	Xe
TF/scf (a)	——	64.29	78.20	98.53	112.79
TF/var (b)	——	20.40	24.80	31.20	35.70
TF/cusp	66.9	87.33	49.13	39.58	37.29
TFD/cusp	19.2	23.99	30.75	32.50	32.84
TFDW/cusp	8.6	19.32	26.81	29.82	30.71
Experim. (c)	——	6.74	19.58	28.80	43.91

(a) Self-consistent results (Schey et al.,1965)
(b) Variational results (Lenz, 1932; Jensen, 1936)
(c) Experimental results (Dehn et al., 1968)

Atomic polarizability

A good method of forming a polarizability (Bruch et al.,
1976) is to use the Hasse-Kirkwood-Vinti approximation
(Hirschfelder et al., 1965) for the atomic polarizability
α in terms of the molar diamagnetic susceptibility given
by

$$\alpha = 16 \frac{[mc^2\chi / Na\ e^2]^2}{Z\ a_o} \tag{19}$$

where a_o is the Bohr radius and the other terms have their
usual meaning. Considering χ as defined by Eq. (15), then
the atomic polarizability in this approximation is given
by

$$\alpha = \frac{4}{9 Z a_o} \langle r^2 \rangle^2 \tag{20}$$

where $\langle r^2 \rangle$ is given by Eq. (16). The values of α for
several atoms are given in Table 8 together with the
TFD (Thomas-Fermi-Dirac) results (Gombás, 1946) and the

Table 8 - Comparison of calculated theoretical polarizabilities ($\alpha \times 10^{24}$ cm^3) for several atoms

Atom Model	Ne	Ar	Kr	Xe	Rn
TFD(a)	43.48	35.76	28.39	24.78	21.22
TF/cusp	470.96	44.58	7.06	3.05	1.70
TFD/cusp	3.88	3.36	2.76	2.06	1.32
TFDW/cusp	0.78	2.18	2.10	1.73	1.15
SCF (b)	0.39	1.98	3.12	5.38	6.88

(a) Thomas-Fermi-Dirac calculations (Gombas, 1946)
(b) Self-consistent results (Fraga et al., 1976)

SCF values (Fraga et al., 1976). The actual results are better than those from the TFD original calculation and are relatively closer to the SCF ones. The TFD values for atomic polarizabilities have been reported by Gombás (1946) and were obtained with limited variational calculations. SCF values are those calculated by means of the use of Hartree-Fock densities. As in the case of the diamagnetic susceptibilities, we can see that although there is an improvement over previous calculations, the results are still not good enough. Moreover, they present an incorrect trend with the increment of atomic number Z. The fact that both diamagnetic susceptibilities and atomic polarizabilities are calculated in the same way is the reason for this anomalous behavior.

Average radial electron density

The radial integrals involving the functions of one electron density $\rho(r)$ as the integrand define useful properties for atoms. The average radial electron density

$$\langle\rho\rangle = \int \rho^2 \, d\tau \tag{21}$$

is an integral which has been shown (Szasz et al., 1975; Hyman et al, 1978; Tal et al., 1980; Sen et al., 1980; Gadre et al., 1983; Sen et al., 1985) to provide rigorous bounds to the total electronic energy E. Recently, a polynomial representation of E according to

$$E = A <\rho>^{2/3} + B <\rho>^{1/3} + C + D <\rho>^{-1/3} + E <\rho>^{-2/3} \qquad (22)$$

within an isoelectronic series of atoms has been found (Sen et al., 1985) to provide a better fit than the conventional Z expansion.

Moreover, it has been shown (Hyman et al., 1978) that $<\rho>$ is an experimentally measurable quantity related to the intensity scattered by an element.

Based on scaling arguments due to Szasz et al (1975), it has been shown numerically that an excellent representation of the total electron-electron repulsion energy E_{ee} for atoms can be written as

$$E_{ee} = 0.3977 \ N^{4/3} <\rho>^{1/3} \qquad (23)$$

Using Eqs. (8) and (21), we have calculated the average radial density for noble gas atoms for the TF, TFD and TFDW models with the cusp condition restriction and N=3, and the results are listed in Table 9. Using these values, it is now possible to test Eq. (23) for the electron-electron interaction energy. For example, the TFDW/cusp value of $<\rho>$ for Ne is 143.581 and then using Eq. (23), we obtain 44.87 for the interelectronic energy that may be compared with the HF value (Clementi et al., 1974) of 41.01. Similar calculations can be done for the other cases.

Table 9 - $<\rho>$ values of several atoms (in a.u.) for the TF, TFD and TFDW models with cusp condition restriction and N=3

Atom Model	He	Ne	Ar	Kr	Xe
TF/cusp	0.4792	153.654	1131.47	11372.1	43096.1
TFD/cusp	0.7165	168.955	1265.25	11735.6	44068.8
TFDW/cusp	0.5713	143.581	1035.54	10332.8	39173.3

Momentum expectation value and inverse momentum expectation value

Recently, there has been a renewal of interest in the momentum density of atoms, a quantity that can be measured (Williams, 1977) by high-energy X-ray Compton scattering experiments.

Gadre et al (1981, 1983ab) have studied the direct and reverse transformation between the electron density ρ and the momentum density Ξ, and Pathak et al (1981) have calculated the peak value of the Compton profile, $J(0)$, for a number of atoms from atomic electron densities.

It has been shown by Pathak et al. (1981) that the momentum expectation value, estimated exclusively from the knowledge of $\rho(r)$, is given by

$$<p> = \pi \left[3/4 \ (3/\pi)^{1/3} \int \rho^{4/3} \ d\tau \right] \qquad (24)$$

where the quantity in large parentheses is the same as $- E_{xc}$ appearing in Eq. (2).

It has been also shown by Pathak et al. (1981) that the inverse momentum expectation value, estimated exclusively from the knowledge of $\rho(r)$, is given by

$$<p^{-1}> = 1/2 \ (3/\pi)^{2/3} \int \rho^{2/3} \ d\tau \qquad (25)$$

Finally, attention is called to the relationship (Pathak et al., 1981)

$$J(0) = 1/2 \ <p^{-1}> \qquad (26)$$

By means of Eqs. (8) and (24), we have calculated the momentum expectation value for several atoms. The results are listed in Table 10. For comparison we have included $<p>$ values calculated through Hartree-Fock densities (Berrondo, 1980; Gadre et al.,1981, 1983, 1987; Pathak et al., 1982; Allan et al., 1985).

Table 10 - Comparison of momentum expectation values for several atoms (in a.u.)

Atom Model	He	Ne	Ar	Kr	Xe
TFD/cusp	2.49	33.65	88.14	277.5	538.0
TFDW/cusp	2.29	32.40	85.81	272.7	545.2
HF (a)	—	35.20	88.70	281.4	—

(a)Hartree-Fock calculations (Gadre et al., 1981,1983,1987;
   Pathak et al., 1981,1982; Berrondo et al., 1980; Allan
   et al., 1985)

By means of Eqs. (8) and (35), we have calculated the inverse momentum expectation value, and with Eq. (30), the peak value of the Compton profile $J(0)$ for several atoms. The results are listed in Table 11 together with the corresponding values calculated through the use Hartree-Fock densities (Williams, 1977; Berrondo et al., 1980;Gadre et al.,1981, 1983, 1987; Pathak et al., 1981).

Table 11 - Comparison of inverse momentum expectation values for several atoms (in a.u.)

Atom Model	He	Ne	Ar	Kr	Xe
TF/cusp	13.13	20.09	17.59	17.82	18.50
TFD/cusp	7.16	11.52	14.20	16.23	16.83
TFDW/cusp	5.22	10.48	13.26	15.54	17.40
HF (a)	—	5.46	10.13	14.48	—

(a)Hartree-Fock calculations (Gadre et al., 1981,1983,1987;
   Pathak et al., 1981,1982; Berrondo et al., 1980; Allan
   et al., 1985)

Final remarks

We have shown an improved way of calculating total atomic energies and several properties associated with them. We believe that this is an improved calculation because we are obtaining good values of atomic energies, in agreement with the results from self-consistent calculations. Moreover, the technique used in this work allows us to obtain the same density as the one obtained from the solution of the differential equation associated with each studied model. In this way, the election of the trial density is no more an arbitrary choice. Also, we have considered the inclusion of an additional constraint to the trial density, namely the nuclear cusp condition restriction, and have shown that both the atomic energies and densities associated with them are improved over previous calculations. In addition to this, the results are obtained without a great numerical effort.

We have calculated some atomic properties with the technique presented here. Our aim is to show how these calculations are done and to stress the point that these values are the result of the solution of particular models and are independent of the trial density used in the minimization procedure. In any case, new models should be developed, but the method of solution will remain the same. Those atomic properties were only given as examples, and we know that they, and many others can be obtained accurately through HF calculations. But it is not so for molecules where HF calculations are computationally very intensive. Using atoms as examples we gain insight into theory and minimization procedures. We can then transfer this experience to the study of more complicated systems such as molecules.

References

Allan, N.L., Cooper, D.L., West, C.G., Grout, P.J., and March, N.H., 1985, *J. Chem. Phys.*, **83**:239-240.
Battacharya, A.K., and Das, S., 1985, *Acta Phys. Polonica*, **67**:803-810.

Berrondo, M., and Flores-Riveros, A., 1980, *J. Chem. Phys.*, **72**:6299-6302.

Bruch, L.W., and Lehnen, A.P.,1976, *J. Chem. Phys.*, **64**:2065-2068.

Clementi. E, and Roetti, C.,1974, *Atomic Data and Nuclear Data Tables*, **14**:177-478.

Csavinszky, P., 1968, *Phys. Rev.*, **166**:53-56.

Csavinszky, P., 1969, *J. Chem. Phys.*, **50**:1176-1181.

Csavinszky, P., 1973, *Phys. Rev. A*, **8**:1688-1701.

Csavinszky, P., 1981, *Phys. Rev. A*, **24**:1215-1217.

Csavinszky, P., 1983, *Phys. Rev. A*, **27**:1184-1186.

Csavinszky, P., 1986, *Int. J. Quant. Chem.: Quant. Chem. Symp.*, **19**:559-565.

Csavinszky, P., and Vosman, F., 1983, *Int. J. Quant. Chem.*, **24**:61-64.

Csavinszky, P., and Vosman, F., 1983, *Int. J. Quant. Chem.*, **23**:1973-1978

Dehn, J., and Mulay, L.N., 1968, *J. Chem. Phys.*, **48**:4910-4912.

Dirac, P.A.M., 1930, *Proc. Cambridge Phil. Soc.*, **26**:376-385.

Donnamaría, M.C., Glossman, M.D., and Alonso, J.A., 1986, *J. Chem. Phys.*, 1986, **85**:6637-6644.

Donnamaría, M.C., Glossman, M.D., Castro, E.A., and Fernández, 1983, *MATCH (Commun. Math. Chem)*, **14**:247-261.

Fraga, S., Karwowski, J, and Saxena, K.M.S., 1976, *Handbook of Atomic Data*, Elsevier, Amsterdam.

Gadre, S.R., and Bengale, R.D., 1983, *J. Chem. Phys.*, **78**:996-999.

Gadre, S.R., and Chakravorty, S.J., 1987, *J. Chem. Phys.*, **86**: 2224-2228.

Gadre, S.R., Gejji, S.P., and Pathak, R.K., 1983, *Phys. Rev. A*, **27**:3328-3331.

Gadre, S.R., Gejji, S.P., and Pathak, R.K., 1983, *Phys. Rev. A*, **28**:462-463.

Gadre, S.R., and Pathak, R.K., 1981,*Phys. Rev. A*, **24**:2906-2912.

Gázquez, J.L., and Parr, R.G., 1978, *J. Chem. Phys.*, **68**:2323-2326.

Ghosh, S.K., and Deb, B.M., 1982, *Phys. Rep.*, **92**:1-44.

Glossman, M.D., and Castro, E.A., 1986, *MATCH (Commun. Math. Chem.)*, **19**:13-17.

Glossman, M.D., and Castro, E.A., 1987, *Z. Phys. D*, **6**:81-82.

Glossman, M.D., and Castro, E.A., 1988, *Bulg. J. Phys.*, 5:459-464.

Glossman, M.D., and Castro, E.A,, 1988, *J. Phys. B: At. Mol. Opt. Phys.*, 21:411-419.

Glossman, M.D., and Castro, E.A., 1988, *S.-Afr. Tydskr. Fis.*, 11:18-19.

Glossman, M.D., and Castro, E.A., 1989, *Z. Phys. D*, 13:89-93.

Glossman, M.D., and Castro, E.A., 1989, *Phys. Rev. A*, 39:4870-72.

Glossman, M.D., and Castro, E.A., 1990, *Chem. Phys. Lett.*, 167:305-308.

Glossman, M.D., Donnamaría, M.C., and Castro, E.A., 1987, *MATCH (Commun. Math. Chem.)*, 22:223-243.

Glossman, M.D., Donnamaría, M.C., and Castro, E.A., 1987, *Few-Body Systems*, 3:95-98.

Glossman, M.D., Donnamaría, M.C., Castro, E.A., and Fernández, F.M., 1983, *Bol·. Soc. Chil. Quim.*, 28:17-24.

Glossman, M.D., Donnamaría, M.C., Castro E.A., and Fernández, F.M., 1984, *Bol. Soc. Quim. Peru*, L:143-149.

Glossman, M.D., Donnamaría, M.C., Castro, E.A., and Fernández, F.M., 1985, *J. Phys. (Paris)*, 46:173-179.

Glossman, M.D., Donnamaría, M.C., and Castro, E.A., 1987, *Acta Phys. Slovaca*, 37:298-315.

Glossman, M.D., Lee, C., Parr, R.G., and Zhou, Z, 1989 (unpublished results).

Glossman, M.D., and Zhou, Z., 1989 (unpublished results).

Goldstein, J.A., and Ruiz-Rieder, G., 1987, *J. Math. Phys.*, 28:1198-1202.

Gombás, P., 1946, *Die Statistische Theorie des Atoms und ihre Anwendungen*, Springer-Verlag, Wien.

Hirschfelder, J.O., Curtiss, C.F., and Bird, R.B., 1965, *Molecular Theory of Gases and Liquids*, Wiley, New York.

Hohenberg, P., and Kohn, W., 1964, *Phys. Rev. B* 136:864-871.

Hyman, A.S., Yaniger, S.I. and Liebman, J.F., 1978, *Int. J. Quant. Chem.*,14:757-766.

Jensen, H., 1936, *Z. Phys.*, 101:141.

Kato, T., 1957, *Commun. Pure Appl. Math.*, 10:151-177.

Kim, Y.S., and Gordon, R.G., 1974, *J. Chem. Phys.*, 60:1842-1850.

Lenz, W., *Z. Phys.*, 1932, 77:713.

Parr, R.G., 1983, *Ann. Rev. Phys. Chem.*, 34:631-656.

Parr, R.G., Donnelly, A., Levy, M. and Palke, W.E., 1978, *J. Chem. Phys.*, **68**:3801-3807.

Parr, R.G., and Ghosh, S.K., 1986, *Proc. Natl. Acad. Sci. USA*, **83**:3577-3579.

Parr, R.G., and Yang, W., 1989, *Density Functional Theory of Atoms and Molecules*, (Edited by Breslow, R., Goodenough, J.B., Halpern, J, and Rowlinson, J.S.), Oxford University Press, New York.

Pathak, R.K., and Gadre, S.R., 1981, *J. Chem. Phys.*, **74**:5925-5926.

Pathak, R.K., and Gadre, S.R., 1983, *Phys. Rev. A*, **28**:1808-1809.

Pathak, R.K., Panat, P.V., and Gadre, S.R., 1982, *Phys. Rev. A*, **26**:3073-3077.

Scheidemann, A., and Dreizler, R.M., 1986, *Z. Phys. D*, **2**:43.

Schey, H.M., and Schwartz, J.L., 1965, *Phys. Rev. A*, **137**:709-716.

Sen, K.D., and Nath., S., 1985, *Theor. Chim. Acta*, **68**:139-142.

Sen, K.D., and Reddy, V.S., 1980, *J. Chem. Phys.*, **81**:5213-5214.

Shih, C.C., and Present, R.D., 1976, *Bull. Am. Phys. Soc.*, **21**:381.

Steiner, E., 1963, *J. Chem. Phys.*, **39**:2365-2366.

Stich, W., Gross E.K.U., Malzacher, P., and Dreizler, R.M., 1982, *Z. Phys. A*, **309**:5-11.

Szasz, L., Berrios-Pagan, I., and Mc Ginn, G., 1975, *Z. Natursforch. Teil A*, **30**:1516-1534.

Tal, Y., and Levy, M., 1980, *J. Chem. Phys.*, **72**:4009-4013.

Tomishima, Y., and Yonei, K., 1966, *J. Phys. Soc. Japan*, **21**:142-153.

von Weizsäcker, C.F., 1935, *Z. Physik*, **96**:431-458.

Wang, W.P., Parr, R.G., Murphy, D.R., and Henderson, 1976, *Chem. Phys. Lett.*, **43**:409-412.

Wang, W.P., and Parr, R.G., 1977, *Phys. Rev. A*, **16**:891-902.

Weinstein, H., Politzer, P., and Srebrenik, S., 1975, *Theor. Chim. Acta*, **38**:159-163.

Wesselow, M.G., 1937, *Zh. Eksp. i. Teor. Fiz.*, **7**:829.

Williams, B.G., 1977, *Phys. Scripta*, **15**: 92-110.

Yang, W., 1986, *Phys. Rev. A*, **34**:4575-4585.

Zhou, Z., and Parr, R.G., 1988 (unpublished results).

23
Formic Acid: Methylamine Complex Studied by the Hartree-Fock and Density Functional Approaches

RONALD A. HILL, JAN K. LABANOWSKI,
DAVID J. HEISTERBERG, AND DUANE D. MILLER

Abstract

A comparison of Hartree-Fock (HF), Møller-Plesset second order perturbation theory (MP2) and Density Functional (DF) results for a formic acid:methylamine complex is presented. The interaction energy for the complex was calculated as a function of the distance between the amine nitrogen and the carboxyl hydrogen at the HF and MP2 level with 3-21G*(*) and DZP basis sets. The Basis Set Superposition Error was estimated. Also, full geometry optimization was performed for the complex at the HF and MP2 level. These results were compared with Local Spin Density (LSD) calculations as implemented in the programs DMol and DGauss, with and without inclusion of gradient corrections to the total energy, using double-zeta basis sets with polarization functions on all atoms. LSD alone, without gradient corrections, substantially overestimates hydrogen bond energy compared with HF and MP2 results. Application of Becke-Perdew gradient corrections in DF calculations for this hydrogen-bonded system closely reproduces the MP2 values.

Introduction

Hydrogen bonding plays a major role in biological recognition and in stabilizing molecular arrangements in biological systems. The proximity of amino and carboxyl groups is essential for a variety of proton transfer reactions, and interaction of these groups is believed to be important for the binding of many ligands with receptors. For some time we have been studying the activity of dopamine agonists and antagonists by synthesizing analogs which have the amine moeity replaced by other charged and uncharged functional groups, and testing their activity in D_1 and D_2 receptor systems (Miller et al., 1988). Based on the known primary sequence of the D_2 receptor (Bunzow et al., 1988) and other evidence, it is likely that the amino group of dopamine interacts with the carboxyl group of an aspartic acid residue in the receptor cavity.

To understand the trends in biological activity of dopamine analogs it is necessary to evaluate the contribution of this interaction to the free energy

---

*To whom the correspondence should be addressed.

of binding of drug to receptor. Unfortunately, the molecular mechanics approach cannot be used due to a lack of parameters for many of the groups represented in our series of dopamine analogs. Semi-empirical approaches are known to perform rather poorly in such systems (Koller and Hadži, 1989). On the other hand, traditional *ab initio* approaches adequately describe hydrogen bonding when large basis sets are used and electron correlation is accounted for (for reviews see: Del Bene, 1986; Hehre, *et al.*, 1986). Applying these methods to real biological systems, however, is impractical even with the world's most powerful supercomputers.

Software for performing calculations on isolated molecules using the Local Spin Density (LSD) variant of DF has only recently become available. Since DF scales as N^3 with the number of basis functions as opposed to N^4 scaling for Hartree-Fock (HF), it has potential for studying larger molecular systems. This might be especially attractive in the area of computer-aided drug design, where researchers are mostly concerned with ground state properties of molecules. Also, the disk space requirements are in general much smaller for DF compared to HF approaches.

It is known that gradient corrections to LSD energies are essential in predicting the strength of chemical bonds. The importance of gradient corrections in studying the hydrogen bond for water dimer was already noticed (J. Mertz, J. Andzelm, unpublished). Typically the LSD geometry optimization is performed first, and then gradient corrections are calculated using LSD densities. This non-selfconsistent approach was found to be satisfactory in calculations on diatomics (Becke, 1986; 1988), polyatomics (Andzelm, this volume), and organometalic species (Ziegler and Tschinke, this volume). We present the first systematic study of a hydrogen bonded system using the LSD approach with and without gradient corrections. The formic acid:methyl amine complex was chosen for this study due to our interest in dopamine.

The amine:carboxyl system was studied previously by *ab initio* methods at the HF level by Hadži and coworkers (Koller, *et al.*, 1984; Hadošček and Hadži, 1989) and by Sapse and Russell (1986). They found that the lowest-energy form of the isolated $CO_2H\cdots NH_2$ couple is the neutral one (i.e. without proton transfer from carboxyl to amino group). The reported binding energy for a bifurcated hydrogen bond in an acetic acid:methylamine complex was -11.7 and -7.5 kcal mol^{-1} for 6-31G and 6-31G* basis sets (Sapse and Russell, 1986). For a quasi-linear geometry the binding energy in formic acid:methylamine was -17.6, -14.8, -14.0 and -11.8 kcal mol^{-1} for STO-3G, 4-31G, 6-31G and 4-31G** basis sets and the estimated basis set superposition errors of these values were 6.0, 2.2, 2.0 and 2.0 kcal mol^{-1}, repectively (Hadošček and Hadži, 1989).

In this study we compared interaction energies calculated by a number of methods at different geometries of the complex. Full geometry optimizations were also performed at HF and Møller-Plesset second order perturba-

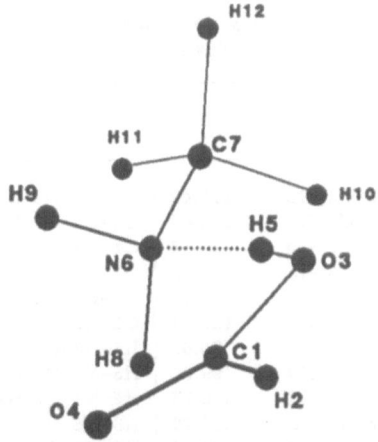

Figure 1: Atom numbering for formic acid:methylamine complex.

tion theory (MP2) levels by CADPAC, and by DMol and DGauss packages using the LSD scheme.

Computational details

Starting geometries of formate and methylammonium were obtained with molecular mechanics using SYBYL modeling software (Sybyl ver. 5.3, Tripos Associates, St. Louis, Missouri), and then optimized at the Hartree-Fock level using a 3-21G basis set (Binkley, *et al.*, 1980; Hariharan and Pople, 1972) augmented with polarization functions only on atoms: C1, O3, O4, N6, H5, H8, H9 (6 d-type functions for C, N and O with 0.8 exponent and p-type functions for H with exponent 1.1). Throughout the remainder of this paper, this collection of basis functions will be refered to as 3-21G*(*). All HF and MP2 *ab initio* calculations were performed with CADPAC (Amos and Rice, 1987) running on a CRAY Y-MP8/864 supercomputer. Geometry optimizations at the HF and MP2 levels were terminated when the largest component of the energy gradient was less than 1.0×10^{-4} Hartree Bohr^{-1}.

The atom numbering we adopted for the complex is shown in Fig.1. The starting geometry for the complex was constructed from formate and methylammonium ions as follows: a linear O3–H5–N6 hydrogen bond with H5 and N6 in the O4–C1–O3 plane was assumed; the length of the hydrogen bond (O4–H5) was set to 1.6 Å; the angle C1-O3-H5 was set to 120 degrees; N6 and C7 were placed in the O3-C1-O4 plane in such a way that C7 was remote from O4; hence, C1, H2, O4, O3, H5, N6, and C7 were all co-planar. The geometry of this complex was then optimized using the 3-21G*(*) basis

Table 1: 321-G*(*) optimised geometry for formic acid: methylamine complex. Bond lengths are expressed in Å and angles in degrees.

Bond lengths		Bond angles		Torsion angles	
C1–H2	1.094	H2–C1–O3	110.42	H2–C1–O3–H5	180.00
C1–O3	1.310	H2–C1–O4	123.44	O4–C1–O3–H5	0.00
C1–O4	1.186	O3–C1–O4	126.13	C1–O3–H5–N6	0.00
O3–H5	0.974	C1–O3–H5	108.05	O3–H5–N6–C7	180.00
H5–N6	1.833	O3–H5–N6	163.35	O3–H5–N6–H8	53.31
N6–C7	1.472	H5–N6–C7	129.74	O3–H5–N6–H9	-53.32
N6–H8	1.005	H5–N6–H8	98.98	H5–N6–C7–H10	-58.80
N6–H9	1.006	H5–N6–H9	99.00	H5–N6–C7–H11	180.00
C7–H10	1.083	C7–N6–H8	110.73	H5–N6–C7–H12	58.83
C7–H11	1.087	C7–N6–H9	110.72	H8–N6–C7–H10	63.32
C7–H12	1.083	H8–N6–H9	104.76	H8–N6–C7–H11	-57.86
		N6–C7–H10	109.12	H8–N6–C7–H12	-179.05
		N6–C7–H11	113.90	H9–N6–C7–H10	179.07
		N6–C7–H12	109.13	H9–N6–C7–H11	57.89
		H10–C7–H11	108.33	H9–N6–C7–H12	-63.30
		H10–C7–H12	107.85		
		H11–C7–H12	108.34		

set with the BFGS optimizer in internal coordinates and also in cartesian coordinates.

To evaluate the dependance of interaction energy on the distance H5···N6, the geometry of the 3-21G*(*) optimised complex was varied by setting this distance to: 0.8, 1.0, 1.2, 1.4, 1.6, 1.833 (the optimal value from 3-21G*(*) calculations), 2.0, 2.2, 2.5, 3.0, 4.0, 6.0, 8.0, and 12.0 Å, while keeping all other bond lengths and valence and torsion angles constant (see Tab.1). The total energy for these rigid molecular arrangements was obtained by HF calculations with 3-21G*(*), and HF and MP2 calculations with DZ basis sets (Dunning and Hay, 1977) with polarization function on all atoms (p-type functions with exponent 1.0 for H atoms and, 6 d-type functions with exponents 0.8 for C and N and 0.9 for O, respectively). The latter set of basis functions will be referred to as DZP throughout this paper. The interaction energy was then calculated for each molecular arrangement as:

$$E_b = E_{FM} - E_F - E_M$$

Here, E_{FM} denotes the total energy of the molecular arrangement and E_F and E_M represent total energies for formic acid and methylamine , respectively, calculated for their geometries in the 3-21G*(*) optimised complex (i.e., identical with their geometries in the corresponding molecular arrangement). No geometry optimization was performed in this case. The binding energies were also calculated (see below) but in this case the total energies E_{FM}, E_F and E_M corresponded to energies resulting from full geometry optimization with an appropriate basis set for the complex, formic acid and methylamine, respectively.

The Basis Set Superposition Error (BSSE) in the interaction energy at each distance was estimated by performing energy calculations for rigid geometries of formic acid and methylamine at their geometry in the molecular arrangement and locating additional basis functions at the appropriate centers such that there were the same number and type as in the corresponding molecular arrangement. Obviously, adding basis functions results in lowering the total energy and, as a consequence, the calculated interaction energy is always smaller when ghost functions are included. Strictly speaking, it is not possible to calculate BSSE for binding energies since the geometries of separate molecules are different from those in the complex. However, we believe that the magnitude of this effect for binding energies should be similar to the BSSE errors for appropriate basis set calculated for the interaction energy corresponding to the H5$\cdots$N6 distance of 1.833 Å because the geometry of this molecular arrangement is similar to the geometry of the HF and MP2 optimized complex, and because the geometries of formic acid and methylamine optimized separately are close to those in the complex.

DF calculations were performed for the same set of geometries by DGauss (Andzelm *et al.*, 1989; Andzelm, this volume) and DMol (DMol, ver. 1.2, Biosym, Inc., San Diego, California) using a CRAY Y-MP8/864 supercomputer. The DGauss program uses the LSD potential by Vosko and coworkers (Vosko *et al.*, 1980), while Dmol incorporates the Barth and Hedin LSD potential (von Barth and Hedin, 1972), and therefore total energies from these programs are not directly comparable. The total LSD energies were included in this paper (Tabs.3, 4 and 5) as a reference rather than for direct comparison.

Basis sets of double zeta quality and polarization functions on all atoms were used with both programs. The DNP (Double Numerical with Polarization) basis set supplied with the program was used with DMol. We made use of default parameters for calculations, but the integration mesh was modified so that integration points extended out from each center by 15 Bohr, the maximum allowed by the program, and the angular sampling frequency was adjusted accordingly. Parameters FASCF and FBSCF (mixing coefficients for charge density and spin density, respectively) were set to 0.4, since this resulted in acceptable convergence of the SCF procedure.

Two LSD-optimized Gaussian basis sets (Godbout *et al.*, 1990) were explored with DGauss in order to test the significance of the BSSE effect. Both sets were of valence double zeta quality with polarization functions (DZVPP), but differed in representation of the core orbitals. The first set (DG-1) had a pattern (621/41/1*), which is a shorthand notation for a DZVPP basis set of 3s, 2p and 1d contractions. The s-type contractions

Table 2: Total energies in Hartrees as a function of H5···N6 distance for different basis sets. The BSSE columns refer to calculations in which the basis functions and centers were the same as for the "complex" but nuclear charges were placed only on real atoms. The ∞ row represents total energies for formic acid and methylamine calculated without ghost functions.

d [Å]	Complex	Formic (BSSE)	Methylamine (BSSE)
HF 3-21G (polarization functions on C1, O3, O4, N6, H5, and H9)			
0.800	-282.2684382	-187.8422810	-94.7409442
1.000	-282.4575668	-187.8410366	-94.7392584
1.200	-282.5359981	-187.8399431	-94.7382043
1.400	-282.5695321	-187.8386934	-94.7376184
1.600	-282.5825630	-187.8373426	-94.7371506
1.833	-282.5860864	-187.8358511	-94.7363435
2.000	-282.5850421	-187.8349464	-94.7357412
2.200	-282.5822218	-187.8340863	-94.7351583
2.500	-282.5771349	-187.8332053	-94.7341689
3.000	-282.5700004	-187.8325430	-94.7320359
4.000	-282.5643668	-187.8323626	-94.7301396
6.000	-282.5628192	-187.8323597	-94.7300686
8.000	-282.5625473	-187.8323597	-94.7300686
12.000	-282.5624482	-187.8323597	-94.7300686
∞		-187.8323597	-94.7300686
HF DZP basis set with polarization functions on all atoms			
0.800	-283.7527918	-188.8154796	-95.2417233
1.000	-283.9414618	-188.8152362	-95.2415791
1.200	-284.0187932	-188.8150373	-95.2414687
1.400	-284.0518493	-188.8148997	-95.2412830
1.600	-284.0653372	-188.8148007	-95.2410510
1.833	-284.0700179	-188.8147040	-95.2408339
2.000	-284.0700479	-188.8146397	-95.2407190
2.200	-284.0685636	-188.8145676	-95.2406083
2.500	-284.0655114	-188.8144744	-95.2405083
3.000	-284.0611919	-188.8143432	-95.2404710
4.000	-284.0566921	-188.8141444	-95.2402647
6.000	-284.0546907	-188.8141076	-95.2401402
8.000	-284.0543721	-188.8141076	-95.2401401
12.000	-284.0542644	-188.8141076	-95.2401401
∞		-188.8141076	-95.2401401
MP2 DZP basis set with polarization functions on all atoms			
0.800	-284.6565704	-189.3480416	-95.6060717
1.000	-284.8413707	-189.3466897	-95.6029886
1.200	-284.9147298	-189.3456710	-95.6003464
1.400	-284.9441343	-189.3449009	-95.5983918
1.600	-284.9547199	-189.3443330	-95.5970781
1.833	-284.9570022	-189.3438335	-95.5961221
2.000	-284.9557851	-189.3435425	-95.5956277
2.200	-284.9531825	-189.3432568	-95.5951443
2.500	-284.9489466	-189.3429380	-95.5946002
3.000	-284.9434190	-189.3425901	-95.5940407
4.000	-284.9379820	-189.3422248	-95.5933728
6.000	-284.9358250	-189.3421672	-95.5931791
8.000	-284.9354910	-189.3421672	-95.5931790
12.000	-284.9353711	-189.3421672	-95.5931790
∞		-189.3421672	-95.5931790

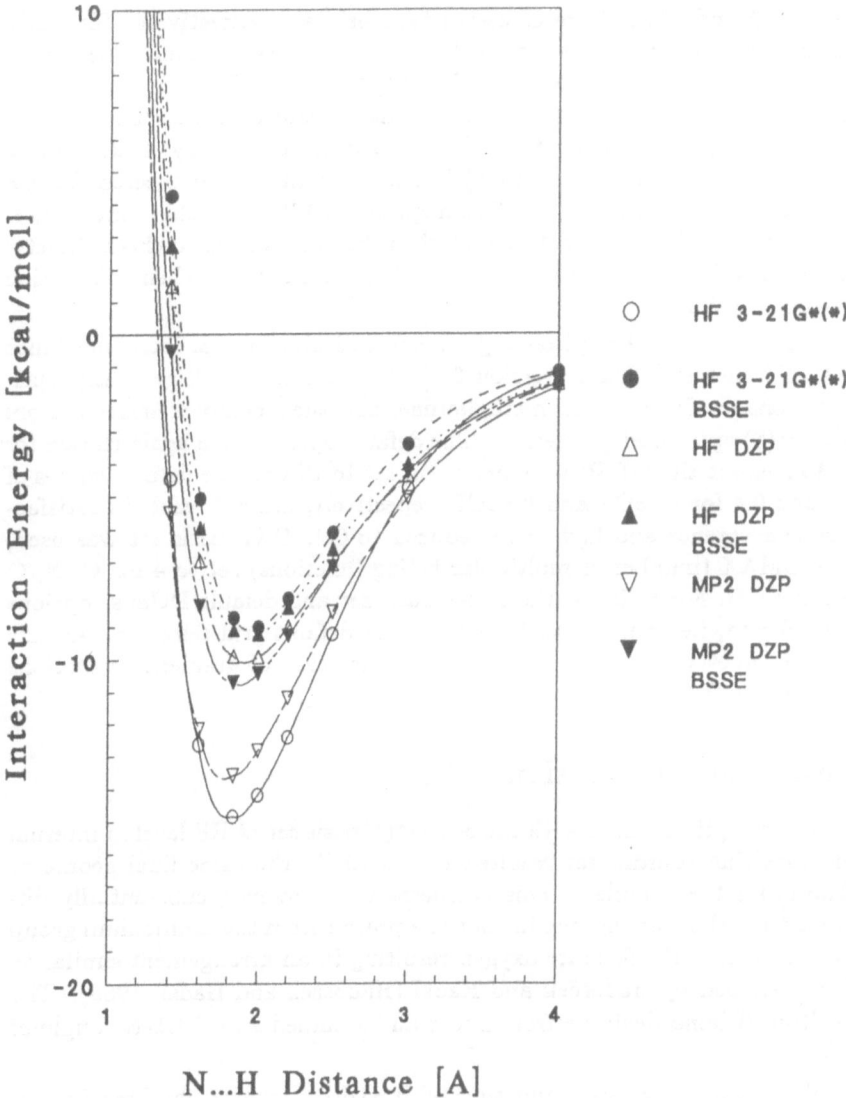

Figure 2: Dependence of interaction energy on the H5···N6 distance obtained by HF and MP2 methods. The effect of Basis Set Superposition Error is shown for 3-21G*(*) and DZP basis sets.

have 6, 2 and 1 primitive Gaussian type orbitals, respectively. Similarly, the p-part contains contraction of 4 and 1 primitive Gaussians. The second basis set (DG-2) had a pattern (721/51/1*) and differed in the number of primitive Gaussian functions representing core orbitals. The method of optimizing the LSD contracted basis sets introduced by Andzelm and coworkers (Andzelm *et al.*, 1985) is based on the one pioneered for HF basis sets by Huzinaga *et al.* (Huzinaga *et al.*, 1984). At this time neither DF program allows the inclusion of ghost basis functions to check directly for BSSE effects. We plan to perform these calculations when this option becomes available.

Quite recently, the option of geometry optimization was introduced into both DGauss and DMol (version 2.0). We performed full geometry optimizations for formic acid, methylamine, and their complex starting from 3-21G*(*) optimized geometries. The default options were again chosen for DMol, except the MESH parameter was set to FINE. The default values of 0.7 and 0.4 for FASCF and FBSCF, repectively, did not result in satisfactory convergence and had to be reduced to 0.2. DNP basis set was used, with LMAX (number of multipolar fitting functions) set to 4 for C, N, O and 3 for H. For DGauss, the DG-1 basis set and default DGauss options were chosen, i.e. triple zeta fitting set and default grid selection. At this time, geometry optimization is without inclusion of gradient corrections (Becke, 1989).

Results and Discussion

Geometry optimizations with the 3-21G*(*)basis set at HF level in internal and cartesian coordinates resulted in essentially the same final geometry (Tab.1) for the complex. This geometry was, however, substantially different from the starting one in that the proton from the ammonium group moved towards the formate oxygen resulting in an arrangement similar to that described by Hadoŝček and Hadži (Hadoŝček and Hadži, 1989). The resulting H-bond deviated from linear and assumed an O3-H5-N6 angle of 163°.

The total energy as a function of H5···N6 distance obtained by HF calculations with 3-21G*(*) basis sets, and HF and MP2 calculations with DZP basis sets, are reported in Tab.2. These data are presented in terms of interaction energies in Fig.2. For those curves which were corrected for superposition errors (labelled BSSE), E_F and E_M refer to formic acid and methylamine energies calculated with ghost basis functions as described above. In the plot, the energy zero corresponds to $E_F + E_M$ for isolated formic acid and methylamine molecules at their geometries in the 3-21G*(*) optimized complex.

The higher-level *ab initio* results suggest that the optimium H5···N6 distance in this arrangement is about 1.8 Å and the corresponding inter-

Table 3: Results of calculations with different DF approaches. Total energies (in Hartrees) of the molecular arrangement derived from 3-21G*(*) optimised complex as a function of H5···N6 distance (in Å).

$d_{H5\cdots N6}$	DMol[1] LSD[3]	DGauss[2] LSD	DGauss[2] Becke-Stoll[4]
0.800	-283.2732442	-283.1679315	-285.3870636
1.000	-283.4445384	-283.3463971	-285.5740960
1.200	-283.5101007	-283.4159256	-285.6521537
1.400	-283.5345387	-283.4419731	-285.6866404
1.600	-283.5415423	-283.4494973	-285.7011342
1.833	-283.5408469	-283.4486787	-285.7074617
2.000	-283.5376894	-283.4456232	-285.7088223
2.200	-283.5333614	-283.4412407	-285.7078829
2.500	-283.5273672	-283.4349614	-285.7051159
3.000	-283.5205093	-283.4275672	-285.7022478
4.000	-283.5149494	-283.4214682	-285.6977108
6.000	-283.5130614	-283.4196324	-285.6954231
8.000	-283.5128124	-283.4191288	-285.6951405
12.000	-283.5127422	-283.4190123	-285.6950408
HCOOH	-188.4479132	-188.3915076	-189.8236360
CH_3NH_2	-95.0648210	-95.0274799	-95.8713837

$d_{H5\cdots N6}$	DGauss[2] Becke-Perdew[6]	DGauss[5] LSD	DGauss[5] Becke-Perdew[6]
0.800	-285.4167163	-283.1922663	-285.4433321
1.000	-285.5982520	-283.3711538	-285.6251044
1.200	-285.6710343	-283.4403778	-285.6973920
1.400	-285.7009482	-283.4662743	-285.7270324
1.600	-285.7114990	-283.4735820	-285.7374377
1.833	-285.7141582	-283.4724816	-285.7400187
2.000	-285.7134722	-283.4693001	-285.7393342
2.200	-285.7106762	-283.4647722	-285.7366041
2.500	-285.7061422	-283.4585133	-285.7321756
3.000	-285.7019023	-283.4512311	-285.7279409
4.000	-285.6975518	-283.4457046	-285.7238110
6.000	-285.6956007	-283.4440900	-285.7219708
8.000	-285.6952987	-283.4435581	-285.7216556
12.000	-285.6951907	-283.4434194	-285.7215466
HCOOH	-189.8148552	-188.4070742	-189.8317654
CH_3NH_2	-95.8803125	-95.0363224	-95.8897596

[1] DNP basis set was used
[2] DG-1 basis set was used
[3] Version 1.2 of DMol was used
[4] LSD with gradient corrections
(Becke, 1986; Stoll et al., 1978).
[5] DG-2 basis set was used
[6] LSD with gradient corrections
(Becke, 1988; Perdew, 1986).

Table 4:   Comparison of energies and optimized geometries of formic acid:methylamine complex calculated by MP2, HF, DGauss (DG-1 basis set) and DMol. Bond lengths in Å and angles in deg. Total energy in Hartrees.

	MP2 DZP	HF DZP	DGauss LSD	DMol LSD
Formic acid:methylamine complex				
Energy	-284.9616978	-284.0713031	-283.4593640	-283.5472272
C1–H2	1.097	1.088	1.118	1.116
C1–O3	1.332	1.309	1.314	1.315
C1–O4	1.225	1.191	1.230	1.229
O3–H5	1.013	0.971	1.086	1.096
H5–N6	1.692	1.874	1.501	1.492
N6–C7	1.473	1.462	1.459	1.458
N6–H8	1.017	1.002	1.030	1.033
N6–H9	1.017	1.002	1.030	1.033
C7–H10	1.090	1.084	1.103	1.103
C7–H11	1.094	1.088	1.107	1.107
C7–H12	1.090	1.084	1.103	1.103
H2–C1–O3	110.55	111.08	112.20	112.20
H2–C1–O4	123.46	123.12	122.38	121.98
O3–C1–O4	125.99	125.80	125.41	125.81
C1–O3–H5	107.30	110.57	105.52	105.28
O3–H5–N6	171.73	175.23	165.19	165.49
H5–N6–C7	122.59	121.39	126.75	126.89
H5–N6–H8	103.64	103.23	99.18	99.18
H5–N6–H9	103.65	103.23	99.33	99.19
C7–N6–H8	110.30	110.79	112.38	112.45
C7–N6–H9	110.30	110.80	112.36	112.44
H8–N6–H9	104.82	106.15	103.90	103.64
N6–C7–H10	108.81	109.30	109.69	109.50
N6–C7–H11	113.66	113.58	113.70	114.03
N6–C7–H12	108.81	109.30	109.75	109.50
H10–C7–H11	108.70	108.36	108.14	108.13
H10–C7–H12	108.02	107.76	107.27	107.33
H11–C7–H12	108.70	108.36	108.08	108.14
H2–C1–O3–H5	180.00	180.00	180.00	180.00
O4–C1–O3–H5	0.00	0.00	0.00	0.00
C1–O3–H5–N6	0.00	0.00	0.00	0.00
O3–H5–N6–C7	180.00	180.00	180.00	180.00
O3–H5–N6–H8	54.60	55.24	52.55	52.75
O3–H5–N6–H9	-54.66	-55.17	-53.30	-52.80
H5–N6–C7–H10	-58.73	-58.86	-58.73	-58.70
H5–N6–C7–H11	180.00	180.00	180.00	180.00
H5–N6–C7–H12	58.74	58.86	58.88	58.72
H8–N6–C7–H10	63.61	62.36	62.78	63.03
H8–N6–C7–H11	-57.66	-58.78	-58.43	-58.25
H8–N6–C7–H12	-178.92	180.00	180.00	180.00
H9–N6–C7–H10	178.92	180.00	180.00	180.00
H9–N6–C7–H11	57.65	58.77	58.34	58.28
H9–N6–C7–H12	-63.61	-62.38	-62.84	-63.02

Table 5: Comparison of energies and optimised geometries for formic acid and methylamine calculated by MP2 and HF methods, DGauss (DG-1 basis set), and DMol. Bond lengths in Å and angles in deg. Total energy in Hartrees.

	MP2 DZP	HF DZP	DGauss LSD	DMol LSD
	Formic acid			
Energy	-189.3453762	-188.8150210	-188.3943201	-188.4452214
C1–H2	1.094	1.086	1.114	1.112
C1–O3	1.352	1.323	1.342	1.343
C1–O4	1.215	1.184	1.211	1.211
O3–H5	0.972	0.950	0.988	0.998
H2–C1–O3	109.25	110.48	109.49	109.64
H2–C1–O4	125.55	124.58	125.66	125.33
O3–C1–O4	125.20	124.94	124.86	125.03
C1–O3–H5	106.54	109.19	106.34	105.98
H2–C1–O3–H5	180.00	180.00	180.00	180.00
O4–C1–O3–H5	0.00	0.00	0.00	0.00
	Methylamine			
Energy	-95.5936433	-95.2405088	-95.0295531	-95.0671446
N6–C7	1.468	1.456	1.451	1.456
N6–H8	1.015	1.001	1.025	1.032
N6–H9	1.015	1.001	1.025	1.032
C7–H10	1.091	1.085	1.105	1.105
C7–H11	1.097	1.092	1.114	1.113
C7–H12	1.091	1.085	1.105	1.105
C7–N6–H8	109.62	110.91	110.87	109.60
C7–N6–H9	109.62	110.91	111.00	109.59
H8–N6–H9	106.01	107.39	107.10	105.17
N6–C7–H10	108.87	109.32	109.45	109.28
N6–C7–H11	115.06	114.53	116.24	115.65
N6–C7–H12	108.88	109.32	109.50	109.29
H10–C7–H11	108.16	108.03	107.50	107.74
H10–C7–H12	107.45	107.37	106.20	106.79
H11–C7–H12	108.17	108.03	107.47	107.74
H8–N6–C7–H10	63.59	61.75	62.64	64.28
H8–N6–C7–H11	-57.97	-59.61	-59.34	-57.45
H8–N6–C7–H12	180.00	179.02	178.67	180.00
H9–N6–C7–H10	180.00	-179.01	-178.45	180.00
H9–N6–C7–H11	58.00	59.63	59.56	57.48
H9–N6–C7–H12	-63.58	-61.74	-62.42	-64.26

action energy is around 10 kcal mol^{-1}. The BSSE with MP2 calculations for the DZP basis amounts to about 3 kcal mol^{-1} while for the HF calculations with the same basis set the BSSE is negligible. The explanation for this effect can be found by examining energies and coefficients of the HOMO (Highest Occupied Molecular Orbital) and LUMO (Lowest Unoccupied Molecular Orbital) for formic acid and methylamine with and without ghost basis functions. For both formic acid and methylamine, the energies of the corresponding HOMO's (and lower-lying orbitals) calculated with and without ghost atomic orbitals are similar and the coefficients of the ghost basis functions are very small. Correspondingly, the BSSE is small at the HF level. However, for both formic acid and methylamine respective LUMO's were substantially lower in energy when ghost basis functions were included, and the coefficients of the ghost functions were very significant. Since the excitations to LUMO contribute most significantly to the MP2 correction to the total energy of the molecule, the calculated MP2 energies of formic acid and methylamine with ghost functions are considerably lower than ones without, and, as a result, the calculated interaction energy appears larger if BSSE correction is not included. This situation once again reaffirms the importance of estimating BSSE and the necessity of examining molecular orbital energies and coefficients before accepting values of *ab initio* calculated energies using the supermolecule approach.

Total energies as a function H5$\cdots$N6 distance for various DF approaches are collected in Tab.3 to serve as a reference. These energies were used in calculating interaction energies, compared in Fig.3. At the moment neither DGauss nor DMol allow for the estimation of BSSE, hence $E_F + E_M$ corresponds to calculations by the corresponding method on formic acid and methylamine without additional basis functions at all points on the curve. There are however some indications that BSSE effects should not be large. The curves obtained by DGauss for two different basis sets (DG-1 and DG-2) are very similar. Since these basis sets differ only in the the size of core orbitals, which are mainly responsible for BSSE, one can expect the BSSE to be small in the LSD Gaussian-type calculations.

The LSD calculations by DMol and DGauss yield similar results for optimum of the interaction energy and the corresponding distance, namely, -19 kcal mol^{-1} and 1.6 Å, respectively. Gradient corrections to the LSD approximation were only available in DGauss at this time. The Becke-Perdew corrections for two different basis sets result in very similar values for the optimum of interaction energy and the corresponding distance, i.e. -12 kcal mol^{-1} and 1.8 Å. The Becke-Stoll corrections were only calculated for the DG-1 basis set with the resulting energy and length being approximately -9 kcal mol^{-1} and 2.0 Å. The Becke-Perdew corrections closely reproduce the MP2 results, whereas the Becke-Stoll corrections result in a hydrogen bond which is too long and too weak (Fig.4).

The fully-optimized geometries and corresponding energies for formic

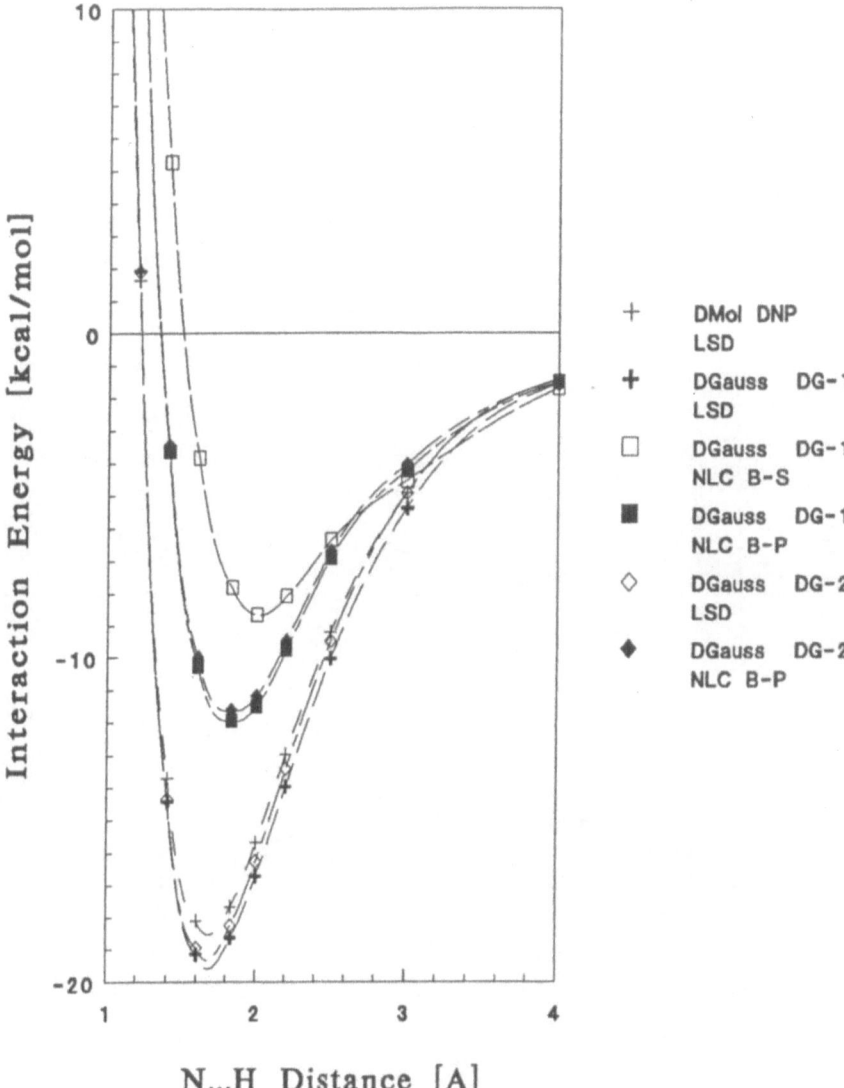

Figure 3: Dependence of interaction energy upon the H5 ⋯ N6 distance calculated by different DF approaches.

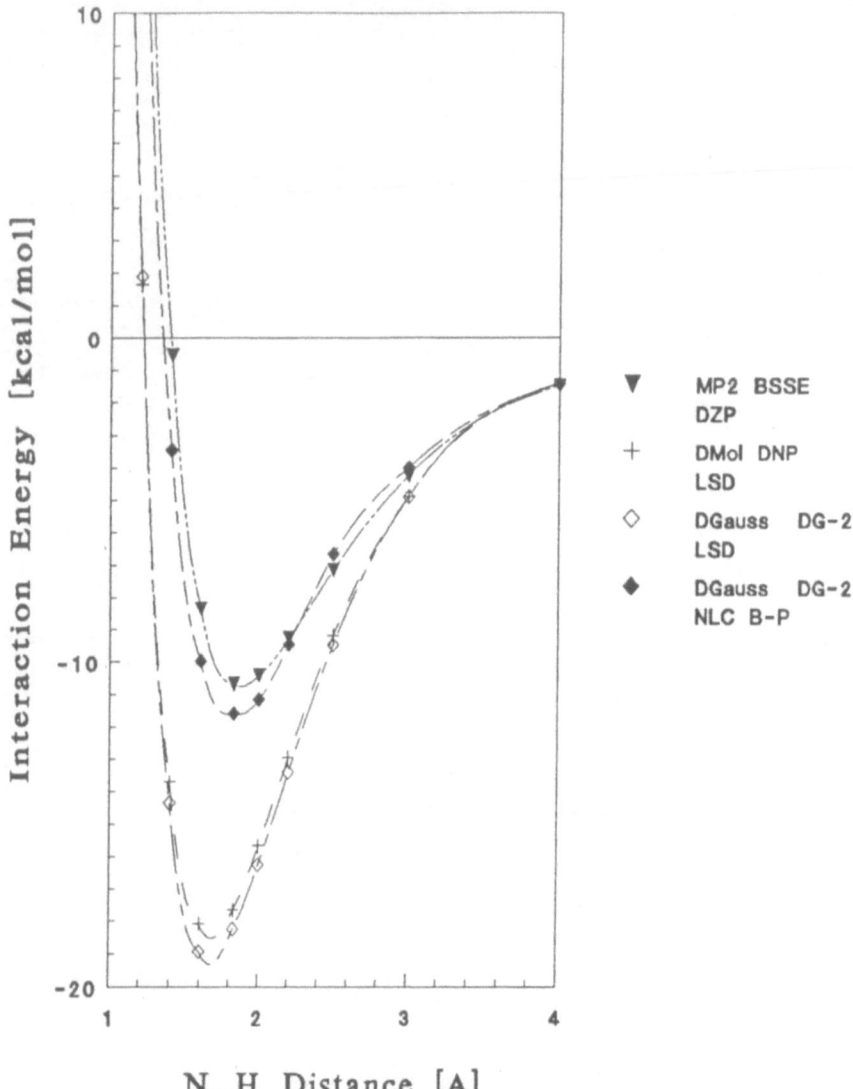

Figure 4: Comparison of interaction energies calculated by MP2 and LSD approaches with and without gradient corrections.

acid, methylamine, and their complex obtained by DMol and DGauss starting from the 3-21G*(*) geometry (without gradient corrections to the LSD in both cases) are compared to those obtained by CADPAC at HF and MP2 levels using DZP basis set in Tabs.4 and 5. The most significant differences are evident in the description of the hydrogen bond. The overall length of the H-bond (O3$\cdots$N6) from LSD calculations is 0.12 Å shorter than MP2 and 0.3 Å shorter than HF, and H5 is 0.2 and 0.4 Å closer to N5 than in MP2 and HF, respectively. Since LSD geometry optimization without gradient corrections by DMol and DGauss yields very similar structures, it is clear that discrepancies between traditional ab initio and LSD can only be remedied by including gradient corrections in geometry optimization. Also, the binding energy is severly overestimated without gradient corrections: -14.23, -9.90, -22.27 and -21.88 kcal mol^{-1} for MP2 DZP, HF DZP, DGauss and DMol respectively. These results suggest that gradient corrections should be used routinely in calculations involving hydrogen-bonded systems. Moreover, it presents little added computational cost.

Computation times for the LSD method compared very favorably to HF and MP2. A single iteration of geometry optimization with DGauss in the case of formic acid:methylamine complex took about 50 seconds of CRAY Y-MP single processor time, while CADPAC, generally considered well vectorized, consumed circa 1100 and 2100 seconds per iteration for HF and MP2, repectively.

Conclusion

Clearly the LSD approximation, without gradient corrections, overestimates hydrogen bond strength compared to HF and MP2 results which serve here as a reference. Also the bond length is considerably shortened. Unfortunately we were unable to find experimental data which would prove or disprove this conjecture; however, geometries of the isolated molecules are similar to those calculated at the MP2 level. Hence the major discrepancy between HF/MP2 and LSD in this case involves only the hydrogen bond, while the strong intramolecular bonds are described in comparable fashion. On the other hand, the Becke-Perdew gradient corrections to LSD dramatically improve the description of the hydrogen bond giving results which are very similar to those obtained by much more computationally expensive HF and MP2 methods.

Acknowledgement

We want to thank Dr. Jan Andzelm from Cray Research, Inc. for performing DGauss calculations for us and for his most valuable comments and discussions. This work was also supported in part by National Institute of Health (NS 17907-0783). The CRAY Y-MP time for CADPAC and DMol calculations was granted by the Ohio Supercomputer Center.

References

Amos, R. D., Rice, J. E., 1987, *CADPAC 4 - Cambridge Analytic Derivative Package, issue 4.0 for Cray X-MP*, Cambridge.

Andzelm, J., Radzio, E., Salahub, D. R., 1985, *J. Comput. Chem.* 6:520.

Andzelm, J., Wimmer, E., Salahub, D. R., 1989, in: *The challenge of d and f electrons: Theory and Computations*, (Eds. D. R. Salahub, M. Zerner), ACS Symposium Series No. 394, p. 228, American Chemical Society, Washington, D.C.

Becke, A. D., 1986, *I. Chem. Phys.* 84:4524.

Becke, A. D., 1988, *Phys. Rev.* A 38:3098.

Becke, A. D., 1989, in: *The challenge of d and f electrons: Theory and Computations*, (Eds. D. R. Salahub, M. Zerner), ACS Symposium Series No. 394, p. 166, American Chemical Society, Washington, D.C.

Binkley, J. S., Pople, J. A., Hehre, W. J., 1980, *J. Am. Chem. Soc.* 102:939.

Bunzow, J. R., VanTol, H. H. M., Grandy, D. K., Albert, P., Salon, J., Christie, M., Machida, C. A., Neve, K. A., Civelli, O., 1988, *Nature* 336:783.

Del Bene, J. E., 1985, *J. Phys. Chem.* 6:296.

Dunning, T. H., Hay, P. J., 1977, in: *Modern Theoretical Chemistry*, (Ed. H. F. Schaefer III), vol. 3, p. 1, Plenum Press, New York.

Godbout, N., Andzelm, J., Wimmer, E., Salahub, D. R., – *to be published.*

Hariharan, P. C., Pople, J. A., 1972, *Chem. Phys. Lett.* 66:217.

Hehre, W. J., Radom L., v. R. Schleyer, P., People, J. A., 1986, Ab Initio *Molecular Orbital Theory* pp. 215–223. Willey, New York.

Hodošček, M., and Hadži, D., 1989, *J. Mol. Struct. (THEOCHEM)* 198:461.

Huzinaga, S., Andzelm, J., Klobukowski, M., Radzio, E., Tatewski, H., 1984, in: *Gaussian Basis Sets for Molecular Calculations*, (Ed. S. Huzinaga), Elsevier, Amsterdam.

Koller, J., Hadošček, M., and Hadži, D., 1984, *J. Mol. Struct. (THEOCHEM)* 106:301.

Koller, J., and Hadži, D., 1989, *J. Mol. Struct. (THEOCHEM)* 200:533.

Miller, D. D., Harrold, M. W., Wallace, R. A., Wallace, L. J., Uretsky, N. J., 1988, *Trends Pharm. Sci.* 9:282, and references therein.

Perdew, J. P., 1986, *Phys. Rev.* B 33, 8822.

Sapse, A. M., and Russell, C. S., 1986, *J. Mol. Struct. (THEOCHEM)* 137:43.

Stoll, H., Pavlidou, C. M. E., Preuss, J., 1978, *Theor. Chim. Acta* 49:143.

von Barth, U., and Hedin, L., 1972, *J. Phys. C.* 5:1629.

Vosko, S. H., Wilk, L., Nusair, M., 1980, *Can. J. Phys.* 58:1200.

24

Electronic and Atomic Structure of Na$_n$Zn Clusters in the Spherically Averaged Pseudopotential Model

J.M. LOPEZ, A. AYUELA, AND J.A. ALONSO

ABSTRACT

Hohenberg–Kohn–Sham density–functional theory is used to obtain the energy of the valence electrons in a spherically averaged external ionic potential for clusters of Na with a single Zn impurity atom. By adding the ion–ion interaction and minimizing the total energy of the cluster with respect to the positions of all ions we obtain the global equilibrium geometry and also the geometries of a few metastable states for a given cluster. Within our model all the clusters with less than five atoms are intrinsically unstable, a fact in good agreement with the experiment. The cohesive energy as a function of the number of atoms of the cluster shows maxima for clusters with 10 and 20 electrons, in agreement with the experimental results, and for clusters with 13 atoms. The last maximum is due to geometrical reasons.

INTRODUCTION

In the past few years it has become possible to study some properties of atomic clusters produced by supersonic bean devices and other techniques. One of the striking results is the evidence for the so called magic clusters. These are much more abundant in the mass spectra than clusters of adjacent sizes (Echt et al. 1981, Kappes et al. 1982, Knight et al. 1984, Begemann et al. 1987, Cohen et al. 1987). The magic numbers of alkali metal clusters –2, 8, 20, 40, 58, and 92– are explained by the filling of electronic shells in a spherically symmetric effective potential intermediate between a three–dimensional isotropic harmonic oscillator and a square well with infinitely hard walls (Katakuse et al. 1986). This potential is obtained by describing the cluster by a homogeneous spherical background of positive charge in which the valence electrons move (Cohen et al. 1987). The electrons then fill shells in the order 1s, 1p, 1d, 2s, 1f, 2p, 1g, 2d, 1h, 3s, etc. Minor features in the spectra are further explained by considering ellipsoidal deformations in open shell clusters (Clemenger 1985). This description is known as the electronic shell model and has also been applied to clusters of noble metals and polyvalent nontransition metals.

Hartree–Fock (Boustani et al. 1987) and Density Functional calculations (Martins et al 1985, Manninen 1986a) have been performed to investigate the geometrical arrangement of the atoms in very small clusters. The study of Na clusters with up to eight atoms (Martins et al. 1985, Manninen 1986a) has shown that at these small sizes the geometries are governed by the electronic structure; in other words, the atoms adapt their relative positions to the shape of the electron density distribution, this one being very similar to that of a collection of electrons in a spherical potential, as the shell model assumes. These calculations give insight about the interpretation of the experimental mass spectra. For instance, they confirm that the clusters Na_2 and Na_8 are magic due to the shell closing effect. The situation is more complicated for larger sizes; in this range Kappes and Schumacher (Kappes and Schumacher 1988) report some differences between "cold" and "warm" clusters. The shell model reproduces the magic numbers (20,40) of the warm (or molten) clusters but features observed at N=19 and 38 for cold clusters may have something to do with geometric effects.

In order to test the validity of the simple jellium model for metal clusters and to shed light on the possible influence of the geometric structure, several experiments have been performed on heteroatomic clusters formed by metallic atoms with different valence (Knight et al. 1985, Kappes et al 1986a, Kappes et al. 1985, Kappes et al 1986b, Kappes et al. 1987, Heiz et al. 1990). If only electronic structure controls the magic numbers then it is expected that abundance maxima will again occur for close electronic shells. Shifts in the electron numbers of experimental abundance maxima with respect to the sequence predicted by the spherical jellium model have been reported in some cases (Kappes et al. 1985). For instance, the appearance of new magic clusters with ten valence electrons. Density functional calculations performed with a modified jellium model for heteroatomic clusters M_xN, with M=Na or K and N indicating a monovalent or divalent impurity, rationalize the change of the magic numbers by a rearrangement of some of the electronic levels induced by the impurity (Baladron and Alonso 1988). In particular, the new magic number for ten electrons arises from an inversion in the relative positions of the 1d and 2s levels. In the model of Baladron and

Alonso the impurity is placed in a hole excavated at the center of a spherical jellium cluster representing the host. The impurity itself is described by the jellium model (with a different density than the host) and, in a more accurate version (Baladron and Alonso 1989), by a point nucleus plus a compensating electronic cloud.

The above examples indicate the necessity to perform calculations beyond the jellium model in the medium size range. Unfortunately the computational difficulties increase greatly in this size range because of the increasing number of electrons involved and some simplifications are needed. It is also desirable to study the competition between the variations of the total energy as a function of size which arise from the filling of electronic levels and the variations arising from geometric structure effects. This competition is not easily revealed in general because geometry and electronic structure are intimately related (Koutecky and Fantucci 1986). The jellium model can be taken as a starting point in which only electronic effects arise, and geometric effects can be introduced by perturbation theory (Upton 1986, Ishii et al. 1986). The complementary point of view is that of Manninen (1986b). This author has minimized the Madelung energy of a system of point-like ions embedded in an uniform electron gas contained in a sphere with radius $R \propto N^{1/3}$, N being the number of atoms in the cluster. Obviously, only geometrical effects arise in such a system. Manninen then states that for alkali clusters the geometrical contribution to the size variation of the cluster energy arising from such model is smaller than the contribution arising from the electronic shell filling effect of the spherical jellium model and that this fact justifies the success of the simple jellium model. The dominance of electronic over structural effects is, however, not obvious from the numerical evidence presented in Manninen's paper.

We have linked these two points of view by considering simultaneously the electronic levels in a "spherically averaged" effective potential (different, however, from the effective potential of the jellium model) and the optimized geometries calculated for ions embedded in a not homogeneous, although spherically symmetric, electronic background. Doing so we are able to study the oscillations of the total energy with size arising from both geometric and electronic effects in a way that is still relatively simple and permits to extend the calculations to a size range where ab-initio three dimensional calculations become very difficult. The model   was   coined   as   SAPS   from   Spherical   Averaged   Pseudo–Potential (Iñiguez et al. 1989).

Previous calculations   for   Na, Mg, Al and Pb clusters (Iñiguez et al. 1989) have shown that the SAPS model is able to appreciate structural effects, beyond the electronic shell effects which form the essence of the predictions of the jellium model. The calculations show   the formation of atomic shells and appreciable reconstruction as the cluster grows. The model has also been   used to study the   immersion of impurities in alkali clusters (Robles et al. 1989, Lammers et al. 1989) and mixing   and segregation effects in   Na/Li and Na/Cs heteroclusters (Lopez at al. 1989, Lopez et al. 1990,   Mañanes et al. 1990).

In this work we use the SAPS model to obtain the equilibrium geometries of $Na_x Zn$ heteroclusters. The variation of the energy as a function of cluster size helps to interpret the experimental mass spectra (Heiz et al. 1990)

THE SAPS MODEL

To calculate the equilibrium geometry of an atomic cluster at T=0 K and the corresponding ground state energy we use the density functional formalism (for a review see March et al 1983). The ground state energy for a given   configuration of the ions, defined by the set of ionic positions $(\bar{R}_I)$, can be written (in atomic units) as a functional of the electron density $n(\bar{r})$:

$$E[n, (\bar{R}_I)] = T[n] + \frac{1}{2} \int d\bar{r} \; d\bar{r}' \; \frac{n(\bar{r}) \; n(\bar{r}')}{|\bar{r} - \bar{r}'|} + E_{xc}[n]$$

$$+ \int d\bar{r} \; V_I(\bar{r}) \; n(\bar{r}) + \frac{1}{2} \sum_{i \neq j} \frac{z_i z_j}{|\bar{R}_i - \bar{R}_j|} . \tag{1}$$

The first term in this equation is the single particle kinetic energy

$$T[n] = \sum_{occ} \langle \psi_i | - \nabla^2 / 2 | \psi_i \rangle, \tag{2}$$

the second term is the classical coulombic electron–electron interaction, the third term is the electronic exchange–correlation energy, the fourth term is the interaction between the electrons and the total ionic potential V_I and finally, the last term is the coulomb interaction energy between point–like ions of charge z_i. The electron density $n(\bar{r})$ is obtained from the occupied single–particle wave functions

$$n(\bar{r})= \sum_{occ} |\psi_i|^2. \tag{3}$$

these being obtained by solving the Kohn–Sham equations (Kohn and Sham 1965)

$$\left[-\frac{\nabla^2}{2} + V_{eff}(\bar{r}) \right] \psi_i(\bar{r}) = \epsilon_i \, \psi_i(\bar{r}). \tag{4}$$

The effective potential V_{eff} in this equation can be written as

$$V_{eff}(\bar{r})=V_I(\bar{r})+\int d\bar{r}' \frac{n(\bar{r}')}{|\bar{r}-\bar{r}'|} + V_{xc}(r) \tag{5}$$

that is, the sum of the Hartree part (the first two terms of the r.h.s. of (5)) and the exchange–correlation piece. The total ionic potential is constructed from the individual ionic potentials

$$V_I(\bar{r})= \sum_i v(\bar{r}-\bar{R}_i). \tag{6}$$

In the present model we approximate the individual ionic potentials by the empty–core model pseudopotential (Ashcroft 1966), which is purely coulombic $(-z/r)$ outside the empty–core radius r_c and zero for $r < r_c$. Here z is the valence of the atom considered. The single parameter of the potential, r_c, is adjusted to reproduce the ionization potential for monovalent atoms, and to give a good overall account of the first ionization potential and the sum of the other ionization potentials for polyvalent atoms. For the elements of interest here $r_c(Na)=1.74$ u.a., $r_c(Zn)=1.15$ a.u.

We use the local density approximation for the exchange and correlation energies, $E_{xc}[n] = \int \varepsilon_{xc}[n(\bar{r})]n(\bar{r})d\bar{r}$, taking for ε_{xc} the Gunnarsson–Lundqvist interpolation formula (Gunnarsson et al. 1976)

$$\varepsilon_{xc}[n] = \frac{-0.916}{r_s} - 0.0666 \ G\left(\frac{r_s}{11.4}\right) , \tag{7}$$

where $r_s = (3/4\pi n)^{1/3}$. In this expression the function G(x) is defined by

$$G(x) = (1 + x^3) \ \ln(1 + 1/x) - x^2 + x/2 - 1/3. \tag{8}$$

Solving the Kohn–Sham equations for the true three-dimensional potential $V_I(\bar{r})$ of eq. (6) is a formidable task in the case of a medium size cluster. To simplify the computations we substitute $V_I(\bar{r})$ in (5) by its spherical average $V_I^{av}(r)$ about the cluster center (in practice we have found that defining the cluster center by the center of ionic charge leads to the most stable cluster). This is equivalent to the minimization of the total energy (1) subject to the constraint of using only spherically symmetric electron densities because such a density distribution will only feel the spherical part of the ionic potential. Since a priori we do not know the stable geometries for small clusters at T=0 K, we calculate the equilibrium geometry by minimization of the total energy (1) also as a function of the ionic coordinates. The process starts by first generating a random ionic configuration. For this geometry the total ionic pseudopotential V_I is spherically averaged to produce $V_I^{av}(r)$ and the Kohn–Sham equations are solved to obtain the ground state electron density and then the total cluster energy from eq. (1). With this electron density frozen (taken spherically symmetric), we now relax the positions $\bar{R}_i$ of the ions in the direction of the net forces acting on them (that is, on the steepest descent direction for the energy). Evidently, this relaxation only affects the last two terms of

(1), that is, the ion-electron and ion-ion interaction energies since the electron density is frozen. For the new geometry ($\bar{R}_i$) we repeat the cycle again, and this process is iterated until convergence is achieved and an equilibrium geometry is found. But, as is usual in geometry optimizations, we must start the whole procedure from many different random configurations of the ions to search for the absolute minimum. In this way we are able to obtain the global equilibrium geometry as well as metastable equilibrium geometries.

RESULTS

A. Geometries

To begin with our presentation of results we report the equilibrium and metastable geometries of Na$_x$Zn heteroclusters. In the range x=3-24 four types of geometries have been found: a) all the Na atoms sharing a common spherical shell with the Zn atom situated at the center of the cluster (type A); b) all the atoms of the cluster sharing a common spherical shell with the cluster center empty (type B); c) the center of the cluster occupied by a Na atom and all the other atoms sharing an external spherical shell (type C); d) the cluster has two atomic shells, the inner one having two or three Na atoms whereas the rest of them are situated on the surface shell (type D). An atomic shell is defined here as a group of atoms at similar distance from the center of the cluster and well separated from other shells. In table 1 we present the type of geometries for the global equilibrium state and the metastable ones. We can observe that the geometry of the ground state corresponds, for the size range studied here, to the type A, that is, the Zn atom is located at the center of the cluster and all the Na atoms form a common external shell. This result is different from the results obtained for Na$_x$Li heteroclusters (Lopez et al. 1990) with the same model. The geometry of the ground state of Na$_x$Li heteroclusters has a dependence with the number of atoms in the cluster. For x < 7 the ground state geometries are of type B (empty clusters), for clusters with x=7-10 the geometry is of type A (the atom located at the center is the Li atom) and for x=11-19 the geometry is of type C. The differences between the

Table   1.-   Global   equilibrium   and   metastable   geometries   of Na_xZn
heteroclusters. A: Sodium   shell with the Zn atom   at the
cluster center. B: empty   cluster. C: Na centered cluster.
D: two or   more Na atoms at the cluster center.

x	Ground state	Metastable (in order of increasing energies)
4–5	A	
6	A	C
7–9	A	B, C
10–14	A	C, B
15–16	A	C, D, B
17–23	A	D, C

equilibrium geometries of Na_xZn and Na_xLi arise from the different
valence of the impurity: z(Zn)=2, whereas z(Li)=1. The presence of the Zn
atom   at   the   cluster   center   leads   to   a   stronger   binding   of   the   1s
electronic   shell   (notice   that   this   electronic   shell   should   not   be
confused with the innermost shell of the inert ion core of the Zn atom,
since core electrons are avoided in our calculation). Pure Na_x clusters
have the center empty for x less than 9, and have one central atom and
the rest of them in an external layer for x between 10 and 19 (Iñiguez
et   al.   1989).   As   a   example   of   the   geometries   obtained   in   our
calculations, we present in Figure 1 the geometries   corresponding to the
ground state and   the two first metastable states of the heterocluster
$Na_{14}Zn$. From the results obtained here we can see that the assumption   of
placing the impurity at the cluster center made by Baladron et al (1989)
in their modified jellium   calculation is correct.

B. **Magic numbers and stability**

In figure 2 we present the binding energy of Na_xZn as a function of the
total number of atoms, n_a=x+1, in the heterocluster. The binding energy
is defined:

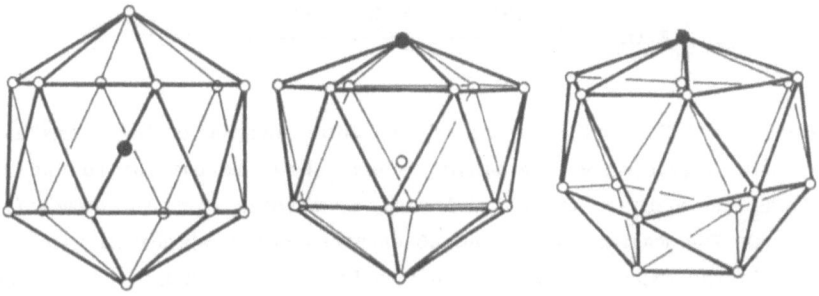

Figure 1.– Global equilibrium and first two metastable geometries of Na$_{14}$Zn. The Zn atom is indicated by the larger black sphere.

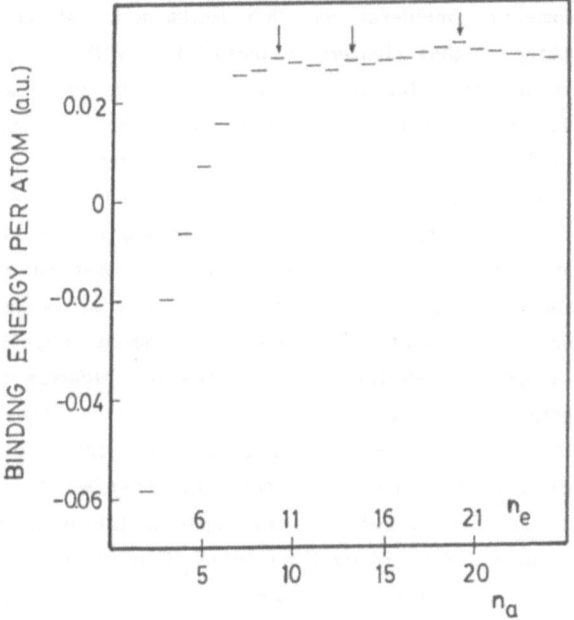

Figure 2.– Binding energy per atom of the cluster Na$_x$Zn versus the number of atoms n$_a$ = x+1. The number of electrons n$_e$ = x+2 is also indicated.

$$E_b(n_a)= \frac{x\,E(Na) + E(Zn) - E(Na_xZn)}{x + 1} \tag{9}$$

where $E(Na)$, $E(Zn)$ and $E(Na_xZn)$ are the total energies of the atoms and the cluster respectively. The binding energy shows maxima for 10 and 20 electrons (9 and 19 atoms) and also a maximum for 13 atoms (14 electrons). The maxima for 10 and 20 electrons are in agreement with the maxima in the experimental mass spectra (Heiz et al. 1990, Kappes et al. 1987) and correspond to the filling of electronic shells. The order of the electronic levels (1s, 1p, 2s, 1d, ...) has suffered a rearrangement with respect to the order obtained in the jellium model for the pure Na cluster (1s, 1p, 1d, 2s, ...). The rearrangement is due to the presence of the Zn impurity at the center of the cluster, where s-type electrons have a sizable probability density. The maximum for 13 atoms is due to geometric considerations. This maximum is related to the very compact geometry of that cluster: an icosahedron with the Zn atom at the center. Some evidence for the enhanced stability of icosahedral 13-atom Li and Na clusters has been obtained in the experiments of Saito and coworkers (Saito et al. 1988, Saito et al. 1989) for clusters generated by the liquid-metal-ion-source technique. However X_{13} is not an abundant cluster in the mass spectra of alkali clusters obtained by vapor expansion techniques. This makes the nature of the 13-atom clusters somehow controversial. Some reasons for the non observability of these clusters in the mass spectra have been recently proposed (Lopez et al. 1990). Specifically, these authors find that the 13-atom clusters are too rigid and that evaporation of atoms at finite temperature can lower their population. From figure 2 we also can notice that the cohesive energy of clusters with less than 5 atoms is negative, and that Na_4Zn is very weakly bound. This prediction agrees with the experimental evidence that Na_xZn clusters with the $x \leq 6$ have not been detected in the mass spectra (Heiz et al. 1990). For a better understanding of this effect we have plotted in Figure 3 the heat of the reaction

$$Na_x + Zn \longrightarrow Na_xZn \tag{10}$$

The reaction is energetically favorable (that is, exothermic) only for $x \geq 5$. As mentioned above, small Na clusters have the center empty.

Furthermore, the most stable position of Zn is at the center of the Na_xZn heterocluster. The negative contribution to the heat of reaction (10) arises from the bonding between the Zn atom and the surrounding Na atoms. However, when x is very small the hole at the center of Na_x is not big enough to accommodate easily the Zn atom. This leads to a forced enlargement of the size of the Na_x cage with a corresponding cost of energy which opposes the introduction of the Zn atom. When x increases the hole at the center becomes larger and the Zn atoms can easily be accommodated. Table 2 gives the calculated cluster radius, R, of Na_x and the radius R_h of the central cavity, defined $R_h = R - r_c(Na)$, where $r_c(Na)$ is the core radius of Na. The atomic radius of the Zn atom is 2.46 a.u. (Kittel 1976), so only starting with Na_5 is the cavity at the center of Na_x large enough to accommodate the Zn atom.

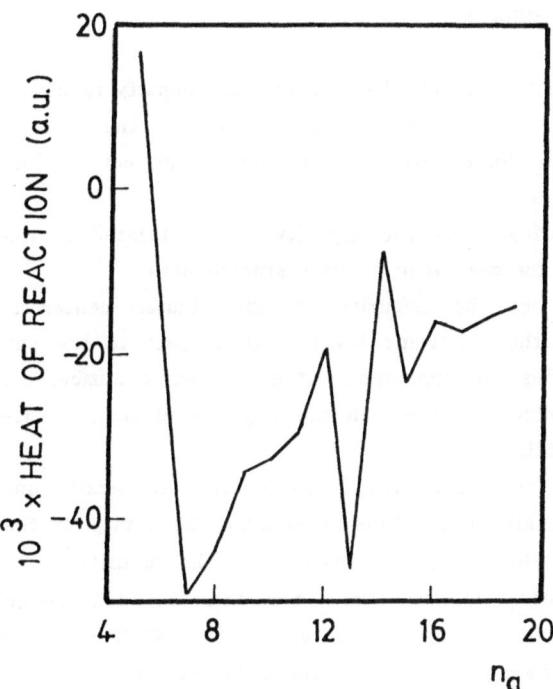

Figure 3.- Heat of the reaction Na_x + Zn ⟶ Na_xZn versus the number of atoms n_a in the final cluster (n_a = x+1).

Table 2.- Radius R of the Na_x cluster, and effective radius R_h of the central cavity $(R_h = R - r_c(Na))$

x	R (a.u.)	R_h (a.u.)
2	2.78	1.04
3	3.48	1.74
4	3.92	2.18
5	4.38	2.64
6	4.72	2.98

SUMMARY AND COMMENTS

In summary, by using the Hohenberg–Kohn–Sham density functional formalism and a convenient computational simplification in the treatment of the ionic potential, we have obtained some interesting conclusions for $Na_x Zn$ heteroclusters:

a) In the equilibrium state the impurity, Zn, is located at the center of the herocluster for the range of sizes studied here.

b) The location of the impurity at the cluster center leads to a rearrangement of the electronic levels with respect to the simple jellium model; this explains the appearance of a new magic number corresponding to 10 electrons, which is absent in the sequence of magic numbers of the simple jellium model.

c) Heteroclusters with less than 5 atoms are intrinsically unstable, and $Na_4 Zn$ is very weakly bound. This is because the cavity at the center of Na_x ($x \leq 4$) is not big enough to accommodate the impurity.

d) Finally we stress that some differences exist between the ground state geometries of $Na_x Li$ or Na_x with respect to $Na_x Zn$. The Zn impurity stabilizes the geometry of type A (impurity at the cluster center) for all the heteroclusters studied here.

Acknowledgments

This work has been supported by grants from Direccion General de Investigacion Cientifica y Tecnica (DGICT, Grant No. PB-86-0654-C02), Junta de Castilla y Leon, and Caja Salamanca.

References

Ashcroft, N. W., 1986, Phys. Lett. 23, 48.

Baladron, C., Alonso, J. A., 1988, Physica B154, 73.

Baladron, C., Alonso, J. A., 1989, Phys. Lett. A140, 67.

Begemann, W., Dreihöfer, S., Meiwes-Broer, K. H., Lutz, H. O., 1987, Physics and chemistry of small clusters. Jena, P., Rao, B. K., Khanna, S. N., (eds.), p. 269, New York, Plenum Press.

Boustani, I.,Pewestorf, W., Fantucci, P., Bonacic-Koutecky, V., Koutecky, J., 1987, Phys. Rev. B35, 9437.

Clemenger, K., 1985, Phys. Rev B32, 1359.

Cohen, M. L., Chou, M. Y., Knight, W. D., de Heer, W. A., 1987, J. Phys. Chem. 91,3141.

Echt, O., Sattler, K., Recknagel, E., 1981, Phys. Rev. Lett. 47, 1121.

Gunnarsson, O., Lundqvist, B. I., 1976, Phys. Rev. B13, 4274.

Heiz, U., Röthlisberger, U., Vayloyan, A., Schumacher, E., 1990, Israel J. of Chem (to be published)

Ishii, Y., Ohnishi, S., Sugano, S., 1986, Phys. Rev. B33, 5271.

Iñiguez, M. P., Lopez, M. J., Alonso, J. A., Soler, J. M., 1989, Z. Phys. D- Atoms, Mol. and Clusters, 11, 163.

Kappes, M. M., Kunz, R. W., Schumacher, E., 1982, Chem. Phys. Lett. 91, 413.

Kappes, M. M., Radi, P., Schär, M., Schumacher, E., 1985, Chem Phys. Lett. 119, 11.

Kappes, M. M., Schär, M., Radi, P., Schumacher, E., 1886a, J. Chem. Phys. 84, 1963.

Kappes, M. M., Radi, P., Schär, M., Yeretzian, C., Schumacher, E., 1986b, Z. Phys. D 3, 115.

Kappes, M. M., Schär, M., Yeretzian, C., Heiz, U. Vayloyan, A., Schumacher, E., 1987, Physics and chemistry of small clusters, Jena, P., Rao, B. K., Khanna, S. N. (eds.). p. 263, New York, Plenum Press.

Kappes, M. M., Schumacher, E. J., 1988, Z. Phys. Chem. 156, 23.

Katakuse, I., Ichihara, T., Fujita, Y., Matsuo, T., Sakurai, T., Matsuda, H., 1986, Int. J. Mass Spectro. Ion Proc. 74, 33.

Kittel, C., 1976, Introduction to Solid State Physics, 5th edition, Wiley, New York.

Kohn, W., Sham, L. J.,1965, Phys. Rev. A140, 1133.

Knight, W. D., Clemenger, K., Saunders, W. A., Chou, M. Y., Cohen, M. L., 1984, Phys. Rev. Lett. 52, 2141.

Knight, W. D., de Heer, W. Clemenger, K., Saunders, W. A., 1985, Solid State Commun. 53, 445.

Koutecky, J., Fantucci, P., 1986, Chem. Rev. 86, 539.

Lammers, U., Mañanes, A., Borstel, G., Alonso, J. A., 1989, Solid State Commun. 71, 591.

Lopez, M. J., Mañanes, A., Alonso, J. A., Iñiguez, M. P.,1989, Z. Phys. D- Atoms, Mol. and Clusters, 12, 237.

Lopez, M. J., Iñiguez, M. P., Alonso, J. A., 1990, Phys. Rev B41, 5636.

Mañanes, A., Iñiguez, M. P., Lopez, M. J., Alonso, J. A., 1990, Phys. Rev. B (to be published)

Manninen, M., 1986a, Phys. Rev. B34, 6886.

Manninen, M., 1986b, Solid State Commun. 59, 281.

March, N. H., Lundqvist, S., 1983, Theory of the inhomogeneous electron gas. New   York, Plenum Press.

Martins, J.-L., Buttet, J., Car, R., 1985, Phys. Rev. B31,   1804.

Robles, J., Iñiguez, M. P., Alonso, J. A., Mañanes, A., 1989, Z. Phys D-Atoms, Mol. and Clusters, 13, 269.

Saito, Y., Watanabe, M., Hagiwara, J., Nishigaki, S., Noda, S., 1988, Jpn. J. Appl. Phys. 27, 424.

Saito, Y., Minami, K., Ishida, T., Noda, T., 1989, Z.   Phys. D-Atoms, Mol. and Clusters, 11, 87.

Upton, T. H., 1986, Phys. Rev. Lett. 56, 2168.

25
Nucleophilic Attacks on Maleic Anhydride: A Density Functional Theory Approach

FRANCISCO MÉNDEZ AND MARCELO GALVÁN

ABSTRACT

The reactivity of maleic anhydride with respect to nucleophiles is studied by using some concepts defined within density functional theory formalism. It is shown that the hard and soft acids and bases principle in a local approach, is useful for distinguish the reactivity of the two type of carbon atoms in the molecule.

INTRODUCTION

Local reactivity criteria permit to distinguish between regions with different behavior inside a molecule. In this context, frontier orbitals (Fukui, 1973) are concepts widely and successfully used within the framework of quantum chemistry. This local quantities are useful reactivity criteria, however they can fail in some cases as was pointed out recently by Dewar (Dewar, 1989).

Density functional theory (DFT) (Parr, 1989; Lundqvist, 1983) has a structure which contains several local quantities that are reactivity parameters. Charge density itself contains information such as where the nucleophilic or electrophilic sites are localized. Moreover, as has been demonstrated by Bader and collaborators (Bader et al, 1984ab; Bader et al, 1985), the laplacian of the charge density has a lot of chemical information.

Fukui function (Parr et al, 1984; Yang et al 1984; Gazquez et al 1987; Lee et al, 1988) is another local quantity that is well defined in DFT. It can give the preferred sites for the initial chemical attacks (Parr et al, 1984) and it is related with the local softness of the molecule (Berkowitz et al, 1988). Besides the application of fukui function to localize reactive sites, it can be seen as a local response coefficient because it is a measure of the charge redistribution during charge transfer processes.

Molecules that contain several reactive sites are optimal candidates for testing local reactivity criteria. Maleic anhydride is one of that candidates because it has different reactive regions with respect to nucleophilic attacks (Flett, 1952; March, 1977; Young, 1975). The objective of this work is to rationalize the reactivity of maleic anhydride with respect to nucleophilic attacks by utilizing the fukui function of the molecule and its relation with local softness.

REACTIVE SITES IN MALEIC ANHYDRIDE

There are two kinds of active carbon atoms with respect to nucleophilic attacks in maleic anhydride: the carbonyl carbons (C_2, C_5) and the carbon atoms located in the α position to carbonyl carbons (C_3, C_4). Experimental evidence shows that, in general, hard nucleophiles interact with the carbon atom in the carbonilic group. On the other hand, soft nucleophiles

Table 1 . Reaction sites of maleic anhydride with respect to some typical nucleophiles.

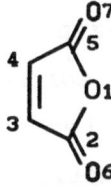

Nucleophile	Attacking point	Hardness[a]
water[b]	C_2	9.5
ammonia[c]	C_2	8.2
methylamine[c]	C_3	7.2
pyrrole[d]	C_3	5.4

a) Values in eV estimated by Pearson (Pearson, 1989: Pearson, 1990).
b) (March, 1977).
c) (Flett, 1952).
d) (Young, 1975).

react with the carbon atom in the α position. Table 1 summarizes this reactive behavior of the molecule by showing some typical reactions of maleic anhydride with nucleophiles.

FUKUI FUNCTION, CONDENSED FUKUI FUNCTION, LOCAL SOFTNESS AND LOCAL HARDNESS.

There are three basic concepts that help us to rationalize the experimental information described above: the Parr and Yang postulate (Parr, 1984), the relation between local softness and fukui function (Berkowitz et al, 1988; Lee et al, 1988; Yang et al, 1986) and the assumption that the hard-hard interaction is mainly driven by electrostatic effects (Pearson, 1973; Klopman, 1974).

Parr and Yang established that the preferred trajectory for a chemical reaction is the one that assures the maximum change in the absolute value of the chemical potential (Parr et al, 1984). In addition to that, fukui function, $f(r)$, is a local quantity that appears in the expression for the chemical potential change:

$$d\mu = \eta dN + \int f(r)\delta\omega(r)dr \qquad (1)$$

η is the global hardness (Parr et al, 1983), dN is the change in the number of electrons and $\delta\omega(r)$ is the variation of the external potential. Equation 1 has two terms, the first is related with the hardness and it is of global nature. On the other hand, the second term is associated with the fukui function, thus, it depends on the local structure of $f(r)$. In fact, the Parr and Yang postulate associates the maximum values of $f(r)$ with the preferred sites for chemical attacks. Therefore, $f(r)$ is a characteristic function of each molecule that contains information of the preferred sites for a chemical interaction; its definition is :

$$f(r) \equiv \left(\frac{\partial \rho(r)}{\partial N} \right)_{\omega(r)} = \left(\frac{\delta \mu}{\delta \omega(r)} \right)_N \quad . \quad (2)$$

It is possible to define (Parr et al, 1984; Yang et al, 1984), as a consequence of the discontinuity of the derivative with respect to the number of electrons (Perdew et al, 1982), three different fukui functions: $f^+(r)$ is the derivative in the direction of anions and it is related with nucleophilic attacks; $f^-(r)$ is related with electrophilic attacks and it is the derivative in the direction of cations; finally, $f^\circ(r)$ is the fukui function associated with free radicals reactions.

In this work we focus the attention on the nucleophilic attacks, thus, the relevant quantity is $f^+(r)$ which can be approximated by (Lee et al, 1988)

$$f^+(r) \cong \rho_{N+1}(r) - \rho_N(r) \quad . \quad (3)$$

$\rho_{N+1}(r)$ and $\rho_N(r)$ are, respectively, the densities of the anion and the neutral specie. $f^+(r)$ calculated through equation 3 is shown in figure 1. The densities were obtained from the minimum of the potential energy surface of the neutral molecule. The minimum was determined by the gradient technique (Pulay, 1977) within the Hartree-Fock-LCAO approximation by employing the MONSTERGAUS program (Peterson, 1988) with a 6-31G internal basis set.

It is evident from figure 1, that the α-carbon is related with greater values of $f^+(r)$ compared with the carbonyl carbon. Thus, according to Parr and Yang

FUKUI FUNCTION FOR MALEIC ANHYDRIDE

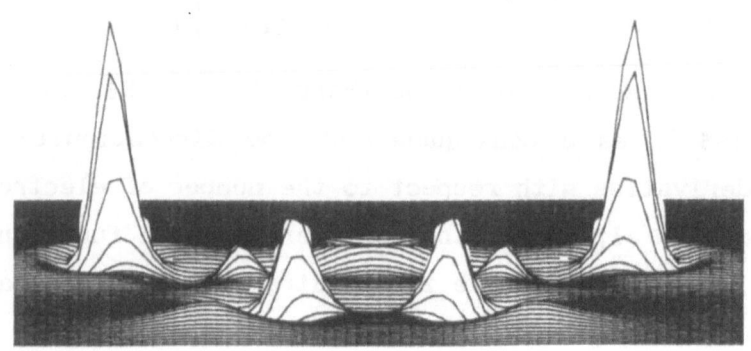

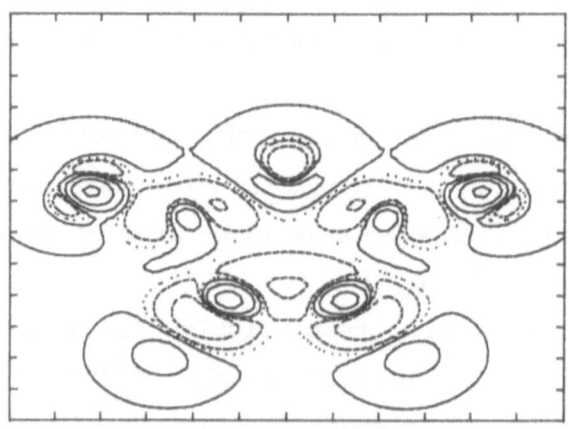

Figure 1.- Surface and contour plots of the fukui function, $f^+(r)$, of the maleic anhydride in a plane which is 0.4 atomic units above the molecular plane and it is parallel to it. The ticks in contour plots are separated by one atomic unit; the dashed lines indicate negative values, the solid lines the positive ones and the doted lines refer to the nodal surface. Positive and negative contours are associated with values that increase as they go away from the nodal surface. The values for the contours are: −0.001, −0.0045; 0.001, 0.0045, 0.01678, 0.0596.

postulate (Parr et al 1984), a nucleophilic attack will take place in the α-carbon. As it was mentioned, this is true for the interaction with soft nucleophiles. However, hard nucleophiles react with the carbonilic group and such a behavior can not be interpreted in terms of the maximum values of the fukui function.

In order to understand the interaction of the maleic anhydride with hard nucleophiles, figure 1 can be seen as the local behavior of the molecular softness. In fact, local softness, $s(r)$ is related in a simple manner to the fukui function (Berkowitz et al, 1988):

$$s(r) = Sf(r) \qquad (4)$$

S is the global softness, that is, the inverse of the global hardness, η. In this context, figure 1 tell us that the carbonyl carbons are harder than the α-carbons. In other words, α-carbons are the softest carbons in the molecule.

Therefore, if the hard and soft acids and bases principle (HSAB) (Pearson, 1973) is invoked in a local sense (Lee et al, 1988; Yang et al, 1986), it is possible to say that the soft regions of the molecule will interact with soft species; in contrast, hard areas will react with hard molecules.

Differences in softness for the two kinds of carbon atoms in maleic anhydride are also obtained through the condensed fukui functions, $f_i^+(r)$ (Lee et al, 1988; Yang et al, 1986). Namely, a reactivity parameter calculated in a similar way as fukui function, but by

Table 2. Net atomic charges and condensed fukui functions for the carbon and oxygen atoms of the maleic anhydride.

Atom[a]	Net Atomic Charge[b]	Condensed fukui function[c]
O_1	-0.6923	0.0529
O_6	-0.4658	0.1677
C_2	0.7426	0.0699
C_3	-0.2183	0.1047

a) See table 1 for atom identification.

b) Obtained from Mulliken population analysis for the neutral molecule.

c) Calculated trough equation 5.

employing gross atomic charges:

$$f_i^+(r) = q_i(N+1) - q_i(N) \tag{5}$$

q_i refers to the gross atomic charge of the i atom in the molecule as is obtained from the Mulliken population analysis (Mulliken, 1955).

Condensed fukui functions for the carbon and oxygen atoms of the maleic anhydride are shown in table 2. It is seen that the behavior of the condensed fukui functions is in agreement with the analysis of figure 1.

In addition to that, the charge distribution in the molecule is such that the carbonyl carbon has a positive character whereas the α-carbon shows a negative net charge. This can be seen in the values of net atomic charges that come from the Mulliken population analysis. Such a distribution of charge indicates that the two kinds of carbon atoms in the molecule have different characteristics from the point of view of electrostatic interactions: carbonyl carbon is more suitable for the interaction with negative species or with the negative areas of an interacting molecule.

DISCUSSION

The results of previous section allow us to say that the analysis of two local quantities, the charge distribution and the fukui function, gives a good description of the reactivity of maleic anhydride with respect to nucleophiles. In a similar manner as in perturbation approaches (Klopman, 1974), there is a distinction between the processes driven by electrostatic interaction and those related with polarization and charge transfer effects. The last type of processes are named, in perturbation theories, orbital-controlled because they are strongly related with frontier orbitals (Klopman, 1974).

In density functional theory approach, the orbital controlled processes are related with fukui function behavior. This is in agreement with the fact that the fukui function reduces to the frontier orbital density in the limit in which the relaxation effects due to charge transfer are neglected (Parr et al 1984; Yang et al, 1984); the frontier orbital related with $f^+(r)$ is the lowest unoccupied molecular orbital (LUMO), $\phi_{LUMO}(r)$, thus,

$$f^+(r) \cong \left[\phi_{LUMO}(r)\right]^2 \tag{6}$$

The resemblance of $f^+(r)$ with the LUMO density is evident by comparing figure 1 with figure 2.

The two approaches, DFT and perturbational, have similarities, however, DFT provides a more structured framework for the analysis of the HSAB principle. Specifically, DFT gives clear definitions of hardness and softness in global and local senses; moreover, both concepts are calculated in a simple manner by employing DFT definitions (Robles et al, 1984; Pearson, 1988; Pearson, 1989; Lee et al, 1988; Gázquez et al 1987).

On the other hand, local reactivity criteria estimated through equation 5, or similar expressions, are independent of orbital transformations and are not attached to an independent particle model (Lee et al, 1988). In addition to that, the densities implied in equation 5 can be evaluated by employing any suitable method.

Finally, we want to mention that previous work (Yang et al, 1986; Lee et al, 1988) combined with the results obtained here, encourage the application of DFT local reactivity concepts to understand chemical reactivity.

LUMO FOR MALEIC ANHYDRIDE

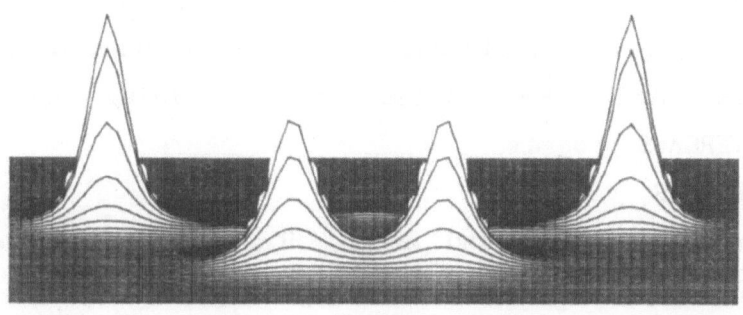

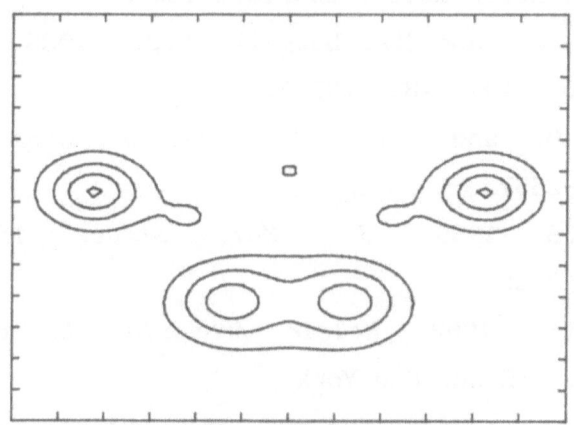

Figure 2. - Surface and contour plots of the orbital density of the lowest unoccupied molecular orbital (LUMO). See figure 1 for the explanation of the figure characteristics. The contour values are : 0.001, 0.0045, 0.01678, 0.596. The scale for the surface plot is the same of the one used in figure 1.

ACKNOWLEDGEMENTS

This work was supported in part by CONACYT through the Programa de Fortalecimiento del Posgrado Nacional. We thank Jose L. Gázquez, Andrés Cedillo and Alberto Vela for many valuable discussions. We also thank Raymond Poirier and Mike Peterson for providing us the MONSTERGAUS program.

REFERENCES

Bader, R. F. W., and Essen, H., 1984a, *J. Chem. Phys.* **80**: 1943-1960.

Bader, R.F.W., Mac Dougall, P.J., and Lau, C.D.H., 1984b, *J. Am. Chem. Soc.* **106**: 1594-1605.

Bader, R.F.W., and Mac Dougall, P.J., 1985, *J. Am. Chem. Soc.* **107**: 6788-6795. 5.

Berkowitz, M., and Parr, R.G., 1988, *J. Chem. Phys* **88**: 2554-2557.

Dewar, M.J.S., 1989, *J. Mol. Struc. (Theochem)* **200**: 301-323!

Flett, H.L., 1952, *Maleic Anhydride Derivatives*, J. Wiley & Sons, New York.

Fukui, K., 1973, *Theory of Orientation and Stereoselection*, Springer-Verlag, West Berlin.

Gázquez, J.L., Vela, A., and Galván, M., 1987, *Structure and Bonding* **66**: 79-97.

Klopman, G., 1974, *Chemical Reactivity and Reaction Paths,* (Edited by Klopman, G.), p. 55-165. John Wiley & Sons, New York.

Lee, Ch., Yang, W., and Parr, R.G., 1988, *J. Mol. Struc. (Theochem)* **163**: 305-313.

Lundqvist, S. and March, N.H., 1983, *Theory of the*

inhomogenous electron gas, Plenum Press, New York.

March, J., 1977, *Advanced Organic Chemistry,* p. 348-348. Mc Graw Hill, New York.

Mulliken, R.S., 1955, *J. Chem. Phys.* **23**: 1833-1840.

Parr, R. G., and Pearson, R. G., 1983, *J. Am. Chem. Soc.* **105**: 7512-7516.

Parr R.G., and Yang, W., 1984, *J. Am. Chem. Soc* **106**: 4049-4050.

Parr, R. G., and Yang, W., 1989, *Density Functional Theory for Atoms and Molecules,* Oxford, New York.

Pearson, R. G., 1973, *Hard and Soft acids and bases,* (Edited by Pearson, R.G.), Dowden, Hutchinson and Ross, Inc., Stroudsburg, Pennsylvania.

Pearson, R. G., 1988, *Inorg. Chem.* **27**: 734-740.

Pearson, R. G., 1989, *J. Org. Chem.* **54**: 1423-1430.

Pearson, R. G., 1990, *Coordination Chemistry Reviews* **100**: 403-425.

Perdew, J. P., Parr., R.G., Levy, M., and Baduz, J. L., 1982, *Phys. Rev. Lett.* **49**: 1691-1694.

Peterson, M. (Dept. of Chemistry, Univ. of Toronto), and Poirier, R. A. (Dept. of Chemistry, Memorial University of Newfoundland) *MONSTERGAUSS Program,* 1988 Version.

Pulay, P., 1977, *Applications of Electronic Structure Theory,* (Edited by Schaefer III, H. F.). p. 153-181. Plenum Press, New York.

Robles, J., and Bartolotti, L. J., 1984, *J. Am. Chem. Soc.* **106**: 3723-3727.

Yang, W., Parr, R. G., and Pucci, R., 1984, *J. Chem. Phys.* **81**: 2862-2863.

Yang, W., and Mortimer, W. J., 1986, *J. Am. Chem. Soc.* **108**: 5708–5711.

Young, D. W., 1975, *Heterocyclic Chemistry*, chapter v, Longman Group Limited, London.

26
"Poor-Man's Self-Consistency"

A. Pisanty, C. Amador,
and M.A. Martínez-Carrillo

Abstract. Self-consistent LSDF potentials resulting from LMTO calculations are used to start embedded-cluster calculations of electronic structure of ordered and moderately disordered solids in a Lloyd multiple-scattering formalism. This is done as a shortcut to remedy the difficulties of attaining full self-consistency in this approach. We report results obtained with this technique on the interplay between the local chemical environment and the magnetic moment of a nickel atom in the nickel-platinum system near the 3:1 composition ratio.

I. Introduction

Self-consistent calculations of the electronic structure within density functional theory are nowadays standard for almost any kind of solid. Bulk elements, bulk ordered alloys, surfaces, interfaces, "sandwiches", etc. can be treated successfully in this way. However, the study of disordered phases still remains a largely unsolved problem. Several methods have been devised in the past to compensate for the mathematical simplifications gone with the loss of crystalline periodicity; on one hand trying to recover the Bloch potential (i.e. VCA, ATA, CPA) and on the other trying to describe the solid by a model of few atoms, i.e. by means of cluster calculations.

The vast majority of calculations done on these solid systems rely on the use of density functional theory and on its simplest implementation, the local density approximation. The accuracy of this particular approach (DFT-LDA) is well known as are also its flaws. So, the main problem with the construction of the electronic potential for a disordered system stems from the lack of periodicity, and not from the treatment of many-body effects (as they are well treated by DFT-LDA). In this regard cluster calculations have proven useful, their main fault being the necessarily limited size of the cluster.

Two different approaches has been widely used in the past. In one of these, the electronic structure for a free cluster is calculated self-consistently. Since these free clusters are finite, the resulting electronic structure consists of a discrete set of levels. To compare with the electronic structure of the extended solid a broadening of these levels is required. There are two main problems with this approach: the first is the inevitable arbitrariness in the broadening process, the second the poorly met boundary conditions (as a usual calculation includes only a central atom and its two nearest neighbours layers in free space). This latter problem has even led some workers to argue that it is not the central atom in a cluster the one that represents the best the typical bulk atom, but an external one.

Alternatively, one can use embedding boundary conditions. In the simplest approach of this kind, due to Lloyd (Lloyd et al., 1972), the potential outside the region occupied by the finite cluster is a negative constant and the electrons are described as free, with plane-wave wavefunctions, outside the cluster. Thus, the resulting electronic structure consists of a continuum. The boundary conditions in this approach are a better representation of the solid, but one has to pay a price: the self-consistency process is not simple, as the constant potential acts as a source or a well of electrons.

The problem can be tackled thoroughly as in the work of Zeller and Dederichs (Zeller et al., 1979). But the resulting technique requires the use of extensive computational resources, far more costlier than the ones required by a cluster calculation in free space. Additionally, the lack of a self-consistency procedure renders the Lloyd approach poorly useful in cases in which a considerable charge transfer is present.

We propose an approximate, pragmatic scheme that can improve the results of cluster-in-condensed matter calculations. Our method is very simple: we take the atomic sphere potential from a fully converged band structure calculation — in this case using the LMTO method (Andersen, 1975; Skriver, 1984) — as the potential inside the muffin-tin sphere of an atom instead of the free-atom charge density used in the Mattheiss construction (Loucks, 1967). With this potential we start the multiple-scattering process by calculating the corresponding phaseshifts, and proceed in the usual way (Keller, 1972; Keller et al., 1980). The rest of this paper is organized as follows. In the next section we present the multiple-scattering formalism, noting that it can be regarded as a two step task: first, the construction of the monoelectronic potential, second, the calculation of the electronic propagator and from this, the calculation of the density of states. In section III, we present our results, first for the pure metals, as a means of calibration and validation of this method, and then, for the investigation of the nickel magnetic moment for different environments of the nickel atom, representing the disordered phase of Ni_3Pt. Section IV is devoted to present our conclusions.

II. The Cluster-in-Condensed-Matter Method

The cluster-in-condensed-matter method starts with the calculation of the potential. Space is divided according to the muffin-tin prescription: atomic spheres centered on the atomic positions define the regions where the potential is spherically symmetric, and an interstitial region where the potential is a constant. The atomic positions correspond to those of the particular lattice (if crystalline) or those of whatever model is being applied (if amorphous).

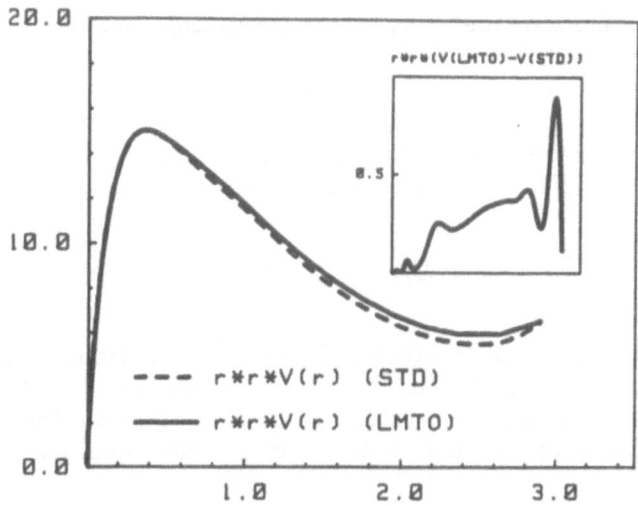

Figure 1. Muffin-tin potential obtained by the Mattheiss construction (broken line) and self-consistent potential from the LMTO calculation (full line). The inset shows the difference between these two potentials.

In the usual implementation of this method, the potential is built up from atomic charge densities calculated for free atoms. Usually these charge densities come from a DFT-LDA (Hohenberg et al., 1964; Kohn et al., 1965) — or LSDF (Kohn et al., 1965; von Barth et al., 1972; Rajagopal et al., 1973), when required — self-consistent calculation, so the use of a specific exchange-correlation functional is implied. These charge densities are then added for each point of a grid in real space. The resulting charge density (a first approximation for the density of an atom in the solid) is used to obtain an electrostatic potential via Poisson's equation. To this Coulomb potential an exchange-correlation contribution is added. Finally, the interstitial constant potential is fixed by averaging over this region. This procedure can include as many atoms as necessary for convergence.

In the present modification of the method, this procedure is changed by taking the potential obtained from a fully converged LMTO calculation. In Figure 1 plots of both potentials are shown.

The next step is the construction of the cluster on which the multiple-scattering calculation will be performed. This cluster consists of a central atom and, usually, its two nearest neighbour layers, in the positions specified by the lattice. The effect of the potential within each sphere is represented, in a partial wave analysis, by the phaseshifts, $\delta_L^i(E)$, where i is the number of the site, $L = (l, m)$ is the angular momentum index, and E is the energy.

The phaseshifts are used to construct the reactance matrix

$$k_l^i(E) = - \tan \delta_l^i(E)/\sqrt{E} . \tag{1}$$

The structural information on the cluster is contained in the free-electron propagator, $G^+(E)$, whose elements are obtained from

$$G_0^{+ij}{}_{LL'} = 4\pi \sum_{L''} C_{LL''}^{L'} i^{l''} h_{l''}^+ \left(\sqrt{E} |\mathbf{r}_{ij}| \right) Y_{L''}(\mathbf{r}_{ij}) , \tag{2}$$

where h_l^+ is a spherical Hankel function, $C_{LL''}^{L'}$ is a Gaunt number and Y_L a spherical harmonic. For the cluster, the electronic propagator is obtained by combining the structural information (from the free electron propagator) with the electronic potential information (from the reactance matrix) as

$$\mathbf{G}^+ = \mathbf{G}_0^+ + \mathbf{G}_0^+ \mathbf{k} \mathbf{G}^+ . \tag{3}$$

This equation is solved for $\mathbf{G}^+$. Finally, The density of states is obtained as

$$n(E) = -\frac{1}{\pi} \operatorname{Im} \operatorname{Tr} \mathbf{G}^+(E) . \tag{4}$$

To make the analysis of the results easier we have defined the multiple scattering ratios (MSR) as follows (Keller et al., 1980; Pisanty et al., 1980). First we consider a particular cluster, that consisting of only one atom — the so called "single-site approximation". The partial density of states for this case has a single, structureless peak centered at the resonance energy, while, for a cluster consisting of several atoms, the density of states curve presents a complex structure. As long as the multiple-scattering effects are treated exactly *for the cluster*, all the differences between these two cases arise from the bonding or multiple-scattering effects. Thus we define the MSR (r_L^i) as the ratio of the partial density of states for the atom in the cluster to the single-site density of states,

$$r_L^i(E) = \frac{\operatorname{Im} G^{+ii}{}_{LL}(E)}{\operatorname{Im} g^{+ii}{}_{LL}(E)} = \frac{n_L(E)}{n_L^0(E)} . \tag{5}$$

By comparing these quantities for different atoms in the cluster we can see the amount in which the bonding effects enhance or suppress states of a certain character at each energy. Then, by comparing these quantities for different kinds of states in the same, or different atoms, we can analyze the effects of hybridization and bonding. For instance, the effect of bonding for the d-type electrons is to split the density of states curve in bonding and antibonding sub-bands; and the comparison of s- and d-MSR's for the same atom reveals the simultaneous enhancement of these states for

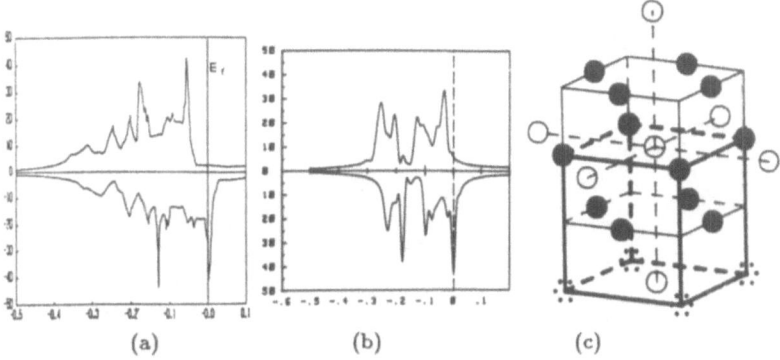

Figure 2. Local density of states for pure nickel. a) Full self-consistent LMTO calculation. b) Cluster calculation employing the same potential: central atom. c) Cluster of nineteen atoms: Heavy lines mark the fcc unit cell. Energies are in Rydberg units.

energies below the center of the d-band and the suppression of s-like states for energies above the center. These effects are often hidden in the complex DOS curves.

III. Results

We have applied the present approach to the study of the magnetic properties in the nickel-platinum alloys (Pisanty et al., 1990). The problem we addressed is mainly whether a stable magnetic moment exists or not, what its value is and, more importantly, how it depends on the local chemical environment, i. e. on the short range order.

Nickel and platinum form alloys in the whole composition range. Near the ratios 3:1, 1:1, and 1:3, ordered phases exist at low temperatures. Near the pure element limit, a disordered phase is stable at all temperatures as it is for the high-temperature region at all compositions. Pure nickel and the 3:1 compound are ferromagnetic. The disordered phase (quenched from a high-temperature solution) is also ferromagnetic up to a concentration of 60% at. Pt. The magnetic disordered state is more stable than the ordered one as the Curie temperature is always higher for the former, and as it exists up to 60% at. Pt, whereas the ordered and annealed alloys are nonmagnetic at Pt concentrations above 40%. The structures of all the phases can be derived by substitution of the face centered cubic, including the 1:1 ordered compound if tetragonal distortion is allowed for. The ordered phases 3:1, 1:1, and 1:3 have the $L1_2$, $L1_0$, and $L1_2$, structures, respectively. The disordered phase is in the A1 structure.

This phase diagram has been studied by means of the tetrahedron approximation to the cluster variation method (Dahmani et al., 1985). This method can successfully reproduce the phase diagram but a further approximation is needed: it is assumed that, in order to have a magnetic moment, a nickel atom must have at least two nickel atoms as nearest neighbours, all three being nearest neighbours. This particular assumption

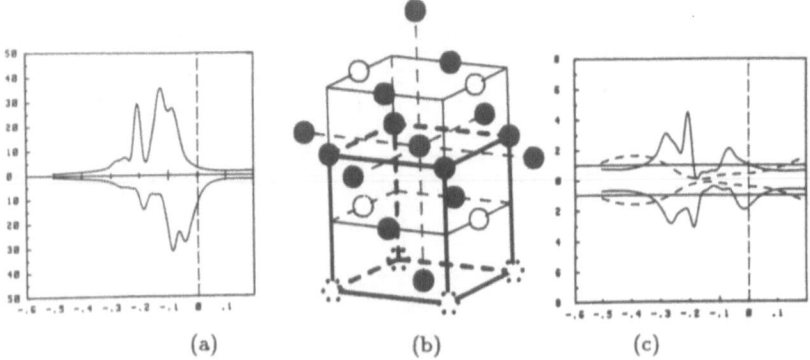

Figure 3. Results for nickel in ordered Ni_3Pt from a cluster calculation using the LMTO potential (cluster 1). a) Local density of states. b) Cluster of nineteen atoms: Heavy lines mark the fcc unit cell. c) Multiple scattering ratios. Energies are in Rydberg units.

will be investigated in the following.

In order to validate the present approach we first report results for the pure nickel metal. In Figure 2, we present our results for the local density of states curve; in figure 2a, we show the LMTO self-consistent calculation, and in figure 2b, the local density of states of the central atom in the cluster calculation performed on the cluster of nineteen atoms depicted in Figure 2c. It is apparent from the figures that the present method produces a very good approximation to the density of states curve as obtained by a band structure calculation. All of the features of that curve are reproduced by the cluster calculation: peaks, valleys, widths and heights. Also, the value of the magnetic moment is well reproduced.

Then, we shift our attention to the 3:1 compound. In Figure 3a we present the local density of states for the nickel atom resulting from the use, in the cluster method, of the LMTO self-consistent potential obtained for the Ni_3Pt ordered phase. In Figure 3b, we depict the cluster of nineteen atoms used in this calculation, and in Figure 3c, the corresponding multiple scattering ratios. Although the local density of states obtained in this fashion lacks the full complex structure of the corresponding LMTO one (not shown), the salient features again are the same, as is the resulting magnetic moment. The MSR curve reveals that the spin polarization can be ascribed almost completely to the d-electrons and that the states are almost the same for the spin up and spin down electrons in the low energy range (states mostly induced by the bonding with platinum) but are shifted in the high energy range, where the magnetization comes from. That is, there is a non-uniform exchange shift in the d band.

Our next step is to model the disordered phase. We perform calculations on several clusters that differ from the one described in that platinum atoms occupy positions substituting nickel atoms in the ordered structure. In the first two of such clusters (referred in the following as cluster 2 and 3), the overall composition is the same: $Ni(Ni_4Pt_8)Ni_6$, but the local arrangement

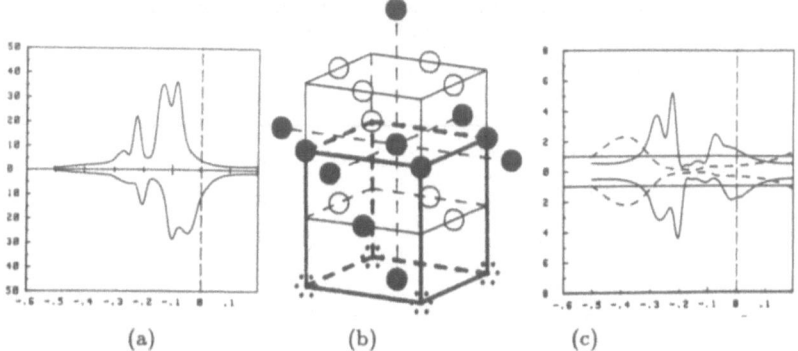

Figure 4. Results for nickel in disordered Ni_3Pt from a cluster calculation using the LMTO potential (cluster 2). a) Local density of states. b) Cluster of nineteen atoms: Heavy lines mark the fcc unit cell. c) Multiple scattering ratios. Energies are in Rydberg units.

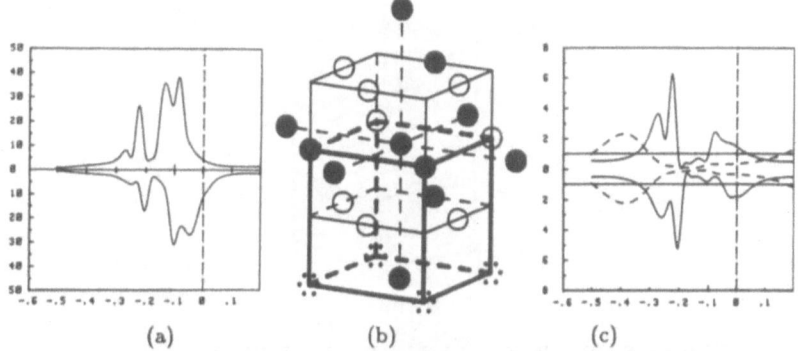

Figure 5. Results for nickel in disordered Ni_3Pt from a cluster calculation using the LMTO potential (cluster 3). a) Local density of states. b) Cluster of nineteen atoms: Heavy lines mark the fcc unit cell. c) Multiple scattering ratios. Energies are in Rydberg units.

of atoms is different. In cluster 2 the "ring" of three nickel atoms that are nearest neighbours exists, while in cluster 3 that arrangement is not present. In Figure 4 we present the density of states, the cluster and the MSR curve for cluster 2, while Figure 5 corresponds to cluster 3. Both the density of states and the MSR of these two clusters are identical, revealing that these quantities are not strongly affected by the small modification of the local structure. Also, the magnetic moment for the nickel atom is practically the same in these clusters, and slightly larger than in the cluster representing the ordered structure (see Table 1). For the MSR curve the same considerations apply as in the case of cluster 1, except that the occupied spin up states are further enhanced, thus increasing the magnetization.

The following clusters (4 and 5, Figures 6 and 7) are even richer in platinum: $Ni(Ni_2Pt_{10})Ni_6$. With only two nickel atoms in the nearest neigh-

Table 1. Magnetic moment on the central nickel atom for the different clusters studied in disordered Ni_3Pt

Cluster	$M\ (\mu_B/\text{atom})$
1^a	0.60
2	0.68
3	0.67
4	0.25
5	0.25
6	0.19

^{a} Ordered phase.

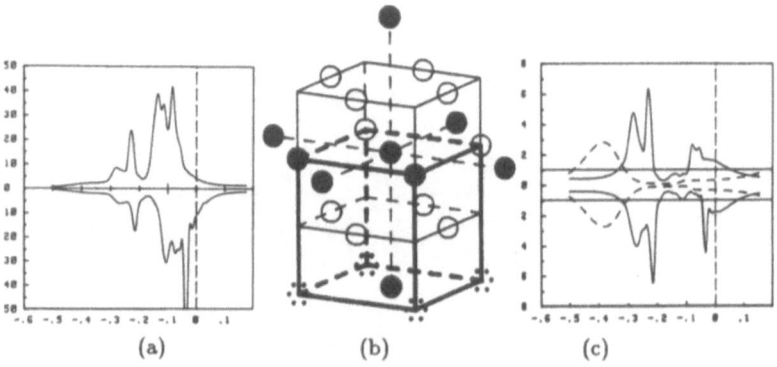

(a)   (b)   (c)

Figure 6. Results for nickel in disordered Ni_3Pt from a cluster calculation using the LMTO potential (cluster 4). a) Local density of states. b) Cluster of nineteen atoms: Heavy lines mark the fcc unit cell. c) Multiple scattering ratios. Energies are in Rydberg units.

bour layers, the magnetization drops to a 0.25 μ_B value. Again, there is no difference in magnetization between these two clusters that differ only in the relative position of one platinum atom but have the same composition. The density of states curves for spin up and spin down in the same cluster look different, but the MSR curves reveal that these differences are due to the relative position of the band centers, as the state enhancements and suppressions are similar for both spins and differ only in the high-energy region. Additionally, it is apparent that the down spin states are further enhanced (while up spin states are not) thus decreasing the magnetization.

Our last cluster (6, Figure 8) corresponds to a central nickel atom with only platinum nearest neighbours, its composition being $NiPt_{12}Ni_6$. The magnetic moment on this nickel atom is only 0.19 μ_B. Again, the smaller magnetization can be tracked to the MSR curve which shows a sharp enhancement of spin down states in the low energy region.

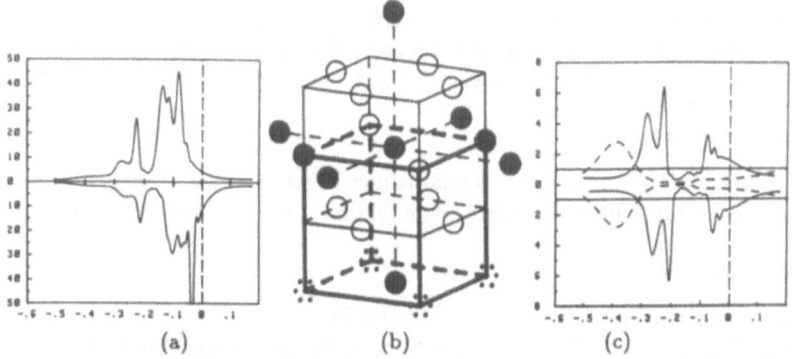

Figure 7. Results for nickel in disordered Ni_3Pt from a cluster calculation using the LMTO potential (cluster 5). a) Local density of states. b) Cluster of nineteen atoms: Heavy lines mark the fcc unit cell. c) Multiple scattering ratios. Energies are in Rydberg units.

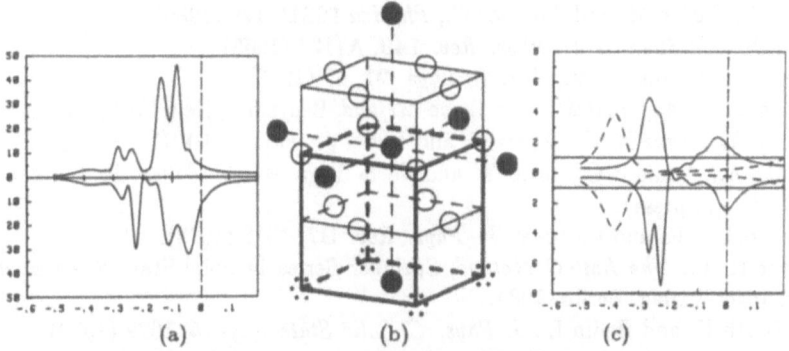

Figure 8. Results for nickel in disordered Ni_3Pt from a cluster calculation using the LMTO potential (cluster 6). a) Local density of states. b) Cluster of nineteen atoms: Heavy lines mark the fcc unit cell. c) Multiple scattering ratios. Energies are in Rydberg units.

IV. Concluding Remarks

The magnetic moment on a nickel atom in the modeled nickel platinum alloy near the composition ratio 3:1 shows a strong dependence on the local environment or short range order. This dependence is not controlled by the precise location of the neighbours, but by the overall cluster composition. Thus, the requirements on the local environment of a nickel atom are not as stiff as proposed by Dahmani et al. (Dahmani et al., 1985), and magnetic moments exist even if the three atoms ring is not present. However, there is a sharp dependence of the nickel magnetic moment on the number of nickel nearest neighbours.

The proposed prescription for a partial self-consistent treatment of clusters in condensed matter works well and allows us to reach meaningful conclusions in a realistic approach. We thus conclude it can be useful, even if not definitive, to use our "poor-man's self-consistency" in cluster models for electronic structure if means for extensive Green's function calculations (of the type of Zeller and Dederichs) are not available.

Acknowledgments

We would like to acknowledge partial support from Comité Universitario de Supercómputo and from Facultad de Química, both at UNAM.

References

Andersen O. K., *Phys. Rev.* **B12**, 3060 (1975).

Dahmani C. E., Cadeville M. C., Sanchez J. M. and Morán-López J. L., *Phys. Rev. Lett.* **55**, 1208 (1985).

Hohenberg P. and Kohn W., *Phys. Rev.* **136**, B864 (1964).

Keller J., in *Computational Methods for Large Molecules and Localized States in Solids*, Herman F., McLean A. D. and Nesbet R. K. (eds.), Plenum, New York, 1972, p. 341.

Keller J., Castro M. and Amador C., *Physica* **102B**, 129 (1980).

Kohn W. and Sham L. J., *Phys. Rev.* **140**, A1133 (1965).

Lloyd P. and Smith P. V., *Adv. in Phys.* **21**, 69 (1972).

Loucks T. L., *Augmented Plane Wave Method*, Benjamin, New York, 1967.

Pisanty A., Orgaz E., de Teresa C. and Keller J., *Physica* **102B**, 78 (1980).

Pisanty A., Amador C., Ruiz Y. and de la Vega M., *Z. Phys. B: Condensed Matter* (in press).

Rajagopal A. K. and Callaway J., *Phys. Rev.* **B7**, 1912 (1973).

Skriver H. L., *The LMTO Method*, Springer Series in Solid-State Sciences **41**, Springer-Verlag, Berlin, 1984.

Von Barth U. and Hedin L., *J. Phys. C: Solid State Phys.* **5**, 1629 (1972).

Zeller R. and Dederichs P. H., *Phys. Rev. Lett.* **42**, 1713 (1979).

27
Density Functional Calculations on Nitro Compounds (Geometries)

PATRICK K. REDINGTON AND JAN W. ANDZELM

Abstract

It is becoming generally accepted that multiconfiguration self-consistent field (MCSCF) is the minimum level of ab initio molecular orbital theory required to accurately treat nitro compounds. A good example is nitromethane where SCF calculations incorrectly predict the ground state to be a triplet. MCSCF calculations are prohibitive except for small molecules. Density functional theory may offer a computationally viable alternative. A set of small nitro compounds including, HNO_2, FNO_2, $HONO_2$, NH_2NO_2, and CH_3NO_2 is used for testing this hypothesis. Optimized geometries from density functional calculations are compared with SCF, MCSCF, and experimental results. The geometries from the density functional calculations are of MCSCF quality and they are obtained with much less computational effort.

Introduction

Molecules containing nitro groups are difficult to treat using molecular orbital theory. For example, an SCF calculation incorrectly predicts the ground state of nitromethane to be a triplet. MCSCF calculations correctly give the ground state as a singlet (Marynick et al., 1984). The problem arises because the nitro group cannot be properly described using a single determinant wave function. This notion is reinforced by the fact that valence bond calculations can also be used to obtain a good representation of nitro molecules (Marynick et al., 1985).

As has been pointed out (Kleier et al., 1984), the nitro group is similar to ozone in that they both possess a three-center, four pi-electron system. Recent work (Fournier et al., 1989) has shown that density functional theory (DFT) can provide a good description of ozone. Therefore the question naturally arises–can DFT provide an accurate description of nitro molecules?

In most cases the nitro molecules that we are interested in are far too large for MCSCF or valence bond calculations. If DFT can be shown to provide a sufficiently accurate description of nitro molecules then its reduced computational burden makes DFT a viable alternative.

In order to characterize the performance of DFT a set of small nitro molecules, for which various types of experimental data are available, was assembled. SCF, MCSCF, and DFT calculations were done, using the standard 6-31G** basis set (Hariharan et al., 1972; Francl et al., 1982). The emphasis in this work is on optimized geometry and the results are compared below.

Computational Methods

Gaussian 86 (Frisch et al., 1984) was used to do the SCF calculations. GAMESS (Dupuis et al., 1980; Schmidt et al., 1987) was used for the MC-SCF calculations. DGauss (Andzelm et al., 1990) was used for the density functional calculations. Complete geometry optimization was carried out for each calculation.

The MCSCF calculations were done using the full optimized reaction space option, also know as CASSCF. Three occupied and three virtual orbitals were included. Symmetry was not used for the geometry optimization.

The DFT calculations were done applying the VWN (Vosko et al., 1980) exchange-correlation potential, the triple zeta fitting set, and the default grid selection. In DGauss this is approximately 1000 grid points per atom. The recently developed approach for analytical gradient calculation (Fournier et al., 1989; Andzelm et al., 1990) was also used in the DGauss calculations.

Computational Results

In this section we summarize the results of our calculations. Comparisons are made between the SCF, MCSCF, DFT, and experimental results.

Optimized geometries

The optimized geometries for each of the test molecules are presented in the following tables. Comparisons are made between each of the standard geometrical parameters. Additionally a parameter ΔR is given. This parameter is used to measure the overall agreement between two sets of molecular geometries. The equation for ΔR is

$$\Delta R = \frac{\left[\sum_{i=1}^{N} (x^T + x_i^E - x_j^R)^2 \right]^{1/2}}{N}, \tag{1}$$

where

$$x^E = A x^O. \tag{2}$$

A is the Euler transformation matrix, x^T is a translation, and N is the number of atoms to match. The superscript R refers to the reference geometry, which is held fixed. The molecular geometry that is to be matched is designated by the superscript O. Rigid body motions (translations and Euler rotations) and are applied to this geometry to bring it into as close alignment as possible with the reference geometry. Computationally we do this by minimizing the value of ΔR where the Euler angles and the translations are the optimization variables (Redington, 1990).

Nitryl hydride (HNO$_2$)

Table 1: Nitryl hydride geometrical parameters (Angstroms and degrees)

Parameter	SCF 6-31G**	MCSCF 6-31G**	DFT 6-31G**
N-H	1.020	1.010	1.061
N-O	1.186	1.221	1.220
O-N-O	128.4	127.7	128.4
H-N-O-O	180.0	180.0	180.0
ΔR	1.345e-02	0.0	1.016e-02

No experimental geometry is available for nitryl hydride so the MCSCF calculation was used as the reference. The DFT calculation is seen to be in good agreement with the MCSCF results. The major disagreement is in the N-H bond length where DFT predicts it to be longer. Too long heavy atom-hydrogen bond lengths is a well known feature of DFT and is seen in the other test cases as well.

Nitryl fluoride (FNO$_2$)

The MCSCF calculation gets the N-F bond distance too short, although it is an improvement over the SCF result. The DFT calculation does much better. The N-O bond distances are also interesting. MCSCF has them slightly different while DFT gets them the same. Perhaps if the active space were made larger this discrepancy would diminish.

Nitric acid (HONO$_2$)

The major problem with the DFT calculation is that the O-H bond lengths are too long. Both the MCSCF and DFT calculations show different bond lengths for the nitro oxygens. This is as expected because the OH on one side makes them nonequivalent. This is also reflected in the slightly different O-N-O bond angles of 116 and 115.

Table 2: Nitryl fluoride geometrical parameters (Angstroms and degrees)

Parameter	SCF 6-31G**	MCSCF 6-31G**	DFT 6-31G**	Exp[a]
N-F	1.344	1.395	1.447	1.47
N-O	1.163	1.172	1.188	1.18
N-O	1.163	1.194	1.188	1.18
O-N-O	133.0	134.4	135.1	136
F-N-O-O	-180.0	-180.0	180.0	0.0
ΔR	2.641e-02	1.414e-02	4.295e-03	0.0

[a]Reference (Mirri et al., 1968)

Table 3: Nitric acid geometrical parameters (Angstroms and degrees)

Parameter	SCF 6-31G**	MCSCF 6-31G**	DFT 6-31G**	Exp[a]	Exp[b]
N1-O1	1.332	1.375	1.398	1.41	1.405
N1-O2	1.188	1.198	1.216	1.22	1.206
N1-O3	1.172	1.179	1.201	1.22	1.206
H1-O1	0.951	0.952	0.986	0.96[c]	0.96[c]
O1-N1-O2	116.0	116.2	115.2	115	116
O1-N1-O3	114.9	114.3	114.2	115	114
O2-N1-O3	129.1	129.5	130.6	130	130
H1-O1-N1	105.3	103.4	101.5	90[c]	102
H1-O1-N1-O2	0.0	0.0	0.0	0	0
ΔR	2.745e-01	2.752e-01	2.830e-01	3.432e-02	0.0

[a]Reference (Stern et al., 1960)

[b]Reference (Cotton et al., 1980)

[c]assumed

Nitramide (NH_2NO_2)

Table 4: Nitramide geometrical parameters (Angstroms and degrees)

Parameter	SCF 6-31G**	MCSCF 6-31G**	DFT 6-31G**	Exp[a]	Exp[b]
N-N	1.353	1.374	1.378	1.427	1.381
N-O	1.191	1.216	1.222	1.206 (assumed)	1.232
N-H	0.996	0.997	1.023	1.005	1.007
H-N-H	117.9	114.5	118.0	115.2	120.9
O-N-O	126.9	126.8	127.8	130.1	132.7
N-N-O	116.5	116.6	116.1	114.9	113.6
H-N-N	111.0	109.0	110.5	109.3	109.7
H-N-N-O	24.5	27.8	24.5	26.5	22.5
ΔR	1.705e-02	9.710e-02	1.370e-02	1.065e-01	0.0

[a] Reference (Tyler, 1963)

[b] Reference (Sadova et al., 1977)

Our optimized geometries are consistent with those reported in the calculations of Saxon and Yoshimine (Saxon et al., 1989). The two reported experimental geometries are quite different (Tyler, 1963; Sadova et al., 1977). The calculations are in better agreement with the geometry of Sadov et. al (Sadova et al., 1977) so we have taken it to be our reference geometry.

Nitromethane (CH_3NO_2)

DFT gives very good agreement with the experimental geometry. The C-H bond lengths are very close to the experimental values. They are still too long, but not nearly as much as with the other heavy atom-hydrogen bonds. These calculations were done at one conformation so that none of the hydrogens are exactly equivalent. The rotational barrier in nitromethane is 6 cal/mole (Tanenbaum et al., 1956) so experimentally the hydrogens are all equivalent. Mckee (McKee, 1989) reports averaged values and our geometrical parameters are in good agreement with his.

Conclusions

DFT gives geometries as good as or better than MCSCF with a far smaller computational effort. The nitromethane calculation took a few minutes of cpu time on a single processor Cray YMP. The MCSCF calculation took several hours on an IBM 3090-400. Molecules such as RDX (trinitraminocy-

Table 5: Nitromethane geometrical parameters (Angstroms and degrees)

Parameter	SCF 6-31G**	MCSCF 6-31G**	DFT 6-31G**	Exp[a]
C-N	1.479	1.477	1.480	1.489
N-O	1.191	1.222	1.225	1.224
N-O	1.191	1.202	1.225	1.224
C-H	1.078	1.076	1.103	1.088
C-H	1.078	1.080	1.099	1.088
C-H	1.078	1.078	1.099	1.088
N-C-H	107.5	108.5	108.5	107.2
N-C-H	107.5	107.1	107.2	107.2
N-C-H	107.5	107.3	108.0	107.2
O-N-O	125.7	125.5	126.6	125.3
ΔR	1.427e-02	1.440e-02	6.277e-03	0.0

[b]Reference (Cox et al., 1972)

clohexane) and HMX (tetranitraminocyclooctane) are far too large to treat with MCSCF, but quite accessible with DFT.

The major error in the DFT calculations is heavy atom-hydrogen bond lengths are predicted to be too long. Basis sets that are optimized for DFT should lessen this error and in general give improved results.

Work making a similar comparison of electronic properties is in progress. In preliminary calculations DFT is giving results comparable to MCSCF theory.

Finally we should mention that DFT was able to correctly predict that the ground state of nitromethane is a singlet. The MCSCF separation was found to be 2.52 eV (Marynick et al., 1984). The DFT values were 3.73 eV and 3.63 eV with the gradient correction.

References

Andzelm, Jan and Wimmer, Erich, 1990. Cray Research, Inc., to be published in this volume.

Cotton, F. Albert and Wilkson, Geoffrey, 1980. *Advanced Inorganic Chemistry*. John Wiley & Sons, New York, Fourth edition. page 244.

Cox, A. Peter and Waring, Stephen, 1972. *J. Chem. Soc., Faraday Trans.*, 2:1060–1071.

Dupuis, M. and Spangler, D. and Wendoloski, J.J., 1980. GAMESS. National Resource for Computations in Chemistry Software Catalog, University of California: Berkley, CA. Program QG01.

Fournier, R. and Andzelm, J. and Salahub, D.R., 1989. *J. Chem. Phys.*, 90(11):6371–6377.

Francl, M.M. and Pietro, W.J. and Hehre, W.J. and Binkley, J.S. and Gordon, M.S. and DeFree, D.J. and Pople, J.A., 1982. *J. Chem. Phys.*, 77:3654.

Frisch, M.J. and Binkley, J.S. and Schlegel, H.B. and Raghavachari, K. and Melius, C.F. and Martin, R.L. and Stewart, J.J.P. and Bobrowic, F.W. and Rohlfing, C.M. and Kahn, L.R. and Defree, D.J. and Seeger, R. and Whiteside, R.A. and Fox, D.J. and Fleuder, E.M. and Pople, J.A., 1984 *Gaussian* 86. Carnegie-Mellon Quantum Chemistry Publishing Unit, Pittsburgh PA, 1984.

Hariharan, P.C. and Pople, J.A., 1972. *Chem. Phys. Lett.*, 66:217.

Kleier, Daniel A. and Lipton, Mark A., 1984. *J. Mol. Struct. (Theochem)*, 109:39–49.

Marynick, Dennis S. and Ray, Asok K. and Fry, John L. and Kleier, Daniel A., 1984. *J. Mol. Struct. (Theochem)*, 108:45–48.

Marynick, Dennis S. and Ray, Asok K. and Fry, John L., 1985. *Chem. Phys. Lett.*, 116(5):429–433.

McKee, Michael L., 1989. *J. Phys. Chem.*, 93(21):7365–7369.

Mirri, A.M. and Cazzoli, G. and Ferretti, L., 1968. *J. Chem. Phys.*, 49(6):2775–2780.

Redington, Patrick K., March 1990. MOLFIT Version 1.0. Hercules Physics Division Technical Report.

Sadova, N.I. and Slepnev, G.E. and Tarasenko, N.A. and Zenkin, A.A. and Vilko, L.V. and Shishkov, I.F. and Pankrushev, Yu. A., 1977. *Zh. Strukt. Khim*, 18:865.

Saxon, Roberta P. and Yoshimine, Megumu, 1989. *J. Phys. Chem.*, 93(8):3130–3135.

Schmidt, M.W. and Boatz, J.A. and Baldridge, K.K. and Koseki, S. and Gordon, M.S. and Elbert, S.T. and Lam, B., 1987. *QCPE Bulletin*, 7:115.

Stern, S. Alexander and Mullhaupt, J.T., 1960. *Chem. Rev.*, 60:185–207.

Tannenbaum, Eileen and Myers, Rollie J. and Gwinn, William D., 1956. *J. Chem. Phys*, 25(1):42–47.

Tyler, J.K., 1963. *J. Mol.Spect.*, 11:39–46.

Vosko, S.J. and Wilk, L. and Nusair, M., 1980. *Can. J. Phys.*, 58:1200–1211.

28
Application of Local Density Functional Theory to the Study of Chemical Reactions

J.M. Seminario, M. Grodzicki, and P. Politzer

Introduction

It is now generally acknowledged that Local Density Functional (LDF) theory (Parr et al., 1989) yields descriptions of molecules and solids in their electronic ground states that are often superior to those obtained by standard ab initio Hartee-Fock (HF) theory, provided that oversimplified approximations to the potential (e.g. the muffin-tin potential) are avoided (Dahl et al., 1984). However it is not clear to what extent this conclusion remains valid when the system under consideration is not in its ground state. With regard to chemically significant problems, it is particularly interesting to determine whether LDF methods are capable of producing reliable descriptions of reaction pathways, i.e. transition states and energy barriers. With very few exceptions, LDF techniques have not previously been applied to the investigation of chemical reactions.

Accordingly, the first part of this contribution is devoted to a systematic comparison of HF, post-HF and experimental data with the results of LDF calculations for a series of symmetry-allowed isomerization reactions. We subsequently look at a symmetry-forbidden process: the cycloaddition of methyleneimine to diazacyclobutane.

The ab initio results are either taken from the literature (Hehre et al, 1986; Whiteside et al) or have been computed with GAUSSIAN 88 (Frisch et al, 1988). All LDF calculations have been carried out with the code DMol (Delley, 1990), which is essentially based on a three-dimensional numerical integration scheme for the matrix elements.

Isomerization Reactions

In comparing LDF, HF and post-HF results, it is most useful to consider reactions that can be handled by the latter at a reasonably high level of sophistication. This requirement is satisfied by isomerization processes of molecules containing just two first-row atoms.

Geometries for these systems have been optimized at the HF/6-31G* level (Whiteside et al), and energy barriers have been computed that include many-body corrections up to fourth-order Møller-Plesset perturbation theory (MP4), with a basis set extended to 6-31G** (Hehre et al, 1986; Whiteside et al). Our LDF results were obtained with a numerical basis set of double-zeta quality including polarization functions (DNP), and using HF/6-31G* structures, since DMol (like most currently available DF codes) does not allow full geometry optimization.

The results are presented in Tables 1-7. A general trend is that the calculated energy barriers are considerably overestimated in the HF approximation, but decrease with an increasing level of many-body corrections. The LDF values are even lower than the MP4. Some of the individual reactions will now be discussed in more detail.

Tables 1-7. Selected local minima and transition states (TS) for isomerization reactions. HF/3-21G results are based on 3-21G optimized geometries; the other results were obtained with HF/6-31G* geometries. Energies are in kcal/mole relative to the ground state.

Table 1.

C_2H_2	Acetylene HC≡CH	Transition State	Vinylidene H $\ \ \ \ C=C:$ H
HF/3-21G	0	64	39
HF/6-31G*	0	49	34
MP2/6-31G**	0	51	49
MP3/6-31G**	0	49	42
MP4/6-31G**	0	48	44
DMOL/DNP	0	45	47

Table 2.

C_2H_4	Ethylene $H_2C=C_2H$	Transition State	Ethylidene $H_3C—\ddot{C}H$
HF/3-21G	0	94	73
HF/6-31G*	0	81	69
MP2/6-31G**	0	81	83
MP3/6-31G**	0	81	79
MP4/6-31G**	0	80	80
DMOL/DNP	0	74	84

Table 3.

CHN	Hydrogen Cyanide H—C≡N	Transition State H $C≡N$	Hydrogen isocyanide H—N≡C:
HF/3-21G	0	68	9
HF/6-31G*	0	51	11
MP2/6-31G**	0	53	19
MP3/6-31G**	0	51	16
MP4/6-31G**	0	49	16
DMOL/DNP	0	44	12

Table 4.

CH$_2$O	Formaldehyde H$_2$CO	Transition State	*trans*-hydroxy-methylene	*cis*-Hydroxy-methylene
HF/3-21G	0	--	47	54
HF/6-31G*	0	101	49	55
MP2/6-31G**	0	90	58	63
MP3/6-31G**	0	94	54	60
MP4/6-31G**	0	89	56	61
DMOL/DNP	0	--	54	58

Table 5.

CH$_3$N	Methylene-imine H$_2$C=NH	Transition State 1 H$_2$C=N	Methyl-nitrene H$_3$C—N:	Transition State 2	Amino-methylene HC—NH$_2$
HF/3-21G	0	88	71	102	27
HF/6-31G*	0	77	78	95	33
MP2/6-31G**	0	83	98	91	41
MP3/6-31G**	0	82	91	92	39
MP4/6-31G**	0	81	92	90	39
DMOL/DNP	0	75	88	75	25

Table 6.

N$_2$H$_2$	*trans*-Diazene	Transition State	Aminonitrene H$_2$N—N:	*cis*-Diazene
HF/3-21G	0	90	9	8
HF/6-31G*	0	89	19	7
MP2/6-31G**	0	78	28	6
MP3/6-31G**	0	83	26	6
MP4/6-31G**	0	78	26	6
DMOL/DNP	0	64	23	4

Table 7.

CH$_5$N	Methylamine H$_3$C—NH$_2$	Transition State	H$_2$C$\cdots$NH$_3$ (complex)
HF/3-21G	0	90	66
HF/6-31G*	0	96	72
MP2/6-31G**	0	90	74
MP3/6-31G**	0	92	74
MP4/6-31G**	0	90	73
DMOL/DNP	0	79	69

Table 1 represents the transition from acetylene to vinylidene. The activation energy decreases with the inclusion of many-body corrections, to the point that the MP4/6-311G** //MP2/6-31G* transition state is only 1 kcal/mole higher than the vinylidene (Hehre et al, 1986). It has therefore been argued that vinylidene may be a saddle point, due to the degenerate rearrangement of the two hydrogens. The LDF results seem to confirm this suggestion, placing vinylidene 2 kcal/mole above the transition state, and it gains additional support from the fact that vinylidene has not yet been detected experimentally. A similar situation is observed in the isomerization of ethylene to ethylidene (Table 2). While a local minimum exists for ethylidene on the HF potential surface, both MP4 and LDF results suggest that it corresponds to a saddle point.

Table 3 deals with the transformation of hydrogen cyanide to the isocyanide. Both iso-mers have been characterized experimentally, with a difference in energy of 10.3 kcal/mole (Maki et al., 1981), but the barrier height has not yet been measured. Both the LDF and the HF results are close to this experimental value, whereas the post-HF calculations overestimate the energy difference by about 50%. The LDF barrier is smaller than the MP4, in accordance with the general trend mentioned above.

Another system for which experimental information is available is formaldehyde (Table 4). The energy difference to the next stable isomer, trans-hydroxymethylene, is 54 kcal/mole, as measured by ion cyclotron double-resonance spectroscopy (Pau et al., 1982). This value is well reproduced by the many-body and LDF methods, while the HF result is 10% smaller.

Ring Dissociation of 1,3-Diazacyclobutane

The reactions that have been discussed so far are all symmetry-allowed. Now we shall present results for a symmetry-forbidden process, the dissociation of 1,3-diazacyclobutane to methyleneimine:

$$\text{(ring structure)} \longrightarrow 2\ H_2C=NH \tag{1}$$

This reaction is structurally analogous to the cycloaddition of two ethylene molecules to form cyclobutane.

Reaction (1) can be analyzed computationally by choosing the angle α, defined in Figure 1, as the reaction coordinate. As α decreases, the distance d_1 increases, leading eventually to the desired reaction products. For each α, a full geometry optimization must be performed in order to determine the correct transition state; the details of this procedure have been described earlier (Grodzicki et al., 1990). The molecular geometries along the

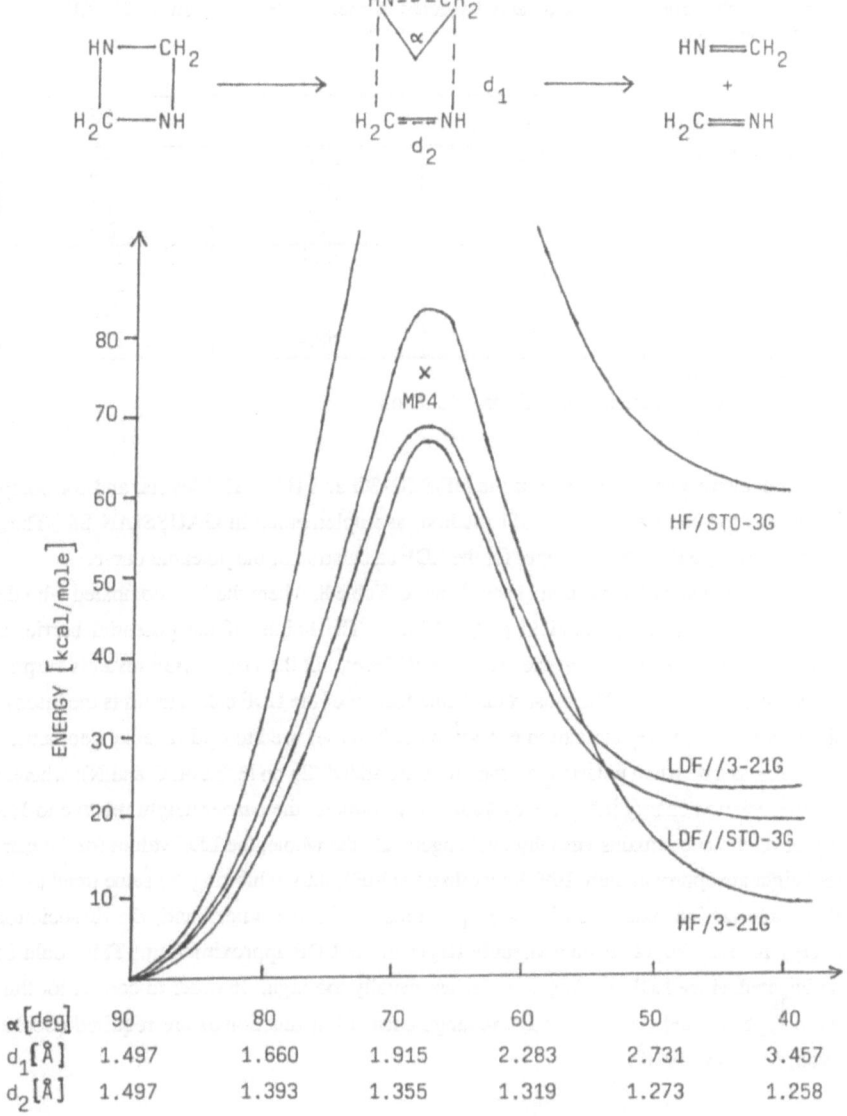

Figure 1. Potential curves for the ring decomposition of 1,3-diazacyclo-butane. For each value of the reaction coordinate α, the optimized distances d₁ and d₂ are given. Structures corresponding to the key points on the potential curve are displayed on the top.

Table 8. Activation energy, E_{act}, and dissociation energy, ΔE, for reaction (1). All energies are given in kcal/mole relative to 1,3-diazacylocbutane.

Table 8.

Methods/ Basis set	Geometry optimization	E_{act}	ΔE
HF/ STO-3G	STO-3G	151.0	59.8
HF/3-21G	3-21G	84.2	9.2
HF/6-31G*	3-21G	78.3	4.1
MP4/6-31G	3-21G	76.0	2.3
LDF/DN[a]	AM1	66.8	19.2
LDF/DN[a]	STO-3G	67.9	20.7
LDF/DN[a]	3-21G	68.5	24.7
LDF/DNP[b]	3-21G	68.4	25.8

[a]DN = double numerical.

[b]DNP = double numerical + polarization function.

potential curve were optimized at the HF/STO-3G and HF/3-21G levels, and for comparison, by the semi-empirical AM1 method, as implemented in GAUSSIAN 88. These geometries were then used as input for the LDF calculation of the potential curve.

The LDF results for reaction (1) are listed in Table 8, where they are compared with the HF and MP4 values (Grodzicki et al., 1990). The height of the potential barrier is considerably overestimated at the HF/STO-3G level, but the HF/6-31G* result is surprisingly close to the MP4. The most remarkable feature of the LDF calculations is the insensitivity of the calculated activation energy to the basis set used to optimize the geometries. The same is true when polarization functions are added (2p on H, 3d on C and N); whereas the dissociation energy increases by about 50 kcal/mole, the barrier height relative to 1,3-diazacyclobutane remains virtually unchanged. On the whole, the LDF values for the barrier height are approximately 10% lower than the MP4, thus exhibiting the same trend as for the isomerization reactions discussed previously. On the other hand, the dissociation energy for reaction (1) is considerably larger in the LDF approximation. This could be anticipated, since LDF binding energies are usually too high. In order to correct for this, more sophisticated (i.e. nonlocal) exchange-correlation functionals are required (Becke, 1988; M. Levy, 1990).

Summary

Our comparative analysis shows that LDF methods are consistent with the most sophisticated many-body techniques currently available, even for the description of chemical reactions. Furthermore, with regard to the discrepancies that do exist between the results obtained by these types of approaches, it is not obvious that the many-body values are necessarily the more accurate ones, since in the few instances for which experimental data are available, these slightly favor the LDF. When it is considered that the LDF requirements of CPU time and memory are orders of magnitude smaller than for the various many-body methods, it seems evident that LDF theory represents a practical first-principles route to the study of large molecules and chemical reactions that are beyond the scope of even HF theory.

Acknowledgement

We greatly appreciate helpful discussions with Dr. Jane S. Murray. We also thank the Office of Naval Research for its support through contract # N00014-85-K-0217, the San Diego Supercomputer Center for a generous grant of time on the CRAY Y-MP8/864, and Biosym Technologies, Inc. for the use of DMol.

References

Becke, A., 1988, *J. Chem. Phys.* 88:1053.

Dahl, J. P., and Avery, J. (eds.), 1984, *Local Density Approximations in Quantum Chemistry and Solid State Physics*, Plenum Press, New York and London.

Delley, B., 1990, *J. Chem. Phys.* 92: 508; Code "Dmol" available from Biosym Technologies Inc.

Frisch, M. J., Head-Gordon, M., Schlegel, H. B., Raghavachari, K., Binkley, J. S., Gonzalez, C., Defrees, D. J., Fox, D. J., Whiteside, R. A., Seeger, R., Melius, C. F., Baker, J., Martin, R., Kahn, L. R., Stewart, J. J. P., Fluder, E. M., Topiol, S., and Pople, J. A., 1988, *GAUSSIAN 88*, Gaussian Inc., Pittsburgh, PA.

Grodzicki, M., Seminario, J. M., and Politzer, P., 1990, *Theor. Chim. Acta* (in press); 1990, *J. Chem. Phys.* (in press).

Hehre, W. J., Radom, L., Schleyer, P.v.R., and Pople, J. A., 1986, *Ab Initio Molecular Orbital Theory*, Wiley Interscience, New York, section 7.6.

Levy, M., 1990, this volume.

Maki, A., and Sams, R., 1981, *J. Chem. Phys.* 75:4178.

Parr, R. G., and Yang, W., 1989, *Density Functional Theory of Atoms and Molecules*, Oxford Univ. Press.

Pau, C. F., and Hehre, W. J., 1982, *J. Phys. Chem.* 86:1252.

Whiteside, R.A., Frisch, M. J., and Pople, J. A. (eds.), *The Carnegie-Mellon Quantum Chemistry Archive*, Carnegie-Mellon Univ., Pittsburgh.

29
Pauli Principle for Heliumlike Atoms

D.-R. Su

The layout of the periodic table of elements is governed by the Pauli exclusion principle. For heliumlike atoms in their ground state, for example, the two electrons must be of the opposite spin with zero probablility for the two electrons to have their spins pointing in the same direction. There are then no possibilities for the heliumlike atoms to exhibit paramagnetism in their ground state. In the density-functional theory (DFT), which is a topic of this book, the spin coupling is in some cases taken into consideration. Developments in the theory allow us to account for magnetism of many-electron atoms. For such a system, the magnetic properties of the ground state should be introduced so that the violations of the Pauli principle are unavoidable. The Pauli paramagnetism is "free electron" in nature and is treated in a similar way to the single particle approximations of DFT. Thus the parallel spin case might then arise as an artifact resulting in violations of the Pauli principle. This paper explains them briefly, and calculations for several heliumlike atoms are presented as illustrations. The final conclusion is that this error within DFT treatment is only substantial for H^- ion. In the light of these results the generally accepted notion of no violations for heliumlike atoms other than H^- is still valid within experimental error (Bethe and Salpeter, 1957; Goeke et al, 1987). Magnetic effects are expressible only as spin density defined as follows (Izuyama et al, 1963; Su and Wu, 1975)

$$m(r) \equiv \psi^+(r)\vec{\sigma}\psi(r) \tag{1}$$

This kind of definition, different from the other group, is convenient to have a simple z-component of $m(r)$

$$m(r) \equiv m_z(r) = n^+(r) - n^-(r) \tag{2}$$

for spin up (down) electron density $n^+(n^-)$. Combined with the total electron density at r

$$n = n^+ + n^- \tag{3}$$

it yields the simple expressions for n^+ and n^-

$$n^+ = (n + m)/2 \tag{4a}$$

$$n^- = (n - m)/2 \tag{4b}$$

Single particle approximation in DFT is used to calculate the energy functional $E_0[n, m]$ of the generalized Thomas-Fermi-Dirac model (TFD),

$$
\begin{aligned}
E_0[n, m] = \int \nu(r)n(r)dr + \mu_B H \int m(r)dr \\
+ A_K \int [(n+m)^{5/3} + (n-m)^{5/3}]dr \\
- A_E \int [(n+m)^{4/3} + (n-m)^{4/3}]dr
\end{aligned} \tag{5}
$$

Here we incorporated the Fermi concept of "electron sea" spin up (n^+) and spin down electrons (n^-) just like in the Pauli paramagnetism. Note (Su, 1987) that the Fermi momenta should be

$$
k_F^{\pm} = (6\pi^2 n^{\pm})^{1/3} \tag{6}
$$

instead of the usual $k_F = (3\pi^2 n)^{1/3}$. A factor of 1/2 will appear in the kinetic energy functional, etc., if (6) is not used. The coefficients in (5) are given as

$$
A_K \equiv 3^{3/5}\pi^{4/3}/20 \tag{7a}
$$

$$
A_E \equiv 3^{4/3}\pi^{1/3}/8 \tag{7b}
$$

In (5) H denotes the uniform magnetic field exerted. In the original TFD, the external electrostatic potential $\nu(r)$ represents the total potential, including the part of the electron-electron (hartree) term. Here, according to DFT, the $e - e$ term is excluded and the above model is called generalized Thomas-Fermi-Dirac model to distinguish it from the original TFD. Furthermore from examination of the external electrostatic term (the first term) in (5), (Su, 1987) yields that DFT is a general response theory. In usual linear response theory (Cottam and Maradudin, 1984) the interaction is linearly proportional to the external source, $\nu(r)$. In this case, $n(r)$ is the response. This means that every $\nu(r)$ determines one unique $n(r)$ (ν-representability). In case of DFT, $n(r)$ is an independent variable and other terms in the energy functional are functionals (or dependent variables) of $n(r)$. Consequently, DFT can be classified as a general response theory (Su, 1987). The same idea is applied to the second term in (5). For the uniform magnetic field H only a single component $m = m_z$ of the vector $m(r)$ will appear, i.e., only $m(r)$ responds to H. So only $m(r)$ of $m(r)$ should be considered as independent variable, but not $m_x(r)$ or $m_y(r)$. It stems from the above that the theory should involve functionals of only two variables $n(r)$ and $m(r)$.

Now we introduce the spin coupling or dipole-dipole term in the Breit equation in Pauli approximation for heliumlike atoms (Bethe and Salpeter, 1957), viz.

$$
H_{d-d} = -\frac{8\pi}{3}\mu_B^2(\vec{\sigma}_1 \cdot \vec{\sigma}_2)\delta(r_1 - r_2) \tag{8}
$$

between two electrons. When deriving H_{d-d}, it is obvious that it arises from the $e-e$ coulomb interaction $|r_1 - r_2|^{-1}$ near the contact moment $r_1 \approx r_2$ (Coulomb hole conjecture). The contribution of H_{d-d} is positive (repulsive) for spins oriented antiparallely and negative (attractive) for spins in parallel. Inclusion of H_{d-d} introduces a source inducing parallel spin orientation which drives the system towards lower energy values. It results in an energy shift $4.80\mu_B^2$ a.u. (Bethe and Salpeter, 1957) for He atom whose spins are antiparallel. It can be shown (Su and Liu, 1988) that within the DFT approach it has a contribution of only $-3.5 \times 10^{-4}\mu_B^2$ a.u. This means that the DFT result incorporates a component which corresponds to two electrons in the He atom having spins parallel; i.e., a violation of the Pauli principle. This will be made clear below. As can be seen from the form of H_{d-d} in (8), the contribution is due to two parallel spins (dipoles) approaching to each other (Fermi hole conjecture). Many experimental attempts have been reported recently (Greenbery and Mohapatra, 1989) for possible violations of the Pauli principle. These reports focused on metals containing large number of electrons, say 40A through Cu for 272h. However no success has been reported yet.

Next we have to express H_{d-d} in (8) in the language of DFT (Su, 1987; Su and Liu, 1988) with only $m(r)$ being an independent variable,

$$E_{d-d}[m] = -\frac{8\pi}{3}\mu_B^2 \frac{1}{2} \int dr m^2(r). \tag{9}$$

We add this functional to $E_0[n, m]$ given above, but in order to obtain a total energy functional we still need a term representing the hartree electron-electron interaction, viz.

$$E_{Hartree}[n] = \frac{1}{2} \iint \frac{n(r)n(r')}{|r-r'|} dr dr'$$

for all the charges. To do that, we shall divide the whole volume of charges into two parts. This introduces several restrictions similar to DFT formalism for solids. Before going any further, we will describe first how to use the above formulations in single particle approximations. We already have derived the total energy functional from (5) and (9)

$$E[n, m] = E_0[n, m] + E_{Hartree}[n] + E_{d-d}[m] \tag{10}$$

Now, we shall apply a variational principle to (10)

$$\frac{\delta E[n, m]}{\delta m} = \mu_B H \tag{11}$$

using the first law of thermodynamics

$$\delta Q = dE - \mu_B H dm \tag{12}$$

for an adiabatic process. In this case we apply the principles of statistical thermodynamics to the variation of the magnetization, δm for m. In this case we need a canonical ensemble with a heat reservoir, or a microcanonical ensemble (Kittel, 1958). As (12) is good only for canonical ensemble we must treat our problem with canonical ensemble inside as a subsystem of the total microcanonical ensemble. The complete microcanonical ensemble consists of two electrons in heliumlike atoms. We choose a portion of it, in the sense of density, with a magnetization m. This part is our canonical ensemble and behaves according to (11). Also (11) determines the magnitude of m or its equivalent. We illustrate this procedure by the following first order theory.

For the two electrons in heliumlike atoms in the ground state, spatial parts of the wave function have the same form (we use atomic units) (Hartree, 1957):

$$|P(r)|^2 = (Z_{eff}^3/\pi)\exp(-2Z_{eff}r) \tag{13}$$

evaluated variationally with the effective charge value

$$Z_{eff} = Z - 5/16 \tag{14}$$

for a heliumlike atom with atomic number Z. For the whole electron system the total density is obtained as

$$\begin{aligned} n_t(r) &= \sum_{}^{occ} |\psi(r)|^2 \\ &= (2Z_{eff}^3/\pi)\exp(-2Z_{eff}r) \end{aligned} \tag{15}$$

Now we assume that there is a part of $n_t(r)$ representing a subsystem. Within this subsystem, the density $n(r)$ and the spin density $m(r)$ are involved in the magnetic entreaties (active magnetic response) and become violations of the Pauli principle. We shall call this subsystem the violation canonical ensemble or the spin hole (SH). It seems plausible to put the densities proportional to $n_t(r)$ globally as,

$$n(r) = (2NZ_{eff}^3/\pi)\exp(-2Z_{eff}r) \tag{16}$$

$$m(r) = (2MZ_{eff}^3/\pi)\exp(-2Z_{eff}r) \tag{17}$$

The proportional constant N varies from 0 to 1. If DFT evaluated N to be zero it would mean the failure of theory in this respect as it does not predict the presence of violation canonical ensemble or SH. On the other hand, if N is unity the whole system or all the electron density is involved in the violation canonical ensemble, i.e., is SH. The other proportional constant M varies and can assume values from -1 to 1. If it is ± 1, the two electrons in heliumlike atoms have their spins oriented in parallel. If it is zero, no magnetic effects occur and the two electrons have spins antiparallel.

Substituting (16) and (17) into (10) the total energy functional is obtained as a function of N and M

$$E[N, M] = \int \nu(\mathbf{r})n(\mathbf{r})d\mathbf{r} + 2\mu_B H M + E_{Hartree}[N] + G[N, M] \quad (18)$$

$$
\begin{aligned}
G[N, M] = &\frac{3}{5}A[(N + M)^{5/3} + (N - M)^{5/3}]/2^{2/3} \\
&- \frac{3}{4}B[(N + M)^{4/3} + (N - M)^{4/3}]/2^{1/3} \\
&- \frac{1}{2}CM^2
\end{aligned} \quad (19)
$$

$$A \equiv 1.21Z_{eff}^2 \quad (20)$$

$$B \equiv 0.450Z_{eff} \quad (21)$$

$$C \equiv \frac{4}{3}Z_{eff}^3 \mu_B^2 \quad (22)$$

The variational principle (11) is changing to

$$\frac{\delta E[N, M]}{\delta M} = 2\mu_B H \quad (23)$$

As a consequence we obtain the folowing algebraic equation:

$$
\begin{aligned}
&A[(N + M)^{2/3} - (N - M)^{2/3}]/2^{2/3} - B[(N + M)^{1/3} \\
&- (N - M)^{1/3}]/2^{1/3} - CM = 0
\end{aligned} \quad (24)
$$

Since Bohr magneton is small ($\mu_B = (276)^{-1}a.u. = (2\alpha)^{-1}$) we have:

$$A \sim B \gg C$$

Further DFT is a "ground state theory". The ground states of ferromagnets correspond to full magnetization and saturation result supplies the following relation:

$$M = -N \equiv M_s \quad (25)$$

The negative sign is due to the fact that the spin down electron is the one which has lower energy in the uniform magnetic field. Then (24) becomes

$$A(-M_s)^{2/3} - B(-M_s)^{1/3} + C(-M_s) = 0 \quad (26)$$

Acceptable solutions of this cubic algebraic equation are

$$M_s = 0 \quad (27)$$

	H^-	He	Li^+	Be^{++}	$B^{(3+)}$
$|M_s|$	0.1538	0.0104	2.58×10^{-3}	9.97×10^{-4}	4.85×10^{-4}

Table I: The values of $|M_s|$ for various He-like atoms

or

$$M_s = -[\frac{B}{A} - \frac{B^2}{A^3}C + O(C^2)]^3$$
$$= -(0.051/Z^3_{eff})(1 - 1.2\mu_B^2) \tag{28}$$

A list of values M_s from (28) for various heliumlike atoms is given in Table I. We can see that the effect is significant only for hydrogen negative ion and should be reflected in experimental measurements (Goeke et al, 1987) and the Kohn's concepts on DFT.

One may ask a question why the vanishing value of M_s given in (27) is not discussed. To answer this question, the author is investigating the hyperfine interaction of He-like atoms,(Su, 1990)

$$H_F = \sum \frac{8\pi}{3}\mu_B(\vec{\sigma}_i \cdot \vec{\mu})\delta(r_i) \tag{29}$$

Here μ is the nuclear magnetic moment. In DFT it is represented as:

$$E_F[m] = \frac{16\pi}{3}\mu_B \int m(r)\mu\delta(r)dr$$
$$= \frac{16\pi}{3}\mu_B\mu m(r=0) \tag{30}$$

The total energy functional in (10) becomes

$$E_1[n, m] = E_0[n, m] + E_{Hartree}[n] + E_{d-d}[m] + E_F[m] \tag{31}$$

Similarly to (19), the functional $G[N, M]$ in (31) for $E[N, M]$ becomes

$$G_1[N, M] = G[N, M] + \frac{16}{3}\mu_B\mu M \tag{32}$$

The variational principle (23) leads to the following algebraic equation

$$A[(N + M)^{2/3} - (N - M)^{2/3}]/2^{2/3} - B[(N + M)^{1/3}$$
$$- (N - M)^{1/3}]/2^{1/3} - CM + F = 0 \tag{33}$$

for the coefficient

$$F = \frac{16}{3}Z^3_{eff}\mu_B\mu \tag{34}$$

Now, the orders of magnitude are:

$$A \sim B \gg C \gg F \gg C^2$$

For the ground state, the "saturated" solution in DFT of (33), $M = -N \equiv M_{s1}$

$$A(-M_{s1})^{2/3} - B(-M_{s1})^{1/3} + CM_{s1} - F = 0 \tag{35}$$

The acceptable solutions of (35) for $(-M_{s1})^{1/3}$ are found from Cardan's rules, and after expanding by the orders of magnitude of the coefficients we get

$$(-M_{s1})^{1/3} = B^2 C/A^3 - F/B + O(\mu_B^4) \tag{36}$$

or

$$(-M_{s1})^{1/3} = B/A - B^2 C/A^3 + F/B + O(\mu_B^4) \tag{37}$$

None of the solutions are zero if $F \to 0$. Examination of these solutions are out of the scope of this article. The conclusion however is that the vanishing solution such as given in (27) should not exist in this case.

Another interesting problem is: Is choosing proportional constants N and M in (16) and (17), respectively, a correct approach? This transforms the variational principle $\delta E[n, m]/\delta m$ to $\delta E[N, M]/\delta M$. This is a common problem in computational physics. In this way local variations are changed to the global ones. In other words, the "variational principle" is changed to "variations of constants or parameters". We illustrate our answer by $E_1[n, m]$ given in (31) and $G_1[N, M]$ given in (32). When performing the exact variation of $E_1[n, m]$ locally using variational calculus, we get

$$\frac{\delta E_1[n, m]}{\delta m} = \mu_B H + \frac{5}{3} A_K [(n + m)^{2/3} - (n - m)^{2/3}] - \frac{4}{3} A_E [(n + m)^{1/3}$$

$$- (n - m)^{1/3}] - \frac{8\pi}{3} \mu_B^2 m + \frac{16\pi}{3} \mu_B \mu \delta(r) \tag{38}$$

Any extremities here cause infinities at $r = 0$. It results in $m = n^+ - n^-$ to be infinite somewhere in the space. As a consequence by using the condition (11) above, we get finally a saturation solution for the spin density

$$m_s = -[0.8 A_E/(A_K 2^{1/3}) + 32\pi A_E^2 2^{2/3} \mu_B^2/(125 A_K^3) + O(\mu_B^4)]^3$$
$$+ (2\mu/\mu_B)\delta(r)$$

Although it is natural to have a point nuclear spin hole at the origin, however, the first term is constant over the whole space including the region of infinite r. Comparing it to the wave function $|P(r)|^2$ in (13) we see that it is non-physical. We must "smooth" the δ-function for the hyperfine interactions or, alternatively, we can solve this problem by global variation as is popularly done,

$$\frac{\delta E_1[N, M]}{\delta M} = \frac{\delta G_1[N, M]}{\delta M} + 2\mu_B H$$

$$= A[(N + M)^{2/3} - (N - M)^{2/3}]/2^{2/3}$$

$$\quad - B[(N + M)^{1/3} - (N - M)^{1/3}]/2^{1/3} \tag{39}$$

$$\quad - CM + F$$

$$\quad + 2\mu_B H$$

	H^-	He	Li^+	Be^{++}	$B^{(3+)}$
E_{d-d}	-5.1×10^{-3}	-3.5×10^{-4}	-8.6×10^{-5}	-3.3×10^{-5}	-1.6×10^{-5}

Table II. The values of E_{d-d} in μ_B^2 atomic units. (Su an d Liu, 1988)

The extremity condition in this case does not lead to difficulties.

Here we have calculated the portion M of the electrons in heliumlike atoms in the ground state which violates the Pauli principle. This is SH in the Fermi sea which strictly obeys the Pauli exclusion principle. Before we calculated the total energy of electrons, we computed the functional $E_{d-d}[m]$ given in (9) for various heliumlike atoms (Table II). We can see that these are very small corrections but negative. Note that these corrections include only the contributions due to the portion of SH. The whole contributions of H_{d-d} in (8) must take into account the whole electron density. A detailed calculation of energies for various heliumlike atoms, including corrections of SH described here, has been done by Liu(1988). It is concluded that the corrections are very small and nearly all of them are within the experimental error.

The crucial point in the derivation presented here is the variational principle (11) derived from the thermodynamic equation(12). The adiabatic condition for (12) needs some discussion. The exchange of physical quantities between SH and the "heat" reservoir is not only the heat energy. It is illustrated here by the electron-electron interaction term. This term in the whole system is

$$E_{Hartree} = \frac{1}{2} \iint \frac{n_t(r)n_t(r')}{|r - r'|} dr dr'$$

The density here is $n_t(r)$ instead of $n(r)$. Of course $n(r)$ of SH is included inside $n_t(r)$. Thus we are not able to perform the variation of $n(r)$. In other words, we do not know the result of $\delta E[n, m]/\delta n(r)$. In ensemble considerations, the exchange quantities between canonical ensemble and its surroundings are more than heat flow, charge variations. There are possible electrostatic energy flows, exchange energy flows etc. Further investigations are required for future usages. We have done some, but no results have been published.

References

Bethe, H.A. and Salpeter, E.E., 1957, Quantum Mechanics of One- and Two-Electron Atoms (Springer-Verlag, Heidelberg)

Cottam, M.G. and Maradudin, A.A., 1984, Surface Linear Response Functions in V.M. Agranovich and R. Loudon, eds., Surface Excitations (North-Holland, Amsterdam)

Goeke, J., Kesseler, J. and Hanne, G.F., 1987, Circularly Polarized He Radiation for Electron Polarimetry in Phys. Rev. Lett. 59: 1413-1415

Greenberg, O.W. and Mohapatra, R.N., 1989, Difficulties with a Local Quantum Field Theory of Possible Violations of the Pauli Principle in Phys. Rev. Lett. 62:712-714

Hartree, D.R., 1957., The Calculation of Atomic Structures (John Wiley & Sons, N.Y.)

Izuyama, T., Kim, D.J. and Kubo, R.,1963, Band Theoretical Interpretation of Neutron Diffraction Phenomena in Ferromagnetic Metals in J. Phys. Soc. Japan 18:1025-1042

Kittel, C., 1958, Elementary Statistical Physics (John Wiley & Sons, N.Y.)

Liu,J., 1988, Thesis, National Tai-Wan University, Tai-Pei, China (unpublished)

Su, D.R. and Wu, T.M., 1975, Resistivity Anomalies for Ferromagnetic Metals at Curie Points in J. Low Temp. Phys. 19:481-491

Su, D.-R., 1987, Density-Functional Theory Extensions for Iron Group Ferromagnets in Chinese J. Phys. 25:529-534

Su, D.-R, and Liu,J., 1988, Ferromagnetic Effect in heliumlike Atoms in Chinese J. Phys. 26:236-249

Su, D.-R., 1990, Dipole and Hyperfine Interactions in Heliumlike Atoms in Density-Functional Theory : Violations of the Pauli Principle (to appear)

Participants of the Ohio Supercomputer Center Workshop on Theory and Applications of Density Functional Approaches to Chemistry

Columbus, Ohio, May 7-9, 1990

Aileen Alvarado-Swaisgood
Amoco Oil Co.
Exploratory & Catalysis Research
P.O. Box 3011 H-2
Naperville, IL 60566

Carlos Amador
Seccion de Quimica Teorica
Facultad de Quimica
U.N.A.M.
Cd. Universitaria
04510 Mexico, D.F., Mexico

Jan Andzelm
Cray Research, Inc.
655-E Lone Oak Dr.
Eagan, MN 55121

Evert Jan Baerends
Vrije University
Department of Chemistry
Section of Theoretical Chemistry
De Boelelaan 1083
1081 HV Amsterdam
The Netherlands

Axel Becke
Department of Chemistry
Queen's University
Kingston, Ontario
K7L 3N6, Canada

Charles F. Bender
Ohio Supercomputer Center
1224 Kinnear Rd
Columbus, OH 43212-1154

Robert J. Boudreau
Division of Nuclear Medicine
University of Minnesota
Hospital Center, Box 381
Minneapolis, MN 55455

Linda J. Broadbelt
University of Delaware
Chemical Engineering
Colburn Lab.
Newark, DE 19716

Francesco Buda
Ohio State University
Department of Physics
Smith Lab.
174 W. 18th Avenue
Columbus, 43210

Dennis Caldwell
Hercules Aerospace
Science and Technology
Physics Division, P.O. Box 98
Bacchus Works, MS A2
Magna, Utah 84044-0098

Douglas B. Cameron
Northwestern University
Department of Chemistry
2145 Sheridan Rd.
Evanston, IL 60208

Nick Camp
Scientific Software Development
Kendall Square Research Corp.
170 Tracer Lane
Waltham, MA 02154-1379

David A. Case
Department of Molecular Biology
Research Institute of Scripps Clinic
10666 N. Torrey Rd.
La Jolla, CA 92037

Miguel Castro
UNAM
Facultad De Quimica
Cd. Universitaria
04510 Mexico, D.F.
Mexico

Andres Cedillo
Universidad Autonoma
       Metropolotana-Istapalapa
Departamento de Quimica
AP. Postal 55-534
09340 Mexico DF, Mexico

Anne Chaka
Chemistry Department
Case Western University
University Circle
Cleveland, OH 44106

Shih-Hung Chou
Computational Science Center
CTP & C, 201-2E-23 3M Corp.
St. Paul, MN 55144-1000

P. Cortona
Universita di Genova
Departimento di Fisica
Via Dodecaneso, 33
16146 Genova, Italy

Christopher J. Cramer
United States Army
ATTN SMCCR-RSP-C
Aberdeen Prov. Grounds
       MD 21010-5423

Ernest R. Davidson
Indiana University
Department of Chemistry
Bloomington, IN 47405

Piero Decleva
Universita di Trieste
Dipartimento Sciense Chimiche
Piassale Europa 1
I-34127 Trieste, Italy

Bernard Delley
Paul Scherrer Institute
RCA Laboratory
Badener Str. 569
CH-8048 Zurich, Switzerland

Walter E. Ditmars
Chemical Abstracts Service, Dept. 57
2540 Olentangy River Rd.
Columbus, OH 43210

David A. Dixon
E.I. du Pont de Nemours & Co.
Central Research and Development
Experimental Station E328/106C
Wilmington, DE 19898

Robert A. Donnelly
Chemistry Department
Auburn University
Auburn, AL 36849

Douglas Dudis
WRDC/MLBP
Bldg 654
Wright Patterson AFB, OH 45433

Brett Dunlap
Naval Research Laboratory
Code 6119
Theoretical Chemistry Section
Washington, D.C. 20375-5000

Robert A. Eades
Cray Research, Inc.
655-E Lone Oak Drive
Eagan, MN 55121

C. Ehrhardt
Sandos Pharma AG
PKF/ZF/DDG/Molecular Modeling
503/551
CH-4001 Basel, Switzerland

Brad Elkin
Cray Research, Inc.
1211 W 22nd St., #610
Oakbrook, IL 60521

George Fitzgerald
Cray Research, Inc.
655-E Lone Oak Dr.
Eagan, MN 55121

Ken Flurchick
NCSC
3021 Cornwallis
Res. Triangle Park, NC 27709

René Fournier
Department of Chemistry
Iowa State University
Ames, IA 50011

Art J. Freeman
Department of Physics
and Astronomy
Northwestern University
Evanston IL 60208-3112

Marcelo Galvan
Universidad Autonoma
        Metropolitana-Iztapalapa
Departamento de Quimica
AP. Postal 55-534
09340 Mexico D.F., Mexico

John Gladitz
Department of Chemistry
SUNY at Stony Brook
Stony Brook, NY 11794

Michael R. Green
Amoco Chemical Company
Computer Applications & Technology
Mail Code C-6, P.O. Box 3011
Naperville, IL 60566

Steven P. Greiner
Allied Signal Inc.
Engineered Materials Research Center
50 East Algonquin Rd
Box 5016
Des Plaines, Il 60017-5016

Michael Grodzicki
University of New Orleans
Department of Chemistry
UNO New Orleans, LA 70148

James M. Groschus
Chemistry Department
Cornel University
Baker Lab.
Ithaca, NY 14853

James F. Harrison
Michigan State University
Department of Chemistry
East Lansing, MI 48824-1322

Joseph G. Harrison
University of Alabama
Department of Physics
Birmingham, AL 35294

David Hartsough
University of Illinois
Department of Chemistry
471 Roger Adams Lab.
1209 W. California Ave
Urbana, IL 61801

Charles Henriet
Senior Computational Chemist
Cray Research France
7, Rue de Tilsitt
Paris, 75017
France

David Heisterberg
Ohio Supercomputer Center
1224 Kinnear Rd
Columbus, OH 43212-1154

Judith Herzfeld
Brandeis University
Department of Chemistry
Waltham, MA 00254

Ronald A. Hill
Ohio State University
College of Pharmacy
500 W 12th Ave
Columbus, OH 43210

Andrew Hope
Biosym Technologies, Inc.
1515 Route 10, Suite 1000
Parsippany, NJ 07054

Elizabeth Jean Jacob
Chemistry Department
University of Toledo
2801 W. Bancroft St.
Toledo, OH 43606

Cynthia A. Jolly
Ohio State University
Department of Chemistry
120 W. 18th Ave.
Columbus, OH 43210

Scott D. Kahn
University of Illinois
Department of Chemistry
1209 W. California Street
Urbana, IL 61801

William P. Keirstead
Department of Chemistry, B-039
Univ. of California at San Diego
La Jolla, CA 92093

Atsuo Kuki
Chemistry Department
Cornel University
Baker Lab.
Ithaca, NY 14853

Jan Labanowski
Ohio Supercomputer Center
1224 Kinnear Rd
Columbus, OH 43212-1154

William D. Laidig
Procter & Gamble Co.
Miami Valley Labs
P.O. Box 398707
Cincinnati, OH 45239-8707

Donald E. Lauffer
Philips Petroleum Company
Petroleum and Chemical Processes
      Branch, 205 CPL
Bartesville, OK 74004

Zachary H. Levine
Ohio State University
Department of Physics
Smith Lab. #4138
174 W. 18th Ave.
Columbus, OH 43210

Peter C. Leung
Computational Science Center
CTP & C, 201-2E-08 3M Corp.
St. Paul, MN 55144-1000

Mel Levy
Tulane University
Department of Chemistry
6400 Freret St, #2015
New Orleans, LA 70118

Yan Li
Department of Physics
Brooklyn College, CUNY
Brooklyn, NY 11210

J. M. Lopez
Departamento de Fisica Teorica.
Universidad Valladoid
Valladoid, Spain

J. N. Louwen
AKZO Research Lab.
Velperveg 76
6924 BM Arnhem
The Netherlands

Karl W. Luth
Ohio State University
Department of Chemistry, Box 288
140 W. 18th Ave
Columbus, OH 43210

Yitbarek H. Mariam
Clark Atlanta University
Chemistry Department
223 J.P. Brawley Dr
Atlanta, GA 30314

Paul M. Mathias
Air Products and Chemical
7201 Hamilton Blvd.
Allentown, PA 18175

Mark McAdon
Dow Chemical
Central Research, 1702 Bldg
Midland, MI 48674

Charles W. McFarland
McGean-Rohco, Inc.
2910 Harvard Ave.
Cleveland, OH 44105

John McKelvey
Computational Science Los/I+CT
Eastman Kodak Co.
B83 Res. Labs
Rochester, NY 14650-2296

Heidi Melman
BIOSYM Technologies, Inc.
1515 Rt 10, Suite 1000
Parsippany, NJ 07054

Duane D. Miller
Ohio State University
College of Pharmacy
500 W. 12th St.
Columbus, OH 43210

Michael D. Miller
Molecular Systems Department
Merck Sharp and Dohme
Research Labs
RY50sw-100
Rahway, NJ 07065
and
IBM Corporation
300 Executive Drive
West Orange, NJ 07052

John W. Mintmire
Department of Physics
Ohio State University
4138a Smith Laboratory
174 W. 10th Ave.
Columbus, OH 43210

Alfred Z. Msezane
Clark Atlanta University
Department of Physics
223 James P. Brawley Dr.
Atlanta, GA 30314

Matthew Neurock
Dept. Chemical Engineering
University of Delaware
Newark, DE 19716

Louis Noodleman
Scripps Clinics
Dept. Molecular Biology, MB1
10666 N. Torrey Pines Rd.
La Jolla CA 92037

Tetsuro Oie
Abbott Labs.
Dept. 47E, Bldg. AP9A
Abbott Park, IL 60064

Harvey L. Paige
WRDC/MLBT
AMC Box 33811
Wright Patterson AFB, OH 45433

Robert G. Parr
University of N.Carolina
   at Chapel Hill
Dept. Chemistry CB# 3290
Chapel Hill, NC 27599-3290

Robert A. Paysen
Department of Chemistry
Bethany College
Bethany, WV 26032

Mark R. Pederson
Naval Research Laboratory
Complex Systems Theory Branch
Bldg 30, Code 4692
Washington, DC 20375-5000

Melanie J. M. Pepper
Ohio State University
Department of Chemistry
120 W. 18th Ave
Columbus, OH 43210

Alejandro Pisanty
Seccion de Quìmica Teorica
Facultad de Quimica
U.N.A.M.
Cd. Universitaria
04510 Mexico, D.F., Mexico

Russell M. Pitzer
Ohio State University
Deptartment of Chemistry
120 W. 18th Ave
Columbus, OH 43210

Lawrence R. Pratt
Los Alamos Natl. Lab.
Chemical and Laser Sciences Division
CLS-2 MS G738
Los Alamos, NM 87545

Andrew Pudzianowski
Squibb Institute for Medical Research
Macromolecular Modeling
P.O. Box 4000
Princeton, NJ 08543-4000

Tapio Rantala
Department of Physics
SUNY at Buffalo
239 Froncsak Hall
Buffalo, NY 14260

Walter Ravenek
Vrije University
Department of Chemistry
Section of Theoretical Chemistry
De Boelelaan 1083
1081 HV Amsterdam
The Netherlands

Patrick Redington
Hercules Aerospace
Science and Technology
Physics Division
Box 98, Bacchus Works, A2
Magna, Utah 84044-0098

M. Dominic Ryan
Smith, Kline Beecham
Physical and Structural Chemistry
709 Swedeland Rd.
King of Prussia, PA 19406

Michal Sabat
University of Virginia
Department of Chemistry
Charlottesville, VA 22901

Dennis Salahub
Departement de Chimie
Universite de Montreal
C. P. 6128, Succursale A
Montreal, Que. H3C 3J7, Canada

Andreas Savin
Institut fur Theoretische Chemie
Universitat Stuttgart
Pfaffenwaldring 55
D-7000, Stuttgart 80, Germany

Andrew Scheiner
IBM, National Engineering
    & Scientific Support Center
1503 LBJ Freeway
Dallas, TX 75234

Ann Schmiedekamp
Penn State University
Ogonts Campus
1600 Woodland Rd.
Abington, PA 19001

William F. Schneider
Department of Chemistry
Ohio State University
Columbus, OH 43210

Jorge M. Seminario
University of New Orleans
Department of Chemistry
UNO New Orleans, LA 70148

Richard Sicka
Bridgestone/Firestone Inc.
Central Research
1200 Firestone Parkway
Akron, OH 44317

Isaiah Shavitt
Department of Chemistry
Ohio State University
Columbus, OH 43210

Douglas A. Smith
University of Toledo
Department of Chemistry
2801 W. Bancroft St.
Toledo, OH 43606

John D. Snoddy
Lilly Corporate Center
Supercompter Applications
and Molecular Dessign, MC7R7
Indianapolis, IN 46285

Roland Stote
Department of Chemistry
Harvard University
Cambridge, MA 02138

Michael Teter
Corning, Inc., PR-2 Sullivan Park,
Corning, NY 14831

James Tyrrell
Dept. Biochemistry & Chemistry
Southern Illinois University
Carbondale, IL 62901

Robert L. Vance
DowElanco - Applied Technology
2800 Mitchel Dr.
Walnut Creek, CA 94598

C. W. Venkatachalam
Polygen Corp.
200 Fifth Ave.
Waltham, MA 02254

Matt Vernon
Department of Chemistry
Columbia University
Box 962, Hauptmayer Hall
New York, NY 10027

Richard Walsh
Minnesota Supercomputer Center
1200 Washington Ave
Minneapolis, MN 55415

Zhiqiang Wang
Department of Physics
Ohio State University
174 W. 10th Ave.
Columbus, OH 43210

Dean Scott Warren
Department of Chemistry
University of S. Carolina
Columbia, SC 29208

Siqing Wei
Department of Physics
Ohio State University
174 W. 18th Ave
Columbus, OH 43210

George E. Whitewell
Akso Chemicals, Inc.
Livingstone Ave.
Dobbs Ferry, NY 10522

Kenneth Wilson
Department of Physics
Ohio State University
4138 Smith Laboratory
174 W. 10th Ave.
Columbus, OH 43210

Erich Wimmer
Cray Research, Inc.
655-E Lone Oak Dr.
Eagan, MN 55121

Weitao Yang
Department of Chemistry
Duke University
Durham, NC 27706

Alan Todd Yeates
WRDC/MLBP
Wright Patterson AFB, OH 45433

Xiao Cheng Zeng
The James Franck Institute
University of Chicago
5640 S. Ellis Ave
Chicago, IL 60637

Tom Ziegler
Department of Chemistry
University of Calgary
2500 University Drive, N.W.
Calgary, Alberta
Canada, T2N 1N4

Hua Zong
Department of Physics
Ohio State University
174 W. 18th Ave
Columbus, OH 43210